The Calvinist Copernicans

History of Science and Scholarship in the Netherlands, volume 1

The series *History of Science and Scholarship in the Netherlands* presents studies on a variety of subjects in the history of science, scholarship and academic institutions in the Netherlands.

Titles in this series

1. Rienk Vermij, *The Calvinist Copernicans. The reception of the new astronomy in the Dutch Republic, 1575–1750.* 2002, ISBN 90-6984-340-4
2. Gerhard Wiesenfeldt, *Leerer Raum in Minervas Haus. Experimentelle Naturlehre an der Universität Leiden, 1675–1715.* 2002, ISBN 90-6984-339-0
3. Rina Knoeff, *Herman Boerhaave (1668–1738). Calvinist chemist and physician.* 2002, ISBN 90-6984-342-0
4. Johanna Levelt Sengers, *How fluids unmix. Discoveries by the School of Van der Waals and Kamerlingh Onnes.* 2002, ISBN 90-6984-357-9

The Calvinist Copernicans

The reception of the new astronomy in the Dutch Republic, 1575-1750

Rienk Vermij

Koninklijke Nederlandse Akademie van Wetenschappen, Amsterdam 2002

Edita KNAW, P.O. Box 19121, 1000 GC Amsterdam, the Netherlands
edita@bureau.knaw.nl, www.knaw.nl/edita

ISBN 90-6984-340-4

The paper in this publication meets the requirements of ∞ ISO-norm 9706 (1994) for permanence

The investigations were supported by the Foundation for Historical Research, which is subsidized by the Netherlands Organization for Scientific Research (NWO)

Contents

Acknowledgements

The debts I have to acknowledge in presenting this work are manifold. In the first place, I am happy to say thanks to Klaas van Berkel of Groningen University, who initiated the whole project and remained a firm supporter and valuable critic throughout. I should also like to thank the other directors of the project 'Culture in the Dutch Republic', in particular A. Th. van Deursen and S. Groenveld, who adopted my work within their overall project. I also thank the other members – directors and researchers alike – of this project for their support and comments. NWO, which provided the funding, deserves special mention.

Several other people helped me a great deal by reading and commenting on my text, or parts of it, as it developed. At Groningen, I found a very interested and enthusiastic colleague in Eric Jorink, who read and commented on all my chapters. Most contacts, however, were abroad or spread across the whole country. Floris Cohen was willing to read my work and provide valuable feedback. I also owe much to the encouragement and expertise of Moti Feingold. At a later stage, Albert van Helden offered much support. A person engaged in closely related studies was Wiep van Bunge. I owe him thanks, both because he was prepared to read my work and because he was willing to show me his. Equally, Theo Verbeek was always very helpful in putting his enormous knowledge on Cartesianism at my disposal. I owe many thanks to Willem van Asselt for his expert comments on my views on the Voetian-Cocceian debate, although I am afraid that on some points he did not convince me. Marc Wingens was willing to share his knowledge of early modern Dutch universities and university professors with me, and Charles van de Heuvel read a version of my chapter on Stevin. Many other people answered questions or gave hints on various aspects of my work.

I carried out this research mainly while working at Groningen University. Nevertheless, I was glad to find that I was still welcome at the Utrecht Institute for the History of Science, where I could use the splendid library and where my former colleagues Bert Theunissen, Frans van Lunteren, Lodewijk Palm, Kees de Pater, Rob Visser and others were always very supportive and helpful. At a later stage, a fellowship at the Herzog August Library in Wolfenbüttel proved very useful for finishing the work off (although, in fact, I went there for other reasons). I owe quite a lot to the Library's generous support, and to discussions with the other fellows. I am glad that Maastricht University, where I am presently engaged, offered me the opportunity to finish the project.

I now come to a very special debt. The late Professor R. Hooykaas studied the reception of Copernicanism in the Netherlands several years before I started my study. His investigations resulted in several publications, the best known of which is his edition of Rheticus' treatise on the Bible and the motion of the earth – a treatise believed to be lost until Hooykaas identified a copy. However, his death in 1994 left the larger project unfinished and unpublished. I owe much gratitude to Professor Hooykaas' widow, as well as to Floris Cohen, the curator of his papers, for allowing me to study the relevant manuscripts at my leisure. It certainly was of much help to become acquainted with the views of such a renowned scholar and eager student of the subject. Here and there, the reader will find in the notes references to Hooykaas' papers.

Hooykaas' papers have by now been transferred to the State Archives in Haarlem. In time, they will be accessible for scholarly research. Inevitably, there are differences between his approach and my approach. Hooykaas starts right away in 1543 and pays much attention to some early reactions to Copernicus' work from mathematicians living in, or originating from, the territory which later would be the Dutch Republic. In some cases, this concerns detailed investigations which perhaps will be published some day. I have left out this episode altogether and start only with the beginnings of the present Dutch state and the foundation of Leiden University in 1575. On the other hand, Hooykaas pays little attention to events after 1650, not to speak of the eighteenth century. He also pays little attention to the debates on Cartesianism and the ecclesiastical strife, although he does deal with some individual authors, such as Van Nierop and Velthuizen. As will be seen, in my own book this later period takes a central place. Still, this leaves a considerable overlap between his work and mine. Authors like Mulerius, Lansbergen and Blaeu take a prominent place in any case. (For some reason, Stevin is not dealt with in the papers I saw.)

Finally, anybody looking through the references will realise how much I owe to the staff of the various archives and libraries both at home and abroad. These people will have to remain unmentioned here – in most cases, I do not even know their names – but it would be very inattentive to omit them completely.

Some final remarks on names and dates. As for dates, the reader should be alert that during the greater part of the period under review, two different calendars were in use in the Dutch Republic. The provinces of Holland and Zealand adopted the Gregorian calendar as early as 1 January 1583, whereas the other provinces (Groningen, Friesland, Overijssel, Gelderland and Utrecht) stuck to the Julian calendar until 1700-1701. Seventeenth-century names are notoriously capricious in their orthography. Moreover, many scholars latinised their names. There seems to be no consistent way to deal with this. As a rule, I opted for the version under which the person in question appeared to be best known in English. Persons indicated by a latinised name are introduced with their real names.

In Dutch names, prefixes like *de*, *van*, etc. are not capitalised. However, this rule can be superseded by another rule, which says that all names, in the form they are written down, must begin with a capital letter. So, one writes Simon van der Moolen, but if one omits the first name, this becomes Van der Moolen. I have followed the Dutch convention throughout this book, except in the references in the notes, as a reminder that the prefixes do not count in the alphabetical order. Patronyms were frequently used in the seventeeth century, usually in an abbreviated form. One writes Jansz. or Pietersz. for Janszoon ('son of Jan') or Pieterszoon ('son of Pieter'). Here, the full stop serves as an abbreviation mark. In order not to confuse the foreign reader, I omitted it in the text.

Introduction

Writing about the history of the Copernican theory may or may not appear to need special justification. It does not need it in that the subject is of acknowledged relevance in the history of science, and it does need it in that one should explain what one can still hope to contribute to a subject already dealt with quite elaborately in the literature. The question I am posing is not really new, viz. why and how did Copernicanism turn into a received and established scientific theory? Nor do I claim much originality for my method, which comes mainly down to studying the discussions on the system of the world in a given context over a longer stretch of time. My justification lies in scope rather than in a new approach to a particular well-known person or episode.

The context I deal with is the Dutch Republic. There are several reasons for this. For one thing, although the Republic was rather small – small enough for a single researcher to overview it – its place in Europe in the seventeenth century was certainly not insignificant. Its universities were famous and drew students from all over Europe. Science had advanced a long way and produced some outstanding scholars. The importance of the United Provinces vis-à-vis science has long been recognised by the international community of historians of science, and some fine studies on the issue have been published. However, there are only some smaller studies on the reception of the new astronomy.[1] What makes the United Provinces a particularly useful case is the fact that state interference with intellectual developments was minimal, though not absent. There was a lively intellectual debate involving many parties. New developments, such as Cartesian philosophy, found an early audience here. Moreover, as a nominal Protestant country with a largely secularised government, it offers a setting in which Copernicanism has been little studied.

[1] For a textbook on the history of science in the Netherlands, see van Berkel, Palm and van Helden (1999). On the history of Copernicanism in the Netherlands, see Vermij (1993); Hooykaas (1976); Snelders (1989); de Smet (1973).

Strictly speaking, the 'reception' of Copernicanism is an ugly term: it implies a passive attitude on the part of the receiver, who can simply accept or reject the theory offered. Of course, this is hardly ever the case. Ideas are not inert, inflexible objects which can only be passed on. They are created, adapted, transformed and interpreted; their meanings change; people realise new consequences, combine them with other ideas and discover new applications. Nowhere this is clearer than in the history of Copernicanism. The 'new astronomy' was not suddenly discovered by Copernicus; modern ideas on the solar system were the result of a long and fateful debate, wherein the work of Copernicus was only one, be it a very crucial, stage. [2]

It is common nowadays to speak of the 'new' astronomy of Copernicus and his followers, as opposed to the 'old' astronomy of Ptolemy and the Medieval period. However, it took a long time before such a dichotomy became clear. True, Copernicus himself had presented his book *De revolutionibus* as a new Almagest, an emulation of the astronomy of Ptolemy. Henceforth, there were two full-fledged astronomies and two views on the constitution of the universe, both of which were supported by astronomical calculations and astronomical authority. Although this stirred interest in astronomy and cosmological speculation, it did not mean that people were forced to choose between the two. Initially, rather than simply choosing between two complete systems, people framed their own ideas, combining existing theories and picking from either theory the elements which suited them. Indeed, Mulerius spoke as early as 1616 of two 'sects' in astronomy, the peripatetics and the Pythagoreans, as he called them. [3] But his own case clearly illustrates that astronomers were not sectarians. They did not side for or against Copernicus, but for or against certain elements in his (or Ptolemy's) work. The debate centred on technical details, not on the systems as such. Before one can speak of 'the reception of Copernicanism', on should ask how, in the wake of Copernicus' work, such a clear-cut 'Copernicanism' came into being at all.

As will be seen, a clear dichotomy between a 'new' and an 'old' astronomy developed only in the course of the seventeenth century. However, by that time the dichotomy was not so much between the astronomical systems themselves, as between rivalling systems of natural philosophy. The 'new astronomy' became part of a 'new physics'. Ideas on the world system were only a part of these – and, as it turned out, a symbol. The major theme in the

---

[2] It is impossible to mention here all titles on the subject. Elementary works on the history of astronomical theory are Dreyer (1953) and Kuhn (1957). On the reaction to Copernicus' work, see for instance North (1975), Baumgartner (1986) and the various works by Westman. For the reception of the new astronomy in different national contexts, see *Reception*.

[3] Mulerius (1616) preface: '*Duas hodie esse Astronomorum sectas in confesso esse.*'

history of Copernicanism is the transition of Copernicanism from an astronomical theory – which mainly gave rise to technical discussions among specialists – to a constitutive element of a completely new, physical world-view. For this reason, historians studying the discussions on the new astronomy have concentrated on the interaction between mathematical and physical arguments. It is this interaction which gave the heliocentric theory its impact and turned its 'reception' into one of the main episodes of the scientific revolution.

This is a complicated affair, as the views on what mathematics and physics entailed have changed over the course of time. Clearly, 'Copernicanism' is a theory on the constitution of the solar system. It is not a mathematical device for calculating planetary and stellar positions, although such devices made up by far the greater part of Copernicus' book. It is not uncommon to make in the early reception of Copernicus' work a distinction between a 'physical' and a 'mathematical' interpretation. Whereas Copernicus intended his heliocentric model to be a representation of reality, most early readers took it as a convenient mathematical model which was helpful for astronomical calculations. There is no doubt that many readers refused to accept Copernicus' heliocentric theory as a description of reality. But it is important to be aware that the distinction between a description of reality and a mathematical device is not the same as the difference between a 'mathematical' and a 'physical' theory, as these concepts were seen at the time.

Here it may be useful to point out that, in Copernicus' time, there was a well-established division of astronomy into astronomy proper and cosmography. Whereas the former taught the calculatory devices by means of which stellar positions could be calculated, the latter comprised the description of the universe – the various celestial spheres, their positions and their periods of revolution. Cosmography was definitely a part of mathematics. In the Middle Ages, it was generally taught from Sacrobosco's treatise *De Sphaera*. [4] This was a very elementary book which did not contain any calculations. Sacrobosco's book remained popular in the sixteenth century. Various updated editions appeared. The best known was the one edited by Christophorus Clavius, the papal mathematician. But by this time, many new works had been written, attesting to the interest the subject elicited. The authors were invariably mathematicians. A new standard was set by Petrus Apianus in his *Cosmographicus liber* (1524), in which he integrated astronomy and geography: the description of the universe comprehended both the system of the world – from the fixed stars and the planets down to the earth – and the description of the earth

[4] Thorndike (1949).

itself, with its mountains and seas, countries and cities. This model was imitated by many. Apianus' book became particularly well known in the adaptation by Gemma Frisius], from Louvain. Other popular cosmographies from the period were the *Protomathesis* (1532) by the Paris mathematician Oronce Fine, *Elementa doctrinae de circulis coelestibus, et primu motu* (1551) by the Wittenberg mathematician Caspar Peucer, and *Novae questiones sphaerae, hoc est, de circulis coelestibus, et primo mobili* (1563) by Sebastian Theodoricus, who was also from Wittenberg. [5]

Both astronomy proper and cosmography were part of mathematics. But whereas the ontological status of the theories of astronomy proper was doubtful, in the case of cosmography there were no such doubts. The description of the world as offered by cosmography should simply represent reality. This description was not 'physical' as contemporaries understood the term. If one were to define sixteenth-century natural philosophy according to its subject matter (one might also defend that it was the discipline which explained Aristotle's *libri naturales*), one would call it the science of qualities and causes. It was certainly *not* the description of reality: that was left to more humble disciplines such as natural history or mathematics. Natural philosophers were concerned with what was *behind* reality; their task was not to describe but to explain it. Their main problem, with hindsight, is that they regarded reality as basically unproblematic and 'given'. As such, philosophy was strictly speaking not equipped to choose between alternative descriptions of reality.

This does not negate the fact that a person's view of the universe was founded in that person's idea of reality in general. In his unpublished work, Hooykaas looked at this problem in an interesting way, viz. by investigating the different ways scholars have represented the cosmos by models. On the one hand, people built purely mechanical models such as Blaeu's *tellurium*, on which Hooykaas collected a lot of documentation. Another example is the model made by Adriaen Anthonisz and described by Mulerius. On the other hand, there are the alchemical experiments by Cornelis Drebbel, which represent the motion of the earth as an animated, organic process. The latter model, for instance, turns up in the work of Lansbergen. By focussing on such modelling, Hooykaas links the ideas on the system of the world to the general philosophical view of nature.

I have chosen a different approach by concentrating on the way cosmography became part of an established philosophical-physical discourse. Contemporaries clearly regarded the natural philosophy of their time as wanting in explanatory power regarding the many new discoveries made during the six-

[5] On sixteenth-century cosmographical literature: Thiele (1995).

teenth and seventeenth centuries. Galileo's telescopic discoveries in particular demonstrated the shortcomings of the old philosophy. Many independent thinkers sought a new philosophy of nature. By the middle of the seventeenth century, the heliocentric theory was an important touchstone for any philosophical system. Hence, 'Copernicanism' could become an important element in the philosophical debates of the time.

The relative importance of and the interaction between physical and mathematical arguments in the discussions on the Copernican theories are a main theme of the first three parts of this book. Part I presents a study of the first reception of Copernicus' ideas in the Dutch Republic. Various elements from Copernicus' work drew a lot of interest, and his theories certainly affected cosmological thinking. But only a few people openly adopted his central thesis of a moving earth and a central sun. Official scholarship saw too many obstacles. Only people on the fringe of the learned world were ready to draw the more radical conclusions from Copernicus' work. All participants in the debate agree that the importance of Copernicus' theory is in its description of reality. No one tries to neutralise its effects by reducing it to a calculatory device. At the same time, philosophical arguments are completely absent. A description of reality is sought by mathematics only.

Part II focuses on the impact of Galileo's telescopic discoveries. It is now commonly accepted that it was these discoveries which turned the Copernican hypotheses from an esoteric doctrine, defended by only a few specialists, into a serious alternative to the traditional world-view. My investigations amply confirm this. The telescope made people look at the heavens with, almost literally, a different eye. Not just the celestial motions but the very nature of the heavens became a matter of investigation and speculation. Mathematical arguments receded into the background. The constitution of the universe became the subject of a lot of physical speculation. Such speculation is unconnected with traditional philosophical arguments. Among its first proponents are people without any formal mathematical or philosophical education. At universities, the teachers of physics only slowly became aware of the relevance of the new astronomical discoveries.

In Part III, the focus is on the apotheosis of the new approach in the philosophy of René Descartes. In Descartes' philosophy, heliocentric cosmography is a central element, but it is no longer treated in a mathematical way. Cosmography is by now the domain of natural philosophers with often the haziest idea of mathematics and astronomy. In this climate, 'Copernicanism' became a definite and recognisable concept. The earlier mathematicians concocted their own systems. The natural philosophers after 1650 simply lacked the competence to deal so freely with mathematical data. They preferred the

Copernican world system not because of its astronomical merits or demerits, but because of their adherence to a general philosophy of nature.

The emergence of clear 'Copernican' and 'anti-Copernican' camps is thus partly explained by the role the world system plays in more comprehensive debates. The debate is no longer about the details of the various solutions to the problem of the constitution of the universe. The respective world systems are so to speak the battle cries or the banners in a debate between competing philosophical schools. One might of course go further and ask why this philosophical debate, unlike the earlier cosmographical debate, became polarised in this way – though perhaps 'polarised' is too strong a word here, as matters were not entirely black and white. Here, one encounters the religious element, which enters in the wake of the philosophical argument. The hardening of the frontiers in the philosophical, and hence cosmographical, debate in the last resort originates in theological oppositions.

Religion is in effect a second main theme in the history of Copernicanism. The religious controversies accompanying the debate on Copernicanism have for a long time been a main point of attention of historians of science. So far, however, this attention has been unevenly distributed: most has been paid to the reactions within the Roman Catholic Church in general and to the trial of Galileo in particular, and very little has been written on the situation in the Protestant churches. The ideas of individual Protestant thinkers are generally known and some nice work has been done on a few topics, such as the reception of Copernicus' ideas at the Lutheran university of Wittenberg in the sixteenth century. But a detailed study of events in the Dutch Republic, which was officially a Protestant country (although it tolerated large groups of dissenters within its borders) should offer a major contribution to the international literature on the subject. The vehement debates which were waged within the Dutch Reformed Church, especially during the second half of the seventeenth century – debates which were widely known at the time and quite influential in the Protestant world – have so far received only scant attention in the international literature on history of science.

Of course, one might well ask why these theological debates should be much more polarised than the discussions between astronomers or philosophers. The point is, of course, that it would be far too simple to regard these theological debates as just that – the uncommitted and uninterested speculations of individual theologians. Indeed, in recent years historians of science have increasingly come to realise that the religious controversies surrounding the introduction of the new science are in fact often politically rather than religiously motivated. Jacob's work on Boyle and the English Revolution, Shapin and Schaffer's work on the controversy between Boyle and Hobbes, and Redondi's and Bagioli's work on Galileo are cases in point. In the Dutch case, too, the furious debates on the system of Copernicus and the philosophy of Descartes become understandable only within their politico-social context. It was not a case of two parties disagreeing about the new philosophy or their exegetical principles, but of two parties competing for power. Very pointedly, one might say that they were not quarrelling because they had a disagreement, but that they created a disagreement because they had a quarrel. This makes the religious issue both far more complicated and far more interesting than a private dispute between theologians. In Part IV, these developments are explained in some detail.

With some exaggeration, one could say that we can speak of a 'reception' of the idea only after 1650, when Copernicanism had become a clear-cut concept, i.e. one which was a given rather than asking to be defined. It then became a matter of gaining more adherents and dealing with the common objections. Of course, this cannot be isolated from the religious and political relations. The dissemination of the Copernican system in the Dutch Republic implied the undoing of the religious sensibilities concerning the subject which had built up in the foregoing period. This was facilitated on the one hand by the calming down of the political opposition, and on the other by the demise of Cartesian philosophy, which in the eyes of many had been too compromised. Copernicanism remained an element in a general physical world-view, but from now on this was based on the theories of Newton, which gave much less offence. This development will be explained in Part V.

I should point out that, although the story is ordered chronologically and divided by some important points, I do not always follow the exact sequel of events. The division is by generations rather than events. That is, I take it that people are formed in their youth and early adulthood, and generally are little affected by new developments after that time. For instance, Lansbergen's main publications appeared about two decades after Galileo's telescopic discoveries, which were published in 1610. Still, in the order of argument Lansbergen's work comes before Galileo's discoveries, as Lansbergen was born in 1561, had his ideas formed before Galileo appeared on the scene, and in his work shows hardly any awareness of the new discoveries. In a similar manner, Beeckman, Hortensius, and Holwarda are treated as pre-Cartesian philosophers, although Beeckman and Hortensius knew Descartes personally, and Holwarda was over 20 years his junior.

Part 1. A world of order

1. The context of learning

A bourgeois society

Hardly any country was less representative of the European state system than the Dutch Republic – the Seven United Provinces. This republic had been established, more or less by coincidence, during the political turmoil of the late sixteenth century. In the 1560s and 1570s, the various Hapsburg territories in the Low Countries had revolted against their overlord, Philip, King of Spain. The result was a split. The southern provinces – Flanders and Brabant, which up till then had been the more important – in the end remained Hapsburg territory. The northern provinces, in the course of a long struggle, managed to remain free from Hapsburg domination. Sovereignty fell to the States, the representative assemblies. The States were constituted by the delegates of the principal cities and nobles of the respective provinces. There was also a national body – the States-General – but this was simply an assembly of deputies of the various provincial States, with, strictly speaking, no power of its own. The so built republic was at the time a very queer construction, but it survived and even prospered. Its independence was officially acknowledged by the peace treaty of Westphalia in 1648.[6]

In the northern provinces, Holland took the leading role. Not only had it assumed leadership in the war, but it was also by far the wealthiest and most powerful of the seven northern provinces. It was densely populated and highly urbanised. During the Republic, Holland paid more than half of the state's expenses. Its dominance is still reflected in many modern languages, where 'Holland' is the name given to the present Netherlands as a whole. In this book, I will only use it in its original meaning, as a denomination for the medieval county and later province. Although one should not forget that Holland was but one of seven provinces and that all seven were virtually independent, Holland did play a very dominant role.

The Dutch Revolt had been in part a war of religion. Religious strife fig-

---

[6] For the history of the Dutch Republic, see Israel (1995).

ures prominently among its causes. In particular the Calvinists emerged as a powerful social force which took on the religious and, when they got the opportunity, the political leadership. After the breakdown of Hapsburg rule in Holland after 1572, they took the opportunity to establish themselves as the official Church in the province and, later, in the Republic as a whole. Roman Catholicism, which had been compromised by its association with the enemy, was suppressed. This happened with the full support of the magistrates and the political leaders of the revolt. However, if they supported the Calvinist Church, it was not because they advocated the Church itself, but because they badly needed the Calvinists as their allies in the struggle against Spanish tyranny. Under the pressure of events, regents and church councils closely cooperated in defending their liberty. But in fact, many magistrates felt rather lukewarm about the new Reformed faith. As soon as people began to feel that Spanish rule was not likely to return, the alliance between Church and magistracy began to show cracks. When in 1609 a truce with the Spanish was concluded, serious religious tensions came to the surface which brought the country to the brink of civil war. Two parties formed within the Dutch Reformed Church. Opposed to the strictly orthodox Calvinist were the Arminians (also called Remonstrants), who in Holland were backed by a majority of the regents. The conflict is much too complex to rehearse here in detail, but it should be mentioned as it was rather a traumatic experience for the Reformed and clearly left its mark. The struggle ended with a military intervention by the stadholder, after which a national synod (Dordrecht, 1619) declared the Arminians heretical and banned them from the Reformed Church. However, this did not end the strife between Church and state. Tension remained latent and when in 1648 a definite peace treaty with the Spanish had been concluded, it flared up again.

Even if most of the regents regarded their country as decidedly Christian or even Reformed, this did not mean that they wanted religion to interfere with their politics. The domains of religion and politics remained separated. The Dutch Reformed Church was publicly acknowledged as the 'public church' of the country and was actively supported by the government, but it was not the official state church. Nor did it ever comprise more than about half of the population as a whole. Other religious beliefs were tolerated. Freedom of conscience was never in danger in the Dutch Republic. In principle, public worship was the monopoly of the Reformed Church, but if dissidents kept a low profile and gave no offence, the regents would tolerate a lot. Offenders could be fined or even banished, but were never executed. The least tolerated was Roman Catholicism, which was still associated with a past of tyranny and oppression, and was regarded by many people as the church of the Antichrist. Still, Catholics remained quite numerous. In peripheral territories which were added later to the Republic, they formed the majority, but there were also many Catholics in Holland itself. Though officially forbidden, in practice Catholics were not harassed too much. Arminianism after 1619 was actively persecuted for a while, but soon established itself as a respected community. Lutherans, Jews and various groups of Mennonites were more or less acknowledged as independent communities. Their leaders were acknowledged by the government, and they founded schools, seminaries, and so on. Nevertheless, Calvinism had a deep impact on the Dutch nation.

To a modern observer, it might seem that the early Dutch Republic – and particularly Holland – was an exciting place to live: it was busy and teeming with life. When the French scholar Joseph Scaliger arrived in Holland in 1597, he found that more boats put in at Leiden on one Saturday than took the popular route from Toulouse to Bordeaux in a whole month.[7] Holland had connections with places right across the globe. The Hollanders had been sailors since the Middle Ages, and during the revolt, they became a first-rate sea power. Before the sixteenth century was over, they had found their way to both the West and the East Indies. Fighting off the Spanish and the Portuguese, they laid the foundations of a sizeable colonial power. Holland's main port – Amsterdam – quickly developed into a metropolis which rivalled any European capital. But to people like Scaliger, Holland was rather a dull place to live, because the Netherlands as a republic lacked what polite people all over Europe thought vital for a truly civilised existence: a princely court.

The court in early modern Europe was more than just the centre of government. The prince tended to monopolise all aspects of public life. He was the fountain of justice and the heart of society, to whom the eyes of all people aspiring to advancement were directed. Consequently, his court was the place where the gentleman was formed and good manners were cultivated. Moreover, it was a centre of culture. The court attracted and sponsored artists and scholars. On the whole, the legitimation of one's talents or activities could take place only in a courtly setting. In a public sense, one existed only in the presence and with the favour of the prince.[8]

The Dutch, then, had deposed not only a king but a way of life. Nor were there any substitutes for a court in the Netherlands. The Reformation had done away with the great ecclesiastical institutions. The great magnates, who had played a dominant role in the Hapsburg Netherlands, in the end kept on the side of the king. Only the Princes of Orange threw in their lot with the revolt. Their position as stadholders of Holland, however, remained

---

[7] Grafton (1993) 373-374.

[8] The classical studies by N. Elias are still valuable. For a study of the impact of court culture on early modern science, see Biagioli (1993).

ambiguous. They never obtained real princely status; formally they were no more than civil servants and their position was consequently unstable. Some of the later Princes of Orange did try to give their position somewhat more lustre, but these episodes alternated with long periods during which the stadholderate of Holland was abolished altogether.

Curiously enough, the Dutch did not seem to care. Holland's interest was in its ships, its trade, its money. Earlier historians have tried to characterise Dutch life as 'bourgeois'. This has met with opposition, and certainly it should not be taken in too absolute a sense. In the Netherlands at the time, there was no 'bourgeois' mentality such as it existed in the nineteenth century. Seventeenth-century Dutch citizens strove to acquire a respectable status as landowner, government official or squire, as did their equals all over Europe. As for fashion, they followed – be it at a great distance – the foreign courts. The local nobility was held in high esteem. But relatively seen, with respect to other seventeenth-century countries, it is certainly warranted to speak of a bourgeois society. First, bourgeois life was undoubtedly more attractive than elsewhere because it was economically so rewarding. Second, as soon as people turned to higher aspirations, Dutch life, compared to the foreign courts which set the tone, proved very provincial indeed.

This affected the arts and sciences in various ways. There were no court astronomers and no court physicians. Patronage could only come from local governments and officials, from rich merchants and some minor noblemen. Even the richest of them could not compete with the smallest European monarch. But taken together, they could. Riches in the Dutch Republic were spread over a large section of the population. Dutch art had to adapt itself to a merchant's drawing room or an occasional town hall. There was no chance of being awarded a royal commission. There were no statues, no stained glass windows, no spectacular works. But there were many town halls and very many merchants who could afford a simple portrait or two. Painters did not need any princely protection. They could make a living on the market. And they did, as the blossoming of seventeenth-century Dutch painting attests.

Learning, too, was largely democratised. Schools abounded in this urbanised society and the rate of literacy was unusually high. In this country, Scaliger noted with astonishment, even the female servants could read.[9] Most cities had got Latin schools and some of these schools, like those at Deventer and Zwolle, attained a high standard of learning and produced important humanists already in the sixteenth century.[10] In connection with this literacy,

[9] Grafton (1993) 375.
[10] Cameron (1990).

there was a great demand for books. Printing houses were established rather early in the main cities of the Netherlands' provinces, and after the establishment of the Republic, printing underwent an impressive boom. This was enhanced by the fact that the press was relatively free, thanks to the tolerant attitude of the Dutch regents. Eventually, many foreign authors had their works printed in the Dutch Republic. The Republic developed into a cultural entrepot, supplying Europe with books.[11]

In Hapsburg times, the really important or ambitious scholars from the northern provinces – Erasmus, Gemma Frisius and others – normally chose a career elsewhere, in the southern provinces (the scholarly centre of the Hapsburg Netherlands was Louvain University), or further afield in Basel, Rome or elsewhere. But the many Latin schools needed teachers, just as the many cities needed trained officials. They offered places to many lesser, though by no means insignificant scholars. The northern Netherlands was certainly not an intellectual centre by the time of the revolt, but humanist learning had spread widely. After the revolt, the new self-assertion and pride of the young Republic quickly capitalised on this beginning, thus laying the foundations for a blossoming scholarly culture.

Leiden University: general

In Hapsburg times, there had been no university in the northern Netherlands. After the revolt, this quickly changed. Leiden University was founded as early as 1575. When it became apparent that the Republic was to steer an independent course, the other provinces came to feel the need for some centre of higher learning. Because every province jealously guarded its independence against the others, and especially against Holland, they all wanted to have their own university. Friesland was the first to follow suit with a university at Franeker, which was opened in 1585. The province of Groningen opened its university in 1614 (in the town of Groningen), and Utrecht did so in 1636 (in the town of Utrecht). Gelderland was last with a university at Harderwijk, which was opened in 1648. The two remaining provinces – Overijssel and Zealand – were too small to support a full university and had to be satisfied with having an 'illustrious school', which in principle offered the same courses as a university but lacked the right to award degrees. The 'illustrious school' of Overijssel was in Deventer, and that of Zealand in Middelburg. Over time, other towns opened their own illustrious school. Some of them,

---

[11] Gibbs (1971).

notably those in Deventer and Amsterdam, attained a high standard of learning and could compete with a minor university like Harderwijk.

The Dutch universities and illustrious schools did not need to compete for prestige or public funds with scientific rivals in other spheres. Any public patronage of science and learning that there was, was administered through them. Leiden in particular, the university of by far the wealthiest province Holland, grew into a prestigious institution. The Hollanders made no investments in pompous buildings and the outward appearance of Leiden University confirmed all foreign prejudices about Dutch rusticity. But the university was well equipped: it acquired an anatomical theatre in 1593 and a botanical garden in 1594 (the famous Clusius was appointed its supervisor). Neither of these features was standard for a university at the time. The famous Antwerp printer Plantin came to Leiden to establish the university press (1584). In the seventeenth century, Leiden was among the first universities to acquire an observatory (1632), a chemical laboratory (1669) and a physical cabinet (1675).

Some characteristics should be kept in mind when speaking of Leiden University. First of all, it was not dominated by the Church. As explained, Dutch regents did not want the Church to become too powerful. So, when it came to formulating the statutes of the university, a quarrel arose with the foreign Calvinist theologians who had been appointed professors. These theologians wanted an unambiguously Calvinist institution, one modelled after the Genevan and French Reformed schools. This would have entailed a close and institutionalised alliance with the Reformed confession. The Leiden magistrates protested that in these times, the government being Christian, such was not necessary. Their official reply even compared such ecclesiastical supervision, rather bluntly, with the hated Inquisition. A compromise was eventually reached which left the regents' dominance intact.[12] As a consequence, the teaching programme offered at Leiden was shaped by secular rather than theological demands – at least, compared to such universities as Wittenberg or Paris. The theologians soon found reason to complain about metaphysical disputations held in the philosophical faculty, which touched on matters of theology.[13] Friction between Church and regents would remain chronic throughout the seventeenth century.

The theological faculty itself, as about the only higher authority in the field within Holland, gained considerable prestige in ecclesiastical matters. But the theology professors regarded their colleagues from the faculty of arts with suspicion. Consequently, philosophy obtained little backing for its function

---

[12] Dankbaar (1946) 144-149.

[13] Dibon (1954) 64-70; Sassen (1941) 26-29.

as 'handmaid of theology'. Nor did traditional philosophy gain much prestige from other quarters. The ordinary courses of philosophy (logic, physics, etc.) were not seen as a field wherein the university was to obtain much honour. These were just preparatory courses for propaedeutic students, and as such were of little value or interest. Accordingly, the field was relatively neglected. Whereas as a rule, the curators tried to attract renowned scholars to Leiden, most professorships in the above fields remained filled by second-rate men. Looking down the lists, it is striking how many professors changed chairs, or left altogether, after only a year or so. It was only with the appointment of Franco Burgersdijk in 1619 that a professor arrived on the scene who succeeded in giving philosophy some academic respectability. Consequently, philosophical education at Leiden was not very advanced. Its contents remained traditional, be it rather eclectic. [14]

The governors of the university sought prestige from another quarter: they strove to turn Leiden into a bulwark of the new humanist learning. In their view, what really mattered were the literary disciplines, such as rhetoric or history. The curator-president of the university – Jan van der Does (Janus Dousa), a local nobleman who had played an important part in the defence of Leiden against the Spanish in 1574 – was a propagator of the new humanist learning as well as a famous neo-Latin poet. He was also well acquainted with the international learned world, which made him capable of turning Leiden into a truly scholarly institution.

Classical philology was at the core of the humanist programme, and Dousa did his utmost to give Leiden some pre-eminence in this field. In 1578, he succeeded in attracting the famous philologist Justus Lipsius to the university. Lipsius was the university's luminary for more than a decade. At Leiden, he published his international best-seller *De constantia* (1584). In 1591, however, he decided to leave Holland for Louvain and publicly professed his adherence to the Church of Rome. People in Holland were upset, but quickly sought alternatives. This led to the great French humanist Joseph Justus Scaliger coming to Leiden. The curators had achieved this by every means at their disposal. Important government officials, including the stadholder Maurice of Nassau, used their full authority to induce him to come. Not only did the curators give him a very high salary and privileges as regards precedence, they even freed him from all teaching obligations. Scaliger decidedly set his mark upon learning at Leiden. Though freed from regular teaching obligations, he gave private tuition – a kind of masterclass – to advanced and gifted students. In this way, he moulded the cream of seventeenth-century Dutch

[14] Van Bunge (1999) 283-285. Sassen (1941). For the influence of philologists, see Dibon (1954) 71-76.

humanism: Daniel Heinsius, Hugo Grotius, Philip Cluverius, Thomas Erpenius, and others.[15] In short, the main influence on the curriculum and the character of the university in general was the humanist programme, not the exigencies of any higher faculty such as theology or medicine.

Mathematics and the humanist programme

The position of mathematics, as a discipline, was ambiguous. On the one hand, it was mainly a practical discipline. In the Dutch Republic, it was connected with surveying, engineering and navigation. At the universities, too, this practical side was incorporated into the programme. This tradition started at Franeker. The first professor of mathematics there was Adriaen Metius, son of the military engineer Adriaen Anthonisz. His father had sent him to university to study law, but he never took his degree; instead, he applied himself nearly exclusively to mathematical studies. After leaving university, he spent some time as an assistant to Tycho Brahe at Hven, did some lecturing at German universities, and finally became assistant to his father as an engineer. Finally, on the recommendation of the stadholder, in 1598 he became a professor at Franeker University, a position he kept until his death. At Franeker, Metius not only taught ordinary students. He asked and obtained permission to teach in the vernacular so that simple practicians would be able to benefit. Other universities followed suit. In the seventeenth and eighteenth centuries, such courses in the vernacular had become quite common.[16]

On the other hand, mathematics could be seen as a part of humanist scholarship. The term 'humanism' is somewhat ambiguous, as it is used by different people in different ways. In this book, I use it in the sense of a cultural ideal of a search for the lost classical wisdom and knowledge, with the purpose of having them re-established. The humanists understood this search as primarily a textual enterprise: it is by means of texts that classical antiquity speaks to us. A prime scholarly task is to find, correct and edit ancient texts. Mathematical knowledge had been a part of classical learning. Meanwhile, many parts had been lost or corrupted and it was a scholar's task to reconstruct this part of ancient wisdom.

Moreover, many humanists realised that mathematics was a valuable source of knowledge and were ready to accept the discipline as a branch of true learning.[17] Astronomy in particular was deemed worthy of study. It was the

---

[15] Grafton (1988) 69-74.

[16] On Metius, see de Waard in NNBW, I, 1325-1327. On vernacular mathematics courses, van Winter (1988).

[17] Hoppe (1996) offers a judicious analysis.

science of the heavens, of the eternal, unchanging celestial motions. It was here that God's hand manifested itself most clearly in the Creation. Such ideas went back to classical antiquity, but they were taken up afresh during the Renaissance. Melanchthon propagated such ideas at Wittenberg. At Leiden, we find similar feelings expressed in the laudatory poems Janus Dousa Jr. and Gruterus wrote about the works of Lansbergen. [18]

Astronomical interest can be noted at an early time among the humanists of Holland. Jacob Susius was famous for his learning. He came from a distinguished patrician family. His cousin served as president of the Council of Holland and he himself was for some time member of the city government of Zierikzee (in Zealand), which by then was a humanist centre of some note. Jacob was a highly learned man, who wrote Latin verse and had befriended Holland's leading humanists. Moreover, he appears to have been well versed in astronomical questions. He wrote a short tract on how to construct a celestial globe and drew up a star catalogue for the 1584 edition of Apianus' *Cosmographia*. He also collated a manuscript of the ancient astrological poem by Manilius. This collation was used by Scaliger in 1599 when, at Leiden, he revised his earlier (1579) edition of the work. [19] Astronomical interests were also pursued by the humanist scholars Janus Dousa Jr. and Johannes Isacius Pontanus, a correspondent and admirer of Tycho Brahe.

Scaliger's coming to Leiden must have contributed much to the interest in mathematics and astronomy among humanist scholars. Scaliger was quite active in this field. His work on historical chronology, which practically established the subject as a new discipline, had become quite famous. Historical chronology entailed a detailed study of all kinds of ancient calendars, and since these normally have an astronomical basis, one could not complete this study without astronomical knowledge. Scaliger's book on chronology (*De emendatione temporum*) had appeared in 1583. At Leiden, Scaliger took up the subject afresh. He published not only a second, revised edition of *De emendatione* in 1598, but also a completely new work on the subject, *Thesaurus temporum* (1606), an edition of several ancient chronological works.

However, Scaliger's 'mathematics' were based on quite different presuppositions than its modern counterpart. When he came to Leiden, as an act of courtesy he dedicated to the curators a book he had written on what at the time was a highly disputed question: the squaring of the circle. Scaliger pretended to have found the definitive solution, by philological rather than by mathematical means. In fact, as far as mathematical contents were concerned,

[18] Methuen (1996), in particular 393-395. Lansbergen (1591).

[19] Grafton (1993) 443.

he took impossible contradictions for granted. For instance, he bluntly declared the length of a circle arch to be smaller that that of its subtensa. Small wonder that the mathematicians criticised the book severely. Scaliger corrected some minor errors, but insisted on his main point. He was a great scholar and philologist, but he definitely had no feeling for the exact sciences. He laughed away the objections of the mathematician Ludolf van Ceulen – who was well versed in questions of circle quadrature – on the grounds that a fencing master should not deem himself competent to correct a scholar. And when Rudolf Snellius, professor of mathematics, tried to convince Scaliger of his error, Scaliger simply replied: 'You ass. Why should I calculate in the same way as you?' [20]

It does not seem that Scaliger contributed much to the acceptance of Copernican ideas. He definitely rejected Copernicus' ideas on the motion of the earth. Still, he may have stimulated the study of the heavens. He developed a new theory on the precession of the equinoxes, a subject which occupied him greatly during his Leiden years. He corresponded on the subject with Tycho Brahe in Denmark, as well as with the astronomer Mulerius, [21] and his ideas were probably well known among the Leiden humanists. In short, they come down to this. Already in antiquity it had been remarked that the sun's position in the sky at the vernal and the autumnal equinox shifted over long stretches of time. Ancient astronomers had explained this by a very slow motion of the spheres of the fixed stars with regard to the earth. Copernicus, on the other hand, had explained this by a slow change in position of the earth's axis with regard to the fixed stars. Scaliger knew of these solutions and rejected them both. In his view, the points of the equinoxes were fixed to a separate, very slowly moving sphere. The relative position of the earth and the sphere of the fixed stars remained constant, a point Scaliger attempted to prove by collecting ancient testimonies on the position of the polar star. Again, Scaliger entangled himself in contradictions. It is only fair to add that he defended his views only in private and shrunk from publishing a full treatise. His *Diatriba* was published posthumously in 1613. It was not the right thing to do to honour his memory, as it earned him only scorn. His modern biographer describes the whole undertaking as Quixotic. [22]

In the end, Scaliger's blunders could not undo the fact that he had clearly demonstrated the relevance of mathematics to humanist scholarship. Still, the

---

[20] On this episode, see Kästner (1796), I, 487-497; Bierens de Haan (1878) 153, 280-314; Grafton (1993) 378-384.

[21] Scaliger to Mulerius, 19 March 1608 st. Jul., printed in Mulerius (1630) 82-83.

[22] Scaliger (1613). Grafton (1993) 459-488 (3.4: 'Don Quixote's last ride: the attack on precession'). Cf. Riccioli (1651) II, 347.

practical side could not be dismissed. Leiden thereupon saw the rather unique construction of two different programmes. For one thing, the university had a professor of mathematics in Rudolf Snellius. Snellius had been appointed as *extraordinarius* in 1580, at the express request of some students. Up to 1580, he had been a physician at Oudewater. Before settling in Oudewater, Snellius had done some teaching (mainly in philosophy) at German universities and written a number of Ramus-inspired textbooks. For the time being, he became only extraordinary professor. Teaching mathematics was his only duty at Leiden. Only occasionally, when another chair became temporarily vacant, was Snellius asked to deputise. So, during two years he taught Hebrew as well. This indicates that Snellius was not a simple practitioner, but a real scholar. However, it seems that as such he was not taken too seriously by his colleagues, because of his defence of the ideas of the French philosopher Petrus Ramus. [23]

However, when in 1598 Metius started his vernacular mathematics courses at Franeker, the Leiden curators apparently felt that their university was not to be outdone and that Leiden needed something similar. In 1600, the university opened a new centre of mathematical instruction, the *Duytsche mathematique* ('vernacular mathematics courses'). The *Duytsche mathematique* was formally part of the university, but in practice it stood largely on its own and had its own professors, who taught exclusively in the vernacular. The courses were not just for ordinary students, but also for interested laymen. The institution was mainly intended for the training of military engineers, which the Dutch army needed badly in the war with Spain. As it turned out, instead of engineers it trained mainly surveyors. It remained a kind of professional training programme with limited practical purposes, not very distinct from other courses in practical mathematics. Its main asset was that is was backed by the prestige of the University. [24]

The *Duytsche mathematique* was clearly not meant to replace the chair occupied by Rudolf Snellius. On the contrary, Snellius was made an ordinary professor in the end. What was effected was a clear split between two forms of mathematics – a practical and a scholarly. Henceforth, curators chose their mathematics professors from among 'real' scholars – people with a classical training, with a knowledge of ancient mathematical texts. However, this does not mean that learned mathematicians were not interested in practical aspects. The happy marriage between the world of mathematics and that of classical philology is nowhere better illustrated than by the career of Rudolf

---

[23] On Rudolf Snellius, see van Berkel (1983) 273-279.

[24] Westra (1992) 82-89.

Snellius' son and successor, Willebrord Snel van Royen (Willebrord Snellius).[25] Young Willebrord went to study law, just as Metius had done before him. However, like Metius, Willebrord preferred the study of mathematics and soon followed in his father's footsteps. In 1600, he obtained permission to lecture at Leiden on Ptolemy's *Almagest*.

Here, however, the similarity with Metius ends. While Metius, despite his study of law, remained a proponent of practical mathematics, Willibrord aligned himself with the world of scholarship. Shortly after his lectures on Ptolemy, he left Leiden for a grand tour through Europe in order to round off his education. Normally, one visited famous humanist scholars during such a tour, but Snellius choose to visit great mathematicians, like Adriaan van Roomen at Würzburg and Tycho Brahe at Prague. Apparently, mathematics could be studied in the same way as 'real' science. After a sojourn at Paris, where he took up the study of law again, he went home to Leiden and obtained his degree in philosophy. From 1609 onwards, he assisted his father in teaching. Eventually, he succeeded him.

There is more to Snellius than a lawyer who could not resist the siren's song of mathematics. He indulged not only in mathematical problems, but also in humanist studies. He appears to have been rather fond of his knowledge of Greek.[26] He was deeply influenced by the dominant philological school at Leiden and appears to have been one of Scaliger's privileged pupils. As he did with the other students under his direction, Scaliger entrusted him with the edition of a classical text: a reconstruction of the work on the cutting off of a ratio and of an area by the Greek mathematician Apollonius, preserved only in abstract. Snellius based his work on a manuscript he had obtained 'by the liberality and munificence of that illustrious man, Joseph Scaliger'.[27] Willebrord appears to have quite forgiven Scaliger the quarrel with his father, and held him in great esteem. In 1616, seven years after Scaliger's death, Snellius edited a posthumous work by him on numismatics.[28]

This did not mean, however, that Snellius abandoned his interest in the more practical aspects of mathematics. What deserves emphasis is that Snellius did not deal with philological and mathematical studies as two separate disciplines. He combined them, treating them as one. Navigation was a typical

---

[25] On Willebrord Snellius, see principally C. de Waard in *NNBW* VII, 1155-1163. A thorough study is sadly missing, but is currently being prepared by L. de Wreede (Utrecht University). For his cosmological ideas, see Nouhuys (1998) 337-360, 528-537; for his work on refraction (Snell's law), see de Pater (1975) 309-311.

[26] Cf. his undated letter to Aemilius Rosendalius, UBU Hs. VII A 26 f 230, about half of which is written in Greek, apparently as some playful entertainment.

[27] Snellius (1607), ad lectorem.

[28] Scaliger (1616).

field of practical mathematics. But Snellius' book on the subject – *Tiphys Batavus* – has an extensive preface on ancient shipping, quoting among other things a long passage by Scaliger on the subject. Snellius' most impressive combination of mathematical and classical studies remains, however, *Eratosthenes Batavus* (1617). In this book, he sets out to determine the size of the earth by means of determining the length of a meridian. The work is famous as a landmark in the history of surveying, another field of practical mathematics. In it, Snellius demonstrates a new method for triangulating large areas. More importantly, he put his ideas into practice, surveying in person. [29]

However, these aspects only occur in the second part of the book. The first part is wholly devoted to an investigation into Eratosthenes' antique determination of the size of the earth. This part of the work clearly owes a lot to classical philology. Snellius not only has his say about Eratosthenes' method, but he does so after making a close scrutiny of the relevant classical sources. He also has recourse to critiques on Eratosthenes by other ancient writers, such as Hipparchos. It is only after this exercise in textual investigation that he introduces, in the second book, his new method for solving Eratosthenes' problem: determining the length of the meridian by means of his new surveying techniques. And even this happens only after a prolonged discussion on standards of length, included classical ones. The book is not a practical manual on surveying, but a study on a classical problem which demands mathematical as well as philological knowledge. Actually, it appears that for Snellius, science consisted primarily in the reconstruction of the knowledge of the ancients. This doctrine of pristine science is expressly stated in his tract on the comet of 1618. [30]

Willebrord Snellius died in 1626 and was succeeded three years later by his colleague Jacob Gool (Golius), professor of Arabic and eastern languages. [31] Golius was even more of a philologist than Snellius, and would combine the two chairs until his death in 1667. In those days, this combination was not strange. The study of Arabic at Leiden owed most to the brilliant scholar Thomas van Erpe (Erpenius), who held the chair of eastern languages from 1613 to 1624. In his oration, Erpenius had made a great effort to undo the prejudices of the students of classical languages against the study of a language like Arabic. Among other things, he pointed out that several classical texts that had been lost in their original, classical version had been preserved in Arabic translations. Knowledge of Arabic was therefore of vital impor-

---

[29] For the technical details, see Haasbroek (1968) 59-115. See also Delambre (1821) II, 92-110.

[30] Snellius (1619). For his ideas on pristine science, see Nouhuys (1998) 348-351.

[31] On Golius, see Juynboll (1931) 119-183; Brugman (1975) 208, 210, 213; Savage-Smith & Wakefield (1994).

tance for the reconstruction of ancient knowledge in such fields as medicine, mathematics, astronomy, etc. [32] This argument surely must have carried some weight among the Leiden humanists. It may well be that it was the principal reason for Golius to engage in the study of Arabic, for indeed, he had applied himself to mathematics before that time. As he explained in a letter to Vossius, Arabic was 'so to speak a universal language, which cannot be separated from the training in Latin and Greek, if we want that human wisdom and the memory of times past will remain among us, so far as possible, in their integrity'. It was for this reason, he explained, that he wanted to add the chair of mathematics to that of Arabic. [33]

In particular, he studied the works of ancient Greek mathematicians, such as Porphyrius and Apollonius. Voyaging in the Levant, he acquired an important collection of Eastern manuscripts. He was particularly fond of his discovery of an Arabic edition, by Thabit ibn-Korrah, of Apollonius' important work on conics, which was only partially known in Europe. Golius wanted to publish an edition of this work, but, to the dismay of mathematicians all over Europe, his death left the project unfinished. In fact, only a few of Golius' projects ever reached completion. His edition, with Latin translation, of Alfraganus' *Elementa astronomica* was published only posthumously. However, he acquired great fame with his *Lexicon Arabicum* (1654), which was to remain the standard dictionary throughout the eighteenth and into the nineteenth century.

To Golius, too, the study of mathematics and mathematical texts must have implied more than textual criticism. In the Levant, he had made several astronomical observations to determine geographical latitude, in order to check his sources. In Holland he made similar observations, in cooperation with Willem Blaeu. [34] There are also indications that he did some experiments on refraction. Occasionally, he even busied himself with a rather practical subject: he revised a manuscript by Simon Stevin on windmills. [35] Moreover, it was under Golius' professorship that the astronomical observatory of Leiden University was opened in 1633, making Leiden the first university in Europe to have an institutional observatory. Apparently it was used, too – at least initially. Some observations of lunar eclipses by Golius are recorded. [36] By 1646, however, a foreign visitor noted that the observatory had fallen into

---

32 Juynboll (1931) 75-76.
33 Juynboll (1931) 136.
34 Juynboll (1931) 141-142.
35 *PW*, V, 336-337.
36 Pingré (1901), 104, 106, 123, 154 (eclipses from 1635 (2x), 1638 and 1642).

disuse.[37] It seems that astronomy was not Golius' main occupation.[38] Writings on mathematics or astronomy by him are not known at all. Golius wanted to be remembered as a scholar and not as a mathematician, and his wish has been fulfilled.

The tendency to appoint scholars rather than practicians to the chair of mathematics was also followed at some other universities. The school at Harderwijk, the precursor of the later university, appointed Johannes Isacius Pontanus to the chair of mathematics and physics. Pontanus was a real scholar, who did not engage himself in practice at all; in fact, he hardly engaged himself in mathematics. He published abundantly, but only one of his works, a new version of a work by Robert Hues on globes, can be deemed mathematical. The rest are in such fields as history and classical philology. For his understanding of mathematics, it is significant that in the dedication of his edition of Hues, he mentioned two contemporary mathematicians who had surpassed the attempts of former ages, viz. Tycho Brahe and Joseph Scaliger.[39] Apparently, it was Pontanus' scholarship, combined with the fact that he had worked for three years with Tycho Brahe at Hven which qualified him for teaching mathematics.

---

[37] Varenius to Jungius, 20 March 1646: '*Instrumenta astronomica in specula a nemine usurpantur*'. Guhrauer (1850) 374; German translation in Avé-Lallemant (1863) 318. – After Golius' death, the conservator Kechelius was given permission to continue the astronomical observations at the Leiden observatory. It is not clear when he had started them.

[38] Cf. ibid.: '... *Domino Gohlio Arabiae linguae Professionem studio Mathematico magis excolente.*'

[39] Hues (1617), dedication; cf. also pp. 13-15, where he adds some quotes by Scaliger and Tycho on the precession of the equinoxes. On Pontanus, see Bouman (1844) 1, 40-46; Pontanus (1909).

2. Cosmography and classical studies

Leiden professors and the system of the world

Who should we look at if we want to know how Copernicus' theories were taken up at Leiden? The professors of mathematics and physics, as the most involved, might seem to be the most obvious choice. However, a brief review of their work will be rather disappointing.

As explained above, physics teachers had little prestige or influence at Leiden. One of the more important figures during the first period was Adrianus Trutius, who taught both logic and physics from 1582 to 1603. Nevertheless, he has remained rather obscure. He left no writings, except for a number of disputations, of which only a very few deal with physics. A disputation on the world from 1599 deals with traditional scholastic questions and shows no awareness of recent developments in astronomy or cosmography. [40] The same applies to a disputation on the heavens from 1603. This offers, for instance, a criticism of the Pythagorean notion of the harmony of the spheres. Trutius argues that such heavenly music is impossible, as above the sphere of fire there is no air to transmit the sound. This argument appears to be taken from an earlier disputation by Petrus Bertius. [41] The system of the world is touched upon only in Trutius' disputation on the earth. As could be expected, the earth is described as immobile and as being the centre of the world. Trutius briefly dismisses the unnamed opponents of this view: what naturally goes down, cannot move in a circle (*in orbem*). [42] Moreover, one of his theses on ethics contains three *Parerga physica*, one of which speaks of the celestial spheres. [43]

Pieter de Bert (Petrus Bertius) formally taught only ethics (from 1599 to 1607, and from 1615 to 1619, when he was dismissed), but he was also subregent of the State's College from 1593 onwards. In this latter function, too, he

[40] Trutius, disp. Leiden 24 July 1599.

[41] Trutius, disp. Leiden 22 Jan. 1603. Cf. Bertius, disp. Leiden 1600, th.phys. 6.

[42] Trutius, disp. Leiden 25 July 1598, th. 8-9.

[43] Trutius, disp. Leiden 13 Jan. 1599, parerga physica, 1.

appears to have done a lot of teaching. He held a lot of disputations and certainly dealt with physical subjects more often than Trutius did. He held disputations on meteors, the elements, and such like. A disputation from 1600 dealing with the whole of philosophy also contains a few theses on spherics – that is to say, elementary cosmography – wherein Bertius explains the triple motion of the heavens, the place of the earth, and eclipses. [44] Still, most of the time cosmography is dealt with in only a very casual way. A disputation on the elements states that the earth is immobile and at the centre of the universe: 'Because if it were to move, as some people feel it does, with the heavens standing still, an infinite number of absurdities would arise.' However, these absurdities are not specified. Bertius rather feels compelled to defend his own view: the next thesis goes on to explain how the rapid motion of the heavens is possible without them catching fire. [45]

The motion of the earth is further discussed (as in the case of Trutius) in a disputation on the earth, from 1604. This disputation contains a physical, a 'stereometric' and a geographical part. The earth's centrality is stated in the third physical thesis, its immobility in the fourth. Bertius here does mention one proponent of a moving earth, viz. the Italian philosopher Francesco Patrizi in his *Pancosmia*. [46] The earth's immobility is also argued in a disputation from 1605. A circular motion of the earth is held to be impossible, as in that case the clouds would appear to fly in the opposite direction, and objects thrown upward could never fall back into the same place. [47]

Gilbertus Jacchaeus lectured on various subjects from 1603 to 1619, when he, too, was dismissed. He published a textbook on physics, *Institutiones physicae*, which was reprinted several times. It consists of nine books. The fifth, on the heavens, deals not only with the traditional questions from Aristotle and on astral influence, but also with the order and motions of the celestial spheres, which would include cosmography. It is interesting to note what topics a philosopher like Jacchaeus deemed worthy of his attention. Jacchaeus explains that the discovery of precession by the Alexandrines made it necessary to introduce a ninth sphere, besides the eight spheres for the planets and the stars. During the reign of King Alphons, a tenth sphere was introduced to explain trepidation (an alleged irregularity in precession) – 'since a simple body is apt to move only with a single motion, while the firmament is moved with three motions.' [48]

---

[44] Bertius, disp. Leiden 1600, sphaerica.
[45] Bertius, disp. Leiden 1598, th. 17, 18.
[46] Bertius, disp. Leiden 1604, th.phys. 3, 4.
[47] Bertius, disp. Leiden 1605, quaestiones physicae, 2.
[48] Jacchaeus, (1624) 116. I have not seen the first edition.

Jacchaeus emphasised that the motion of the heavens was natural, not forced. Still, the heavens were not moved by an internal principle, but by external agents, which Jacchaeus identified as angels.[49] The apparent difficulty with this – viz. that a natural motion was brought about by an external agent – had been discussed by him in two earlier disputations, both from 1604. The motion was natural with respect not to the active but to the passive principle. The heavens really appear to have been a physical hobbyhorse of Jacchaeus. By 1607, he had a 'seventh disputation' on this subject defended. But his views remain traditionally scholastic. Nowhere does he mention, not even to dismiss them, the new ideas on the world's order.

The only philosophy professor who occasionally showed a real interest in cosmological questions was Johannes Murdison. Originally from Scotland (like Jacchaeus), he had been rector at Middelburg in Zealand before he came to Leiden. He stayed only a few years, from 1599 to 1605. Until 1603 he taught physics as well as logic, thereafter he moved to the chair of logic. During these few years, however, he paid more attention to the system of the world than most other professors, although in most cases his treatment is very cursory and not much different from that of his colleagues. A disputation of 1601, on 'the anatomy of the whole of nature', gives a full exposé of the system of the world according to the traditional world-view. He speaks of the threefold heaven, the circularity of celestial motion, celestial influence, and such like. Cosmography is absent, apart from a remark that he concedes no natural rest to the heavens.[50] That 'the earth does not move' is stated in a proposition to a disputation from 1600.[51] Clearly, Murdison does not transgress the boundaries of traditional learning. However, at some places he appears informed of recent developments and even shows some familiarity with the work of Copernicus himself. I shall return to that below.

These were the ordinary philosophy courses: a rehearsal of standard knowledge, taught to young students by people of little standing. They just sought answers to the traditional questions. Developments which were not covered by these simply remained out of sight. The question is not so much what these people taught about the system of the world, but whether they were saying anything about it at all. It is clear that all of them reject the motion of the earth. But how they regarded Copernicus' innovations in other respects cannot be made out. Most probably, most of them simply took no notice.

The chair of mathematics would seem to be of more relevance. Rudolf Snellius, the first professor in this field, was a serious scholar, well read in

[49] Jacchaeus (1624) 100, 111-112.
[50] Murdison, disp. Leiden 2 June 1601, th. 11.
[51] Murdison, disp. Leiden 18 Nov. 1600, coroll. phys. 4.

ancient and modern astronomical literature. However, he never took a public stance on the Copernican theories and it is simply impossible to ascertain his views on the subject. As for the professors at the *Duytsche mathematique*, they restricted themselves to such subjects as surveying and fortification and can be ignored for our present purposes.

Substantial reactions to the Copernican theories can be expected not from the professors in philosophy or mathematics, but from their colleagues in the trivial arts: rhetoric, history, etc. These people were not just concerned with rehearsing traditional and elementary matters, but strove to revive ancient wisdom. This in principle encompassed the whole of learning. Mathematics and astronomy, as shown above, were vital parts of that. Philosophy too fell under their domain. Lipsius earned much praise for his reconstruction of ancient Stoic philosophy. For our purpose, this is of little interest as philosophy did not bother much about a humble description of the world. It is especially the mathematical interest of the humanists which is important here. As mathematics and astronomy were among the prime concerns of the humanistically inspired scholars at Leiden, they clearly could not afford to ignore Copernicus. They had to take a stance on his theories, and on the system of the world in general.

The people at Leiden were not the first to encounter the problem. By 1575, a kind of general scholarly consensus regarding Copernicus' work had been established. Copernicus' mathematical theories, as laid down in books II-VI of *De revolutionibus*, were generally highly valued. His heliocentric cosmography, however, as laid down in the first book, was not regarded as essential to his work. This way of reading Copernicus' theories had been forcefully advocated by Philippus Melanchthon and other scholars at Wittenberg University. The historian Robert Westman, who has made an elaborate study of these astronomers, has therefore named this the 'Wittenberg interpretation' of Copernicus' theories, playfully alluding to the Copenhagen interpretation of quantum mechanics.[52] Like the Copenhagen interpretation, the 'Wittenberg interpretation' did not remain confined to the university of its origin, but gained widespread allegiance. It is therefore a useful term, which by now is quite current among historians.

It will be my point in the following that most scholars at Leiden, and in the Netherlands in general, were not adherents of the 'Wittenberg interpretation', but diverged from it on essential points. At first sight, the differences may not seem that great. In the Netherlands as elsewhere, people were hesitant about heliocentrism. They, too, picked up some elements of Copernicus' theories while ignoring others, and the elements they picked up here and elsewhere

---

[52] Westman (1975)a 166-167.

were partly the same. Leiden scholars, like those at Wittenberg, regarded the new astronomical theories from a background of classical learning and a search for *prisca sapientia*. All applauded Copernicus' aim of constructing all planetary theories by means of regular circular motions only. But in the end, the Dutch humanists made different choices, and for different reasons. Wittenberg astronomers in the end regarded Copernicus' work foremost as a theory on planetary positions (as seen by us from the earth). That is, they regarded it as pure astronomy, not as a cosmography which showed the order of the celestial spheres. What Wittenberg astronomers ignored was what Copernicus had called the '*symmetria*' of his system, that is, the way its parts were integrated into the whole.[53] Now, as we shall see shortly, this '*symmetria*' was exactly what made Copernicus' theories attractive to Dutch scholars.

One may well wonder why things went differently at Leiden. The fact that Leiden was late to enter the field may have been of influence. Westman has noted that by 1570 there was a new generation, and it showed more interest in the cosmological aspects of the new astronomy than the earlier Wittenberg scholars had.[54] In Germany, they adjusted to what had become the established interpretation. As Leiden was founded only in 1575, the impact of older traditions was probably rather limited there, which left individual scholars more scope to pursue their interests.

A more important factor seems to have been what was said above on the secular character of Leiden University. It appears certain that the 'Wittenberg interpretation' was at least partly inspired by the desire to neutralise any effects Copernicus' ideas might have on the traditional interpretation of the Bible. At Wittenberg, and at many other universities, theology was the main faculty. The arts served mainly to support theological learning and therefore faced strict limits. At Leiden, as we have seen, theological influence was limited. The core programme consisted of classical philology and humanist learning. Within this context, mathematics and astronomy were well-established, valued disciplines. They served as keys to the understanding of the world. In the following, I shall try to sketch how cosmographical thinking was shaped under humanist influence.

Ancient precedence

The frame of reference in which the Leiden humanists received Copernicus' theories was classical learning. The humanist programme entailed first of all

[53] Westman (1975)b 287.

[54] Westman (1975)b 288-289, 338-339.

the recovery of lost classical wisdom. This determined not only the methods used – philological text analysis, mainly – but also the scholars' view of the contents of the discipline. Knowledge was something to be recovered, not invented. This of course might seem a hindrance to the reception of novel ideas. To humanist scholars, however, the Copernican system was not an invention of the sixteenth century, but an age-old idea, which Copernicus had simply brought to light again. This made the idea acceptable in principle. A famous scholar like Gerard Vossis rejected Copernicanism on the grounds that he did not want to deviate from the opinion of classical authority and the ancient Church. But at the same time, the fact that many thinkers of great antiquity had defended heliocentricity, made him hesitant on the subject.[55]

Copernicus himself had pointed out the classical precedence for his theory: the earth's diurnal motion had been defended by the Pythagoreans Heraclides and Ecphantus, and by Nicetas of Syracuse. In his disputation *De mundo* from 1600, the Leiden professor Murdison mentioned these same names. However, he conflated the diurnal and the annual motions: he attributed to these Pythagoreans the view that the earth moved in a circle around a central fire (a view he rejected). As he stated, they had been followed by Copernicus in this.[56]

Paullus Merula, professor of history, appears somewhat better informed. He refers to the question of the motion of the earth in his *Cosmographia generalis*. This large work combined cosmography and geography – a common approach since Gemma Frisius. Moreover, it started with a large and learned overview of the universe (dealing with this subject following the Aristotelian tradition) and the Creation, thus more or less imitating Mercator's *Atlas*. In the context of his cosmography, he speaks about the earth being positioned in the centre of the universe and about its being at rest there, mentioning and refuting the opposite views. Although Merula rejects the motion of the earth, he regards Copernicus' work as valuable in other respects: he praises Copernicus' geometrical determination of the distances between the sun, the moon and the earth.[57]

As defenders of the view that the earth moves in the centre of the universe (that is to say, with a rotation on its axis), Merula offers the same names as Murdison: Heraclides, Ecphantes and Nicetas. Not to be outdone, he adds references to the classical sources and includes Cleanthes of Samos on the list. If he wanted to display his scholarship in this way, the ploy backfired. According to Plutarch, this Cleanthes (not from Samos) criticised the helio-

[55] See his letter to V.F. Plemp, 11 Aug. 1635, quoted by Rademaker (1981) 249. See also de Smet (1973) 18.

[56] Murdison, disp. Leiden 1600 *de mundo*, th. 18.

[57] Merula (1605) 210-216.

centric theory of Aristarchos, but a corrupt version of Plutarch's text turned him into a partisan of the earth's motion. This error is Merula's source.[58]

As defenders of a motion of the earth around the world's centre, Merula mentions Philolaus and Aristarchos of Samos; the former had already been mentioned by Copernicus. Only at the end of his enumeration does he mention the modern authors Caelio Calcagnini and Nicolaus Copernicus, and state that the latter's arguments were respected by Pontus de Thyard, Gemma Frisius and Johannes Stadius. Copernicus' most famous precursor – Aristarchos of Samos – had not been mentioned by Copernicus himself, at least not in his published work. Still, his name turns up nearly immediately in discussions on the system of the world. Aristarchos was considered the originator of the heliocentric theory by Melanchthon in 1549, by Robert Recorde in 1556 and by an ever-growing number of writers thereafter.[59] At Leiden, he – like Merula – was mentioned in some disputations. Murdison mentioned him in a disputation on the elements from 1600. Herein, he mentioned the theories of Aristarchos and his modern followers, primarily Copernicus, and stated that these people were 'wrapped in a cloud of errors'.[60]

Whereas most authors simply copied the list of ancient authorities as given by Copernicus or others, there are also cases of Leiden scholars contributing to this kind of research themselves. For example, the famous humanist scholar Gerard Vossius in his *magnum opus* on gentile theology. As an ancient proponent of the view that the sun is in the centre of the universe, he mentioned not only Aristarchos of Samos but also the Roman king, Numa Pompilius. According to Plutarch, the latter had a temple built which was modelled on the universe: it was round and had a fire burning in the centre.[61] Vossius regarded this as a clear reference to heliocentrism. This suggestion was well received. Not only can it be traced in a number of Dutch dissertations and astronomical textbooks, but it was also used by Isaac Newton in his theological work.[62]

Cosmic harmony: the Capellan system

The Leiden humanists were deeply aware of the significance of mathematics: it was regarded as a key to the understanding of reality. As Mulerius noted,

---

[58] Merula (1605) 214. Wall (1975) 205-206.

[59] Hartfelder (1889) 244. Wall (1975) 218-219.

[60] Murdison, disp. Leiden 5 July 1601, th. 23. Aristarchos is also mentioned in the doctoral disputation by Stembor, disp. Leiden 31 Dec. 1606, thes. astr. 2.

[61] Vossius (1641) II, 311.

[62] Schaffer (1987) 240.

without mathematics one only perceives the shadow of things, 'but mathematics teaches [*cognoscit*] the truth of things'.[63] After all, God had created the cosmos as a harmoniously ordered whole. It was only by mathematics that people could discover the inherent order of the cosmos and appreciate its harmony. At Leiden, nobody stated that Copernicus' theories were to be endorsed only in so far as they served the calculation of stellar positions. Actually, most humanist scholars did not take a great interest in such calculatory work. The astronomical data derived from Copernicus' work concern cosmological dimensions – the distances between the sun, the moon and the earth. People at Leiden rightly felt that Copernicus' significance lay in his new view of the universe, not in his calculatory devices.

Copernicus had argued the case of heliocentricity from 'the principle governing the order in which the planets follow one another, and by the harmony of the entire universe'.[64] He derived his main argument from the orbits of Venus and Mercury. Traditionally, the planets were arranged according to their periods of revolution: the longer a revolution takes, the larger the sphere. Thus Saturn was allotted the largest, outermost sphere, and the moon the innermost. This principle failed, however, in the case of Venus, Mercury and the sun, which all have the same period of revolution (one year). This gave rise to divergent opinions on the exact position of Venus and Mercury – i.e. whether they were above or below the sun – as Copernicus amply demonstrated. This passage was certainly known at Leiden. Johannes Murdison in a disputation *De mundo* from 1600 referred to the divergence of opinion on this point quoting a whole sentence, with due acknowledgement, from *De revolutionibus*.[65]

Now, Copernicus stated, 'either the earth is not the centre to which the order of the planets and spheres is referred, or there really is no principle of arrangement nor any apparent reason why the highest place belongs to Saturn rather than to any other planet.'[66] He concluded, therefore, that Venus and Mercury were moving not around the earth, but around the sun, as had been stated earlier by Martianus Capella, an encyclopaedist from the early Middle Ages. (One should add that from a mathematical point of view, this 'Capellan' system is equivalent to the traditional Ptolemaic system.) Copernicus then introduces his heliocentric system as an elaboration of the Capellan system. Although he gives some additional arguments, his discussion of the system

---

[63] Mulerius (1616), dedication.

[64] Book 1, Chapter 9. Translation by Rosen in Copernicus (1978) 18.

[65] Murdison, disp. Leiden 1600 *de mundo*, th.13. The quote is from *De revolutionibus* book 1, Chapter 10.

[66] *De revolutionibus*, book 1 Chapter 10. Rosen's translation in Copernicus (1978) 20.

as a whole is much shorter. He seems to feel that he has won his cause with the argument from Venus and Mercury.

So, Copernicus finds his main argument for his rearrangement of the cosmos in its presumed harmony; its '*symmetria*'. It was this element the Wittenberg scholars choose to ignore. However, the idea in itself was attractive to many readers. It appears, however, that most of the people who accepted the idea in principle, followed Copernicus only halfway. That is, they agreed that the evidence supporting a sun-centred orbit of Venus and Mercury, which Copernicus had so elaborately argued, made a very strong case. On the other hand, they felt that the heliocentricity of the other planets, or the motion of the earth, remained to be proven. The harmony of the cosmos could be maintained without introducing this element. The publication of *De revolutionibus* did not convince many people of the motion of the earth, but it did cause a revival of the system of Martianus Capella. Scholars one could mention this respect are Jean Pena and Gerard Mercator. At Leiden, too, it drew attention. As I shall argue in the following, it even became the standard view of the universe.

At Leiden, the Capellan system found support not just in Copernicus' arguments, but also from a fascinating antique source: a lavishly illustrated manuscript of Aratus' *Phaenomena*, in a Latin translation by Germanicus Caesar. This manuscript – the *Aratea* – is now one of the treasures of the library of Leiden University. It dates from Carolingian times (i.e. the early ninth century) and is thought to derive from a monastery in northern France. Probably it went astray during the wars of religion. It turned up at Ghent (Flanders), where in 1573 it was bought by Jacob Susius.[67]

Aratus was a Greek poet at the Macedonian court in the third century BC. His *Phaenomena* is mainly a poetic enumeration of the constellations. Most texts offer illustrations of the various constellations as well. The present manuscript, however, contains some illustrations which are unique for this copy. Most prominent is a schema of the system of the world. In this scheme, orbits around the sun have been added for the planets Venus and Mercury, in agreement with the opinion of Martianus Capella. Although these orbits clearly have been superadded to an earlier Aristotelian scheme, they have been integrated into the design as it is now.[68]

---

[67] Obbema (1989) 15, and the facsimile volume. An inscription by Susius on the flyleaf reads: '*Sum Jacobi D.F.P.N. Susii. E pictoris pergula emptus mihi Gandavi Anno a Christo corporato* MDLXXIII, *Mense Januario. Machilina bis capta carenti, και παροικῳ.*' D.F.P.N. should be read as: 'Danielis Filius Petri Nepos' (not 'Pater Nicolai', as Obbema assumes). The manuscript was subsequently owned by Susius, Grotius, Queen Christina of Sweden, and Isaac Vossius, who bequeathed it to Leiden University.

[68] See for a detailed treatment of the astronomical aspects, Eastwood (1983). His dating of the planetary configuration has been corrected by Mostert and Mostert (1990). For the manuscript in general: Obbema (1989); Verkerk (1980) 251. See also Byvanck (1949).

By 1590, the manuscript must have been known to Susius' humanist friends in Holland. In his 1590 dedication of a book of poetry to Janus Dousa, Susius mentions the manuscript and his intention to publish it soon: 'At last, I shall put my hand to the Manilius: which I received to give to you soon, with the Aratus by Germanicus Caesar emended, the one with very old pictures of the stars, if I shall, with your help, find someone qualified to engrave them.' [69]

In 1591, there appeared at Leiden the first part of a long poem *De rebus coelestibus* ('On the heavenly things'). It was written by Janus Dousa Jr. – the son of the curator-president of the university – a bright, very promising young scholar, for whom his father had great hopes. The work offers a poetic exposé of astronomical theory. Five books in all were foreseen. The first book explains the heavens from the point of view of natural philosophy. The other four, which were never published, were to explain the mathematical theory of the motions of the planets and the sphere of the fixed stars. Learned poems like these were not uncommon at the time. Dousa himself referred to the *Sphaera* by the Scottish neo-Latin poet Georg Buchanan, a well-known figure among the Leiden humanists. There are indications that Jan van Hout, secretary to the Leiden town council and a close friend of Dousa Sr., made a Dutch translation (which has been lost) of an earlier version of Buchanan's *Sphaera*. [70]

The part of the poem young Dousa did publish deals with the substance of the heavens and the stars, nature and motions of the heavenly spheres, and such like. He also touches on more cosmographical questions, such as the order of the spheres and the question why the sun is not eclipsed by Venus or Mercury. His ideas appear on the whole rather traditional. Several ancient authorities – Plato and Aristotle – as well as the Bible are quoted. Copernicus is also referred to, as an authority for the distance between the earth and the sun. [71] Near the end, however, there is a rather remarkable illustration: 'Indication of the orbs of Venus and Mercury according to the Egyptian and Pythagorean opinion'. It shows the two inner planets each in a heliocentric orb. Dousa stated quite clearly that he prefers this 'Egyptian' system to the traditional one. [72]

The sources on this system quoted by Dousa are all classical. He referred to a passage in Macrobius' *Somnium Scipionis* to substantiate his claim of an Egyp-

[69] Susius (1590) dedication.

[70] For the relations between Buchanan and Leiden, see Dorsten (1962) 41-44. The existence of Van Hout's translation is argued for by van der Valk (1906). On Buchanan's *Sphaera*, see I.D. McFarlane, *Buchanan* (London 1981) 355-378.

[71] Dousa (1591) 16. Dousa states here that the distance between the earth and the sun is 1111 ('MCXI') earth radii, *'ut ex Copernici observationibus constat'*. To what observations he is referring is not clear.

[72] Dousa (1591) 17.

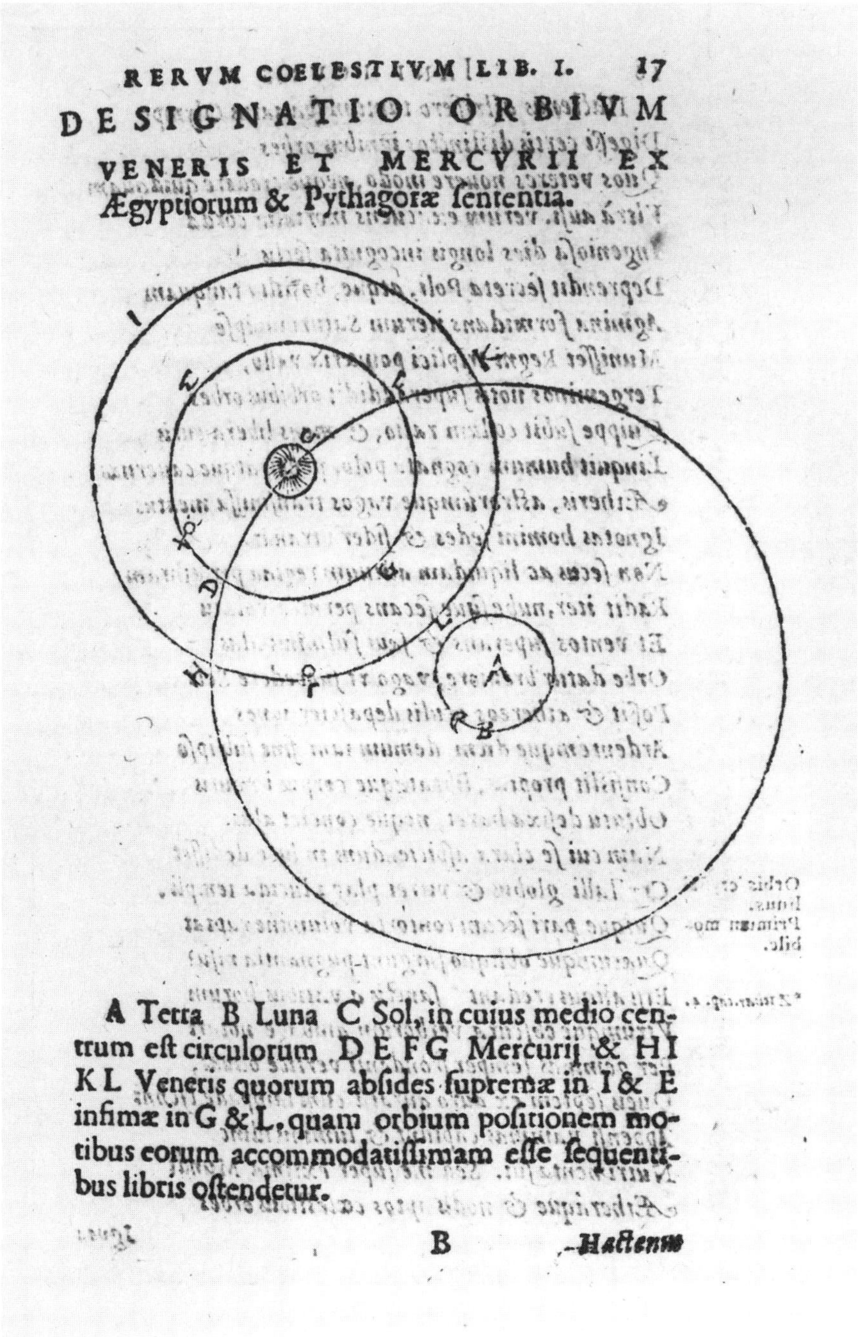

RERVM COELESTIVM LIB. I. 17

DESIGNATIO ORBIVM VENERIS ET MERCVRII EX Ægyptiorum & Pythagoræ ſententia.

A Terra B Luna C Sol, in cuius medio centrum eſt circulorum D E F G Mercurij & H I K L Veneris quorum abſides ſupremæ in I & E infimæ in G & L, quam orbium poſitionem motibus eorum accommodatiſſimam eſſe ſequentibus libris oſtendetur.

B Hactenus

Figure 1: The 'Egyptian system' according to Janus Dousa jr., *Rerum caelestium liber*, 1591. (Utrecht university library.)

tian origin. This claim is rather dubious. Macrobius does speak about some Egyptian ideas on the order of the planets, but one needs a rather tendentious exegesis to read this as though he meant a system like that of Martianus Capella.[73] Dousa appears to be the first to make this identification in print. The name 'Egyptian system' stuck and was commonly used in the seventeenth century. The designation 'Pythagorean', on the other hand, is hardly found elsewhere. The Pythagoreans soon came to be regarded as precursors of Copernicus himself, so that any association with the Capellan system became obsolete. For his attribution of the Capellan system to the Pythagoreans, Dousa referred to Martianus Capella himself, but it is not clear which passage he had in mind.

Although Dousa's poem may not have been widely read, his interpretation of the sources reappears in another work published at Leiden at about this time: Johannes Isacius Pontanus' edition of the works of Macrobius, from 1597. Pontanus was the son of a Dutch ambassador to Denmark. As a youth, he studied at Franeker and Leiden and went on extensive travels. He appears to have been a close friend of Janus Dousa Jr. The latter was among the first to write a poem in Pontanus' *album amicorum*, in 1592.[74] In his edition of Macrobius, Pontanus included an epitaph on Dousa; this epitaph referred to Dousa's poem on the heavens.[75]

In his edition of the *Somnium Scipionis*, Pontanus added a note explaining that Macrobius' understanding of the orbs of Venus and Mercury was equivalent to Martianus Capella's, as explained by Copernicus (whose text, from *De revolutionibus*, he quoted). He appears to be the first commentator on Macrobius to make this identification.[76] Pontantus used the occasion to express his dislike for Copernicus' system and his agreement with the construction Tycho ascribed to the heavens, 'fairly similar to this by the Egyptians, as our text considers it; as Capella considers it as an invention by Pythagoras.'[77] Pontanus was a well-connected scholar and may well have spread his, or Dousa's, ideas further. He stayed with Tycho at Hven for three years and, as he pointed out, he had been instructed in Tycho's system by Tycho himself.[78] (All this

[73] See the discussion in Eastwood (1982) 385-390. The idea that the Egyptians possessed a perfect astronomy is already found in Rheticus, see Hooykaas (1984) 149-150.

[74] Bodel Nijenhuis (1840) 87. The first entry was Lipsius', in 1591. Dousa Sr. and Scaliger inscribed in 1595.

[75] Pontanus (1597) 675.

[76] Eastwood (1982) 390.

[77] Pontanus (1597) 680: '*Quam deinde solertiam aemulus Clarissimus ille Danorum & aevi nostri Atlas, vir inlustris Tycho Brahe, non Soli Coelove, vt Torinensis ille torptentem quietem, sed naturae eorum, quae vt anima nostru in motu semper, conuenientissimum habitum tribuit, assimilem plane Aegyptiorum huic, vti noster sentit; vt Capella, Pythagorae inuento.*'

[78] Pontanus (1597) 680.

happened, by the way, only after the publication of Dousa's poem.) Shortly after 1597, the same interpretation of Macrobius is found, seemingly independently, in the work of another of Tycho's students, Johannes Kepler, who ventured it in his unpublished defence against Ursus. He does not appear to have used Pontanus' edition of Macrobius, so probably he had heard about it either from Tycho or from Pontanus himself.[79]

As stated, Susius was preparing an edition of his Aratus manuscript. He collated it with the Paris 1559 edition of Aratus' poem.[80] Young Dousa appears to have had a keen interest in the project. By 1592 Susius had moved to Liège, an independent prince-bishopric in the southern Netherlands. When Lipsius happened to pass through, Dousa Jr. used the opportunity to inquire after Susius. In particular, he asked whether there was any hope for the 'Germanicus's.[81] But Susius died at Liège at the end of the year with the project still unfinished.[82]

It is not quite clear how Susius' manuscript then got to Leiden, but that it arrived there is certain. It may well have been Dousa Jr. who took care of it in order to edit it in the end. However, young Dousa died in December 1596 at the age of 25, and the Aratea eventually passed to a new rising star in the intellectual firmament, Hugo de Groot (Grotius). Grotius was to acquire his fame mainly in the study of law, but his scholarship was thoroughly embedded in the humanist and philological learning of his age. He started his career by publishing some editions of ancient texts. In 1599, there appeared the *Satyricon* by Martianus Capella, after whom the Capellan system is named, and in 1600 the *Syntagma Arateorum*. Martianus Capella's *Satyricon* or 'Marriage of Philology and Mercury' was a textbook on the seven liberal arts from late antiquity. Grotius' edition was based on an earlier one (Basel 1577) by Bonaventura Vulcanus, who by 1599 was professor of Greek at Leiden. Grotius' contribution consisted mainly of elaborate annotations clarifying the meaning of the text, as regards both grammar and content.[83] It should be noted that it needed some textual interpretation to find the 'system of Martianus Capella' in the original text. Most earlier editions follow a reading which would require intersecting orbits for Mercury and Venus. Vulcanus' 1577 edition, Grotius' main source, was the first to indicate concentric heliocentric orbs for

---

79 On Kepler's interpretation: Eastwood, (1982) 383-385, 390-391. On 390, he argues against Kepler knowing Pontanus' edition.

80 Grafton (1993) 443.

81 Dousa Jr. to Lipsius, 20 Jan. 1592. Lipsius, *Epistolae*, v, 83.

82 Van Leeuwen (1685). Obbema (1989) says he died in 1596. The date 25 November 1592 is confirmed by Lipsius, *Epistolae*, v (nos. 1361, 1368.)

83 Nelson (1952) 38-39, see also 46-60.

them. Grotius notes clarify this further.[84] He compared Martianus with a lot of other classical authors on the mathematical arts. He had also used several other manuscripts of Martianus' book, one of them obtained from Scaliger. For a seventeen-year old student, it certainly was a very fine introduction to the world of scholarship. Scaliger himself contributed a laudatory poem to the book.

The *Syntagma Arateorum*, on the other hand, was certainly more ambitious than a student's paper. The bulk of the work consisted of an edition of Germanicus Caesar's translation of Aratus' *Phaenomena*, based on Susius' manuscript. Grotius also included editions of some minor, related works. All these works he provided with annotations, mainly of a philological nature. Among these were also some emendations originally devised by Susius.[85] Also, all the illustrations from the manuscript in the possession of Jacob Susius were reproduced in engravings by Jacob de Gheyn. Most prominent was a fold-out print of the system of the world according to the Capellan system. In his annotations to this print, Grotius remarked that Venus and Mercury had their spheres around the sun, 'according to the opinion of the ancient astronomers', and referred to his edition of Capella for further explanation. As earlier remarked by Eastwood, the reproduction is not faultless. Some of the captions in the schema have been exchanged, so that what belongs to Jupiter refers to Saturn, and vice versa; something similar occurs in the case of the moon and Mercury. This attests to 'more than a lack of care': the copy 'displays an apparent lack of comprehension of what the schema represents'.[86] One might add that at some places the orbs themselves are drawn entangled in a somewhat puzzling way. It must be De Gheyn who is responsible for this. Until we know on what footing he did his work and who commissioned it (it might have been the author or his father, or the publisher), we cannot say very much about it.

The *Syntagma Arateorum* was really an astronomical text, whereas in Martianus Capella's *Satyricon*, astronomy is only a part, along with geometry, arithmetic and music. Nevertheless, it seems probable that the publication of these two texts in such quick succession, formed in fact one large project, and that what really was at stake was the publishing of the Susius manuscript. The Capella edition may then be regarded as a kind of preparatory work, an exercise to acquire the astronomical skills required.

So it appears that the Leiden humanists thought Copernicus' arguments concerning the orbits of Venus and Mercury quite convincing, largely be-

---

[84] Eastwood (1982) 370-371.

[85] Van Dam (1996) 80-83. Grafton (1993) 443 note 26.

[86] Eastwood (1983) 9.

cause they shared his ideas about the harmony of the structure of the universe. Both Dousa and Grotius were very young (under 20) when they published these works. It is reasonable to assume that their choice of subject was guided by some higher authority. It is possible, however, to regard the age of Dousa Jr. and of Grotius as significant in some other way. As young men, not bound by fixed roles and seeking fame, they were attracted to new and topical subjects. Astronomy apparently was a field that aroused interest. Indeed the *Aratea* appears to have had a marked fascination for the Leiden philologists, for even after Grotius had edited it, another ambitious young student, Johan van de Wouwer (Janus Wowerius), proposed that he should edit it anew. Scaliger quickly steered his student towards more useful undertakings. [87] The fact that these young men, at this stage in their career, were drawn to astronomy and devoted attention to alternative world systems, at least indicates that at Leiden, people felt the impact of Copernicus on the traditional world-view.

Another young student who was impressed by the Capellan system was Gerardus Vossius, by then a student of theology at the State's college; later he was to become a famous historian. In 1598, he graduated as master of arts on a dissertation containing 30 theses under various headings, denoting the various fields of philosophy. [88] Vossius clearly was no mathematician, but some of the theses do attest to an interest in the details of the heavenly spheres, for instance how they cause phenomena like precession and trepidation. Three theses are classified as 'παράδοξα *astrologica*'. The third of these reads: 'There are no spheres of Venus and Mercury'. [89] This does not seem very clear at first. A commentary by Vossius is mainly concerned with a description of the motion of the planets and definitions of deferents and epicycles. It does not argue, or even mention, the thesis itself. It seems safe to assume, however, that Vossius had the Capellan system in mind, where indeed Venus and Mercury do not have spheres of their own; their epicycles are carried by the sphere of the sun. This interpretation is confirmed by Vossius' own testimony of some 40 years later. In his *De theologia gentili*, he mentions that he had defended the Capellan system for his doctoral degree. At this later time, he still did not have any doubt that Venus and Mercury move around the sun, 'as has been firmly shown by Copernicus and Tycho'. [90]

---

[87] Grafton (1988) 71; cf. Grafton (1993) 493.

[88] For a modern edition, with introduction etc., see Vossius (1955). This edition offers also a transcription of a manuscript by Vossius, written for the defence of his thesis. Vossius' dissertation is discussed by Rademaker (1981) 34-40.

[89] '*Nullae sunt Sphaerae Veneris nec Mercurij.*' Vossius, disp. Leiden 23 Febr. 1598. Vossius (1955) 44; see also Vossius' own commentary on this thesis, ibid., 97-98.

[90] Vossius (1641) 1, 469. Both van Straaten in Vossius (1955) and Rademaker (1981) 40, miss the point of the quoted *paradoxon*.

The Capellan system then became something of a standard view among Dutch scholars. It was upheld by many prominent astronomers, as we will see. There is even an indication that at Leiden the system of Martianus Capella was taught in the courses. The library of Utrecht University contains a copy of Cornelius Valerius' popular textbook on physics. Valerius had been a professor at Louvain. His *Physicae, seu de naturae philosophia institutio*, originally from 1566, is an attempt to formulate the Aristotelian world-view anew. Keeping his distance from Thomas Aquinas, he instead seems to have drawn his inspiration from the *Exercitationes exotericae* by Julius Caesar Scaliger (the father of the Leiden humanist).[91] Valerius' textbooks were rather popular and were used at Leiden, too. In cosmology, Valerius defended the geocentricity and immobility of the earth. The present copy of his book, however, from an edition printed in Utrecht in 1613, is interspersed with manuscript notes, both on the pages themselves and on inserted leaves of paper; the hand has been identified as that of the former Utrecht librarian Cornelis Booth, who lived from 1605 to 1678. He matriculated at Leiden University on 3 June 1622. In 1628, he went to Caen, France, to obtain his doctorate in medicine. Probably, Booth used the book during his introductory courses at Leiden. The notes, then, must derive from a physics course given at Leiden in around 1622-1623.[92] The regular professor of physics, Jacchaeus, was suspended from 1619 to 1623 as a result of the Arminian troubles. It is believed that Willebrord Snellius during this time took over the physics courses.[93] Booth's notes therefore probably derive from him.

Copernicus is mentioned in the manuscript notes, but only with respect to the dimension of the universe. Another note deals with the problem of whether the daily rotation is to be ascribed to the heavens or the earth. The latter opinion is ascribed to Anaxagoras (though the annotator probably meant 'Aristarchos'); Copernicus is not mentioned here. Some arguments pro and con are given, but no conclusion is drawn. On the system of the world, however, the annotator is explicit: 'Venus and Mercury are the sun's *σύνδρομοι* [companions], because the centre of the epicycle of Venus and Mercury is the sun itself...'. (The use of the Greek word *sundromoi* indeed supports Snellius as a source.) In order to prevent all misunderstanding, this

---

[91] Sassen (1941) 11-12.

[92] Booth's interest in the subject was apparently older. A copy of Grotius' edition of Martianus Capella, now at Leiden University, bears the name of Booth ('Cornelius Ever: Bootius, Ultraject:') on both flyleaf and title page, as well as an inscription on the flyleaf: '*empt. 17. st. Hag: Comit.* MDCXIX' (*'bought at The Hague for 17 pennies, 1619'). In 1619, Booth was about fourteen years old.*

[93] De Pater (1975) 309.

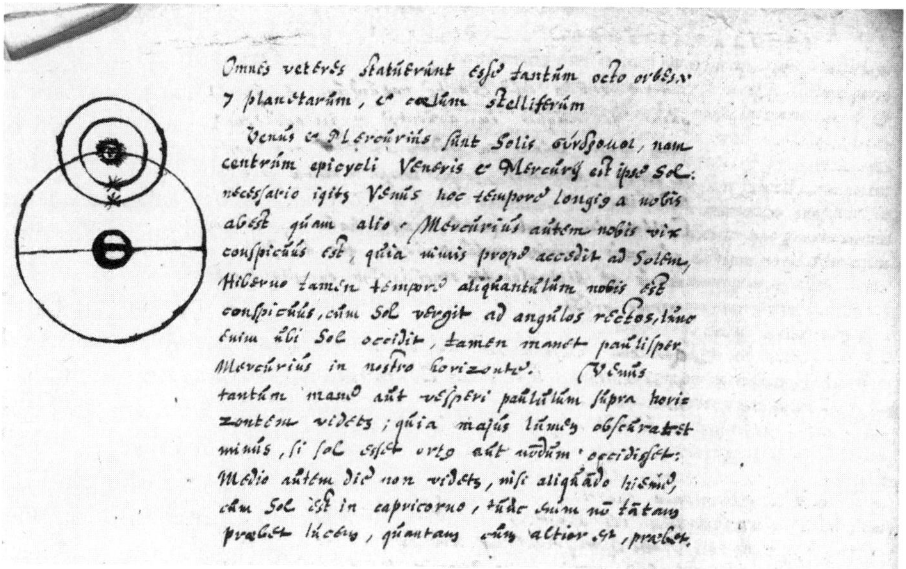
Omnes veteres statuerunt esse tantum octo orbes: 7 planetarum, et caelum stelliferum
Venus et Mercurius sunt Solis [illegible], nam centrum epicycli Veneris et Mercurii est ipse Sol: necessario igitur Venus hoc tempore longius a nobis abest quam alio. Mercurius autem nobis vix conspicuus est, quia nimis prope accedit ad Solem, Hiberno tamen tempore aliquantulum nobis est conspicuus, cum Sol vergit ad angulos rectos, tunc enim ubi Sol occidit, tamen manet paulisper Mercurius in nostro horizonte. Venus tantum mane aut vesperi paululum supra horizontem videtur; quia magis lumen obscuraret minus, si sol esset ortus aut nondum occidisset: Medio autem die non videtur, nisi aliquando hieme, cum Sol est in capricorno, tunc enim non tantum praebet lucem, quantum cum altior est, praebet.

Figure 2: The orbits of the sun, Venus and Mercury as drawn by Cornelis Booth in his copy of the physics textbook by Cornelius Valerius. (Utrecht university library.)

statement is accompanied by a drawing representing the sun, Mercury and Venus in their respective orbits. [94]

To the humanist scholars at Leiden, the relevance of Copernicus' work lay in its drawing attention to and forceful propagation of the principle of cosmic harmony. This is not to say that people were ready to endorse, on Copernicus' authority, the motion of the earth. But the Capellan system appears to have been well received by practically all Dutch scholars who took cosmography to heart. Copernicus' mathematical astronomy, which he developed in order to calculate planetary positions, was not even mentioned by most authors. So far, however, we have mainly dealt with people who were studying astronomy as part of their general education. The search for classical precedence and the belief in cosmic harmony were shared by all, or at least most, scholars at the time. But besides scholars who were interested in astronomy in a more general way, there were those who became active astronomers. One wonders what their stance might have been. Two of these people, who had been educated at Leiden, will be studied in the following chapter.

[94] Utrecht University Library, Hs. 8+ C. 13, Chs. 8 and 13; leaf between 20 and 21, recto; between 14 and 15, verso; between 16 and 17, verso.

3. Humanist mathematicians and Copernicus

Willebrord Snellius

Above, we focused on Willebrord Snellius' ambitions as a humanist scholar and the way he combined mathematical and philological expertise. We will now take a closer look at his ideas about the system of the world. Snellius spoke out most openly on the subject at the very beginning of his career. In his doctoral dissertation (1608) he stated that 'it seems more probable that the earth moves in a one year's course, than that it rests.'[95] This shows sympathy for the Copernican system, but not real conviction. These doubts seem to have persisted. Although the question clearly occupied him, Snellius never again expressed his opinions about the Copernican system, nor did he ever explain what he saw as its merits and demerits.

What he clearly did take for granted was the heliocentric orbit of Venus and Mercury, as propounded by the Capellan system. As we saw above regarding the manuscript notes by Booth, he probably taught this system at Leiden. He openly pondered the question of the planet's orbits in his book on the comet of 1618. Dealing with the question why a comet's tail is always directed away from the sun, Snellius reflects upon the forces hidden in the sun, 'which the oldest physicians called the heart of the heavens'.[96] The sun directs the course of the planets. 'Venus and Mercury circumvent it [the sun] closest as cognates [*laterones*]. And Tycho may have seen right that the other, superior, planets do the same, but with larger circumvolutions, which comprehend also the earth.'[97] That is to say, Snellius takes it as beyond dispute that Mercury and Venus are orbiting the sun, and now suggests that the

[95] Snellius, disp. Leiden 12 July 1608. The first of the theses on astronomy: '*Probabilius videtur terram moveri in orbe annuo, quam quiescere*.' The next thesis denies the real existence of the planetary orbs 'according to the ancient and Ptolemaean hypotheses', because from these hypotheses follows a 'penetration of orbs' (*orbium penetratio*). This is not very clear. The impossibility of a penetration of orbs was a standard argument against the Tychonic system.
[96] W. Snellius (1619) 48.
[97] Ibid. 49.

superior planets – Mars, Jupiter and Saturn – may do the same. Keeping the earth central (and stationary) would result in the Tychonic system. [98]

It deserves notice that Snellius regards heliocentric orbits much less probable for the superior than for the inferior planets. Perhaps this was because Tycho's ideas seemed incompatible with the idea of the universe as consisting of solid spheres. But it may well originate from the common ideas on cosmic harmony. The Capellan system primarily aimed at saving a harmonious order of the celestial spheres. The Tychonic system simplified the planetary motions, but could hardly be said to result in a harmonious whole. Only if the earth too were circling the sun, as in the system of Copernicus, would the heliocentric motion of the superior planets contribute to a harmonious order.

Here, Snellius, Willebrord does not speak about the earth. He had done so, however, a year earlier, in *Eratosthenes Batavus*, where he shortly argued that 'the earth is the midst of the whole world, and as it were the centre'. [99] His arguments are the familiar ones from Ptolemy: the earth is heavy and its centre is the centre of gravity. It should be added that he does not emphasise the point. After having argued that the size of the earth is negligible compared to the heavens, he adds: 'If someone has another opinion on its [the earth's] place, that does not affect my argument. Such people have to admit the same about the size of the annual orbit in which they make the earth rotate, as we assert here on just the size of the terrestrial globe [to wit, that its size is negligible]. However, we prefer to follow this argument [that the earth is immobile], as it is simpler and less complicated to prove the thing we contend.' [100] As a geocentrist, Snellius does not sound very convinced either.

That the matter really engaged Snellius can be seen from a treatise he published in 1612 on the occasion of Scheiner's discovery of sunspots. [101] The work, in the form of a letter to the Leiden curator Cornelius van der Mijle, was published anonymously and has up till now not been recognised as a work by Snellius. However, the work is clearly identical to a tract he appears to have sent to his friend Aemilius Roosendael in 1612. In the accompanying letter, Snellius indicates its contents and that it was written at the request of

---

[98] Snellius' tract on the comet of 1618 is discussed by Nouhuys (1998) 337-360. See 351-352 for the passage quoted in the text. [Snellius' argument that approaching the sun these planets are in apogee, getting away in perigee, is difficult to understand. Perhaps speaking of apogee he hints at their absolute distance from the earth rather than at the position of the centre of their epicycle on the deferent].

[99] Snellius (1617) 11, the title of Part 1, Chapter 11, part 11: *Terram totius mundi esse mediam, & tanquam centrum.*

[100] Ibid., 13.

[101] [Snellius] (1612).

Van der Mijle.[102] The work is mainly a commentary on Scheiner's writings, wherein he disclosed his discovery, and contains no observations of his own. Among other things, Snellius notes that the motion of the earth is still an unsolved problem and suggests that sunspots might offer a means by which one could decide the issue. One should find a sunspot which is not moving, or hardly moving, from our point of view. When the earth moves around the sun, the position of a spot on the sun's surface will accordingly shift from one solstice to the other and back: between summer and winter from north to south, and between winter and summer from south to north, in a way which is symmetric over the two half years. If a sunspot remains visible at the same place on the sun's surface for a whole year, the sun clearly always shows us its same face and we can safely deduce that the earth must be at rest. The argument is rather similar to one later used by Galileo in his *Dialogo*. It may well be that Galileo got the idea from it. Van der Mijle, as Snellius wrote to Roosendael, wanted to send the letter to Welser at Augsburg, so it could have reached Galileo as well.[103]

Nicolaus Mulerius

Nicolaus Mulerius was another Leiden-educated astronomer who appears to have been vexed by the question of the system of the world. But contrary to Snellius, Mulerius extensively put his considerations on paper. He therefore deserves an elaborate treatment.

Nicolas des Muliers, or Nicolaus Mulerius as he Latinised his name, was born in 1564 in Bruges, Flanders. His family moved north because of the religious persecution in the Spanish Netherlands (his mother had been buried alive by the Inquisition). Mulerius studied at Leiden from 1582 to 1589, took a doctorate in medicine and practised for many years as a physician. In 1608, he decided to become rector of a Latin school (without abandoning his medical practice, it seems). When finally he was appointed professor of medicine and mathematics at the newly founded Groningen University in 1614, he was nearly 50. Originally, he was asked to teach medicine and Greek, mathematics having been attached to the chair of history. It was probably at his own request that he could teach mathematics, while Greek went to his colleague for history.[104] Mulerius' teaching of medicine suffered from his many preoccupations. His teaching of mathematics was taken much more seriously.[105]

---

[102] Snellius to Aemilius Rosendalius, 18 Feb. 1612, Utrecht, University Library, Hs. VII A 26.

[103] Snellius (1612) 16-18. Cf. Topper (1999).

[104] *Effigies* (1654) 2, 4, 67.

[105] Huisman (1990) 289.

Indeed, it was mathematics which engaged Mulerius from an early age. In 1594 he tried to obtain a chair at Leiden, not in medicine but in mathematics. In the same year, he approached Scaliger on a rather unimaginative subject, the squaring of the circle.[106] Part of his activity was of a mainly practical nature. In 1595, he published a manual for the use of the astrolabe. He also regularly wrote an almanac for the new year. The oldest known is from 1604, and it is reported that he continued writing almanacs until 1626. They were published in Groningen. These booklets contained a variety of information, some of it purely practical (the times of departure of the market boat to Amsterdam on various dates), some of an astronomical nature, such as the times of old and new moon, the tides and some indications on the position of the planets. Dates were given in both old and new style, and some clues were given to further chronological issues, such as the dates according to the Jewish calendar. They also contained an astrological prognostication.[107] Almanacs like these were printed in large quantities. In a letter from 1619, Mulerius says he expects that another 10,000 copies of the almanac now in press will be printed[108].

However, Mulerius soon concentrated on astronomy and, closely linked with it, chronology. Especially in the latter field, Mulerius' mathematics incorporated more scholarly methods. It appears that he did not think his work inferior to the scholarly work of the humanists. In 1608, he corresponded with Scaliger on the latter's chronological work.[109] Although he greatly admired Scaliger's work, he criticised it on points of detail. Mulerius thought that the Jewish calendar was based on the observations of Hipparchos and Ptolemaios only, and that Scaliger had been wrong to take other authors into account.[110] During the course of his life, Mulerius wrote several essays on chronological subjects. Some of these he published along with his astronomical tables. A diatribe on the Arabic calendar was contributed to the *Opus chronologicum* (1619) by his friend Ubbo Emmius. In 1630 he published an account of the Jewish, Arabic and Turkish calendars.

Mulerius made various astronomical observations. In 1604, he observed a new star in the heavens, known as Kepler's nova.[111] By March 1613, he had obtained a telescope and observed sunspots. A letter has survived in which Mulerius speculates about their nature.[112] (It appears that Metius had drawn

---

[106] Huisman (1990) 287.
[107] Description after Mulerius (1608). On Dutch almanacs, see Salman (1999).
[108] See Huisman (1990) 284-285 on Mulerius' almanacs.
[109] A letter by Scaliger to Mulerius from 19 March 1608 (st.v.) printed in Mulerius (1630) 82-83.
[110] Mulerius (1630) preface.
[111] Mulerius to Onias Geldorpius, 12 Dec. 1604. Leeuwarden, Provincial library, coll. Gabbema.
[112] Mulerius to Wicheringe, 13 March 1613. Leeuwarden, Provincial library, coll. Gabbema. Cf. Huisman (1990) 286-287. Mulerius also refers to these sunspots in Mulerius (1618) Chapter III.

his attention to the phenomenon.) Moreover, Mulerius published several astronomical works. As an almanac maker, he had to calculate the old and new moons from astronomical tables. Perhaps this was how he discovered that some of the current tables by Reinhold were wanting. He thereupon decided to publish a new set of astronomical tables.[113] In 1608 he obtained letters patent for a set of '*tabulae eclipsium triplices*', according to the hypotheses of Ptolemy, Copernicus and Tycho Brahe. The work appeared some years later, in 1611: *Tabulae Frisicae lunae-solares quadruplices*. By then, a fourth source had been added, the Medieval Alfonsine tables (which drew upon Ptolemy's theories). Mulerius' tables deal only with the sun and the moon. Mulerius mainly wanted to make his sources more accessible, but his work was not just one of compilation. All data had been recalculated according to the 'Julian period' introduced by Scaliger. The tables were composed such that the motions of the sun and the moon were related. This made calculating eclipses much easier. A second part, written in 1612, concerning the other planets, never appeared and is only preserved in manuscript.

Having become a professor at Groningen, Mulerius quickly wrote an elementary textbook on astronomy, mainly spherics.[114] The book seems to have had some success and was republished after Mulerius' death in 1649. He also engaged in editing ancient works. He is said to have prepared a Latin translation, elucidated with notes, of Ptolemy's *Tetrabiblos*, the authoritative work on judiciary astrology.[115] However, this never appeared and the manuscript has been lost. What did appear, however, was his new edition of Copernicus' *De revolutionibus*, with many notes and explanations. These notes are mainly of a didactic nature: they make the book more accessible to students. This edition was to remain the standard edition for Copernicus' work for a long time to come and is now probably Mulerius' best-known work. Mulerius seems to have considered it to be more than just a didactic exercise. He gave the book the title *Astronomia instaurata* ('Renewed astronomy'), clearly hinting at the humanistic enterprise of recovering ancient wisdom. Actually, it echoes Scaliger's words in a letter to Mulerius: we owe much to Copernicus 'because he has restored astronomy and torn it away from the hands of the Arabs, in order to return it to Ptolemy, Hipparchos and the ancient Greek authors, its fathers'.[116]

Clearly, Mulerius regarded Copernicus' work as deserving serious study.

---

[113] Mulerius wrote about the reasons to make these tables in Mulerius (1630) 81. See also Mulerius (1611), dedication.

[114] Mulerius (1616).

[115] *Effigies* (1654) 69.

[116] Scaliger to Mulerius, 19 March 1608 st. Jul., in: Mulerius (1630) 82-83.

However, it is striking that he, as a practising astronomer, makes little use of Copernicus' theories when computing stellar positions. As noted, Mulerius' Friesian tables were mainly a compilation of the results of Ptolemy, Copernicus, Tycho Brahe and King Alphonsus. This, though, concerns only the published part, with tables on the motion of the sun and the moon. The unpublished second part of his book, however, with the tables of the planets, is based on the work of Ptolemy and the Alphonsine tables only, as its title indicates: 'Second part of the Friesian tables. Containing the calculus of the various planets, from the sources of Ptolemy and King Alphonsus'. Nevertheless, the volume does contain a table of the moon according to Copernicus, as well as a table of the 'true place' of the several planets at the beginning of 1614 according to Ptolemy, Alphonsus and Copernicus. As a whole, however, the Copernican theories are conspicuously absent. [117]

As a matter of fact, Mulerius felt that astronomy was not about theories which were mere calculational devices. 'In astronomical calculations, two kinds of motion are considered. The first is real motion as it is thought to exist in the heavens themselves, to wit circular and equal, not subject to any intensification or slackening. The other is apparent motion, as it appears to us who are looking on it from afar, according to the judgement of the eyes. This second kind is called anomalous or unequal...' [118]

Now, it is doubtful whether Mulerius meant that astronomical hypotheses could adequately describe reality. He agrees with Ptolemy that it is not appropriate to compare human things (framed models) with divine things (the heavens themselves). [119] But the model had to answer to the criteria which were thought to apply to real motion. It did not suffice that astronomical hypotheses just 'saved the phenomena', they had to do so by motions which were 'circular and equal'. Although their circularity had to be assumed, it could be proved, according to Mulerius, that the motions were equal: 'As they have an internal and eternal principle of their motion, they suffer no delay, and it cannot be that their motions are not equal. So the apparent inequality should, by means of true hypothesis of circles, be reduced to equability.' [120] On various occasions, Mulerius spoke of his conviction 'that all circular motion is equal with regard to its centre.' [121]

This emphasising of the regularity of the celestial circular motions is clearly

[117] Groningen University Library, ms. 106.
[118] In: Copernicus (1617) 164.
[119] Mulerius in Copernicus (1617), a note to Osiander's preface. The reference is to book 13, par. 2 of the *Almagest*, see Ptolemy (1984) 600.
[120] In: Copernicus (1617) 7.
[121] In: Copernicus (1617) 342. See also ibid. 164 and Mulerius' note to Lib. I cap. VII.

inspired by the work of Copernicus. Copernicus' theories were constructed from uniform circular motion only, whereas Ptolemy had had resort to the introduction of the equant, which made the motion in the model irregular. Mulerius' admiration for Copernicus as the restorer of astronomy seems largely based on the latter's abolition of the equant. Indeed, this appears to have been a more general view. Several astronomers referred to the caption of the fourth chapter of the first book of Copernicus' *De revolutionibus* – 'The motion of the heavenly bodies is uniform, eternal and circular or compounded of circular motions' – as an 'axioma'. [122] Mulerius fully agreed with this, not because such motions were useful in dealing with calculations, but because they were true. In this respect at least, Copernicus could be said to be nearer to the truth than Ptolemy.

Mulerius pondered the truth not only of astronomical theories, but also of the cosmographical model of the world. Like other Leiden scholars, he accepted the heliocentric orbits of Venus and Mercury, as in the system of Martianus Capella. For this, he referred to Copernicus, but also to telescopic observations (perhaps he meant the phases of Venus). Mulerius himself seems to regard this view as rather self-evident, as he refers to it only casually. [123]

As stated above, the system of Martianus Capella was attractive because it maintained the order of the heavenly spheres: the larger spheres would move more slowly. Now, Mulerius was prepared to apply the same principle in order to argue for the daily rotation of the earth on its axis. As he argued, if the spheres move faster the nearer they are to the centre, it seems reasonable that the earth, in the very centre, moves fastest of all, while the starry sky, at the periphery, remains immobile. 'In this way, the earth as well as the sun, the moon, the other planets and the starry sky, would each be moved by a single motion. These motions would proceed in a uniform way, from west to east, and be congruent with the size of their own circles.' [124]

The crucial part of Copernicanism is the annual motion – or rest – of the earth. This question really worried Mulerius. In the preface to his *Institutio astronomica*, he stated that he had pondered the various arguments for 25 years, which would include nearly the whole of his mathematical career. In the end, Mulerius was not prepared to endorse the heliocentric theory, mainly, it seems, for religious reasons. His first reference to the issue seems to be in his *Tabula Frisica*, where, however, he does not really discuss the matter. He

---

[122] Rosen's translation in Copernicus (1978) 10. Gingerich (1993), 176-177, 179.

[123] Copernicus (1617) 322: '*Caeterum Veneris & Mercurij situs ad Solem is omnino esse videtur quem Copernicus illis tribuit lib. I. cap. 10. nam id oculorum testimonio confirmari potest, quando stellae istae usu perspicilli oblongi visuntur infra Solem.*'

[124] Mulerius (1616) preface.

simply states that he stands by the concept of a central and stable earth, although he knows that some people prefer the old Pythagorean notion of a moving earth; but 'such is the authority of Holy Scriptures with us, and so moved are we by reverence for it, that against its clear statement, we do not dare to descend to the Pythagorean opinion.' He kept to the vestiges of the best philosophers and of Ptolemy, 'which are in accordance with the sacred words.'[125] In his *Institutio*, he considered the question in more detail. In the end, however, he did not see 'a good and sufficiently valid cause which could force me to diverge from an opinion approved by so many great men.' He admits that although the ancient astronomical hypotheses contain absurdities, Copernicus' system entails other, more serious absurdities. After all, the absurdities of the ancient theories concern only their mathematical hypotheses, which are human figments anyhow.[126]

It seems clear that Mulerius' misgivings about Copernicus' system derive mainly from his religious conviction. Mulerius, not surprising considering his family history, was deeply committed to Calvinism. Along with Ubbo Emmius – his friend and colleague at Groningen – he belonged to the strict wing within the Dutch Reformed Church and abhorred anything that smacked of Arminianism. According to a seventeenth-century biography, he took part, on the orthodox side, in the ecclesiastical controversies surrounding the Leiden professor Vorstius and the Alkmaar minister Venator, who were accused of Socinianism.[127] Both these affairs were quite notorious, the Vorstius case even provoking an intervention by King James of England. No writings by Mulerius on these affairs survive, however, and one should not make too much of a biographical note which, although on the whole fairly reliable, clearly tries to underline his religious orthodoxy. However, there survives a comment by Mulerius on Socinian tendencies within the Mennonite community, apparently written as a kind of warning to the Reformed, which confirms that these tendencies indeed raised Mulerius' concern.[128]

Mulerius' problem with Copernicanism is that it does not conform to the Bible. He was willing to accept the daily rotation of the earth only because this did not seem incompatible with Scripture: 'Indeed, our earth (which does not lack admiration) would be called rightly immobile, as it does not change its place, and because its motion cannot be perceived by sense. (...) So, this

---

[125] Mulerius (1611) 318.

[126] Mulerius (1616) preface.

[127] *Effigies* (1654) 68.

[128] *Notabel verhael van een hoochgeleert ende vermaert persoon* [Mulerius], *seer dienstich tot waerschouwinge van alle Christenen, die enige affectie draegen tot de Christelijcke religie*. ARA, Archief Ned. Herv. Kerk, oudsynodaal archief 1566-1816, nr. 5, 321-326.

hypothesis does not seem to contain anything repugnant to the authority of the Holy Scriptures.' [129] However, Mulerius does not specify his objections to the annual motion. He does not use any theological arguments, nor does he quote any biblical passages. As noted, he speaks of the 'absurdities' the Copernican system entails, but these do not seem to be of a theological nature. The main absurdity Mulerius noted was the immense distance to the fixed stars that the Copernican system demanded. He could accept that the magnitude of the earth is nothing compared to the heavens; after all, antiquity itself recognised this fact. But if even the distance between the sun and the earth were nothing compared to the distance to the fixed stars, the sun, equally, would be no larger than a star of the first magnitude. In a universe of such sizes, but yet not infinite, one would expect to find more suns, each illuminating a part of the world. Now, 'that this deserves to be called absurd and against Christian piety, shall be seen by those who take the defence of Copernicus' opinion to heart.' [130] Moreover, such consequences run counter to Copernicus' own words concerning the central role of the sun, which illuminates the whole world. [131]

This argument probably was inspired by the impact of Galileo's telescopic discoveries, which elicited a lot of speculation about a plurality of worlds in the universe. On the other hand, one did not need Galileo to perceive the further cosmological consequences of Copernicanism. As early as 1608, Mulerius' friend Ubbo Emmius had been deeply shocked by Stevin's suggestion that the earth was a planet like any other. In a private letter to his friend Lubbertus, he commented: 'Good God, how far will this impiety bring us? What better could one do to bring down the whole of our doctrine and religion? For these things demolish the very foundations. Were they true, as our author, I learn, argues vigorously, then Moses would be a liar and the whole of Scripture untrue...' [132] One should add that Emmius had never seen Stevin's work. His information was from hearsay, as he himself admitted, and his source may well have been Mulerius.

This suggests that Mulerius' dismissive stance on heliocentrism was not so much inspired by its incompatibility with specific biblical passages, as by its incompatibility with the 'Christian cosmos'. Apparently, Christian and Aristotelian elements had become inextricably linked in his views (and in those of others) on the order in the universe, the Creation, the role of the heavens, and

---

129 Mulerius (1616) preface.

130 Mulerius (1616) preface.

131 Copernicus (1617) 64.

132 Emmius to Lubbertus, 29 Sept. 1608. Emmius, *Briefwechsel*, II, 51-52. See also ibid. 54, 55, 60-62. Cf. Dijksterhuis (1943) 331-332.

so on. This made it difficult to eliminate traditional physical and cosmographical notions. That it not to say that Mulerius was completely unwilling to accept changes in the physical view of the heavens. So, he regarded comets as eternal celestial bodies.[133] But a plurality of worlds clearly was more than he could swallow. It is not clear whether he thought the annual motion of the earth contrary to specific biblical passages, but it was clearly incompatible with what he regarded as 'Mosaic philosophy'.

However, he did not reject it lightly. Apparently, he considered the arguments which supported heliocentrism strong. As he accepted the Capellan system as well as the daily rotation of the earth because of their conformity with cosmic harmony, one might surmise that it was this very '*symmetria*' of the Copernican system which made him waver on the question of heliocentrism, too. Moreover, Mulerius did not regard the issue as a theological one. Despite his vigilance in dogmatic orthodoxy, he did not go so far as to accuse the followers of Copernicus of heresy. 'This dispute is learned and sharp,' he said, 'but without hatred.'[134]

It is evident that humanist-educated scholars were very interested in cosmographical questions. The tool they found most apt for penetrating the secrets of the cosmos was mathematics, largely because they regarded the cosmos itself as mathematically ordered. Hence, they were very interested in Copernicus' work – not because of his mathematical devices, I keep stressing, but because of his view of cosmographical reality. They adopted several of his innovations. The annual motion of the earth, however, proved too great a leap. None of the authors dealt with so far openly endorsed the heliocentric motion of the earth. Some people, like Snellius, appear to have had serious doubts, but none of them made a public stance in favour of heliocentrism.

The Leiden humanists, it would seem, were too much the defenders of tradition to allow themselves such a radical break with the past. However, in the hands of people with a different education, who had less respect for classical learning, their arguments and considerations could lead to different conclusions. In the following two chapters, I shall argue that this is exactly what happened. Some people, taking the arguments of the humanists out of their original context, became open defenders of the motion of the earth.

---

[133] Nouhuys (1998) 360-368, 216-218 and 537-542. Jorink (1996).
[134] Mulerius (1616), preface: '*Erudita & acris sed sine odio contentio haec est.*'

4. Simon Stevin and the tradition of practical mathematics

Mathematical practitioners and the system of the world

Apart from state patronage offered by the universities, all arts and sciences in the Dutch Republic had to adapt to the demands of the local market. In order to survive, they had to be useful for Dutch magistrates, merchants or burghers. The preoccupations of such people were to a large degree practical. The science most likely to gain support was practical mathematics. In particular, surveying – which was much in demand because of the many reclamation projects being carried out at the time – was a respected occupation. Professional surveyors had to be certified by the provincial authorities. They had to pass an exam and take an oath before they could start their job. Some of these persons, such as Gerard Dou, became well-known mathematicians.

After the outbreak of the revolt, mathematics enjoyed considerably more prestige. Dutch towns were much in need of new fortifications and the continuous warfare encouraged them to keep up with new developments in military engineering. This created opportunities for many mathematicians.[135] But it was the development of overseas trade which made practical mathematics a matter of public interest. Quite a number of mathematicians came to earn a living as a schoolmaster, teaching navigation techniques. As the study of navigation implied a lot of astronomical knowledge, this subject is of special importance to our purpose [136].

The Dutch knew how much they owed to their shipping and took great pride in it. Hence, the science of navigation became a fashion as well as a need. There was an enormous increase in the production of maps and atlases. Indeed, seventeenth-century cartography was completely dominated by Dutch – more in particular, Amsterdam – printing houses. Partly, of course, the increased demand for maps had to do with their use for sailing and shipping.

---

[135] Westra (1992).

[136] Davids (1985) is an outstanding study on the development of navigation techniques in Holland. On teaching see 312-316.

But the precious globes, the voluminous atlases and the splendid, decorated wall-maps, all produced in large quantities, were surely never intended to be taken out to sea. Most of these must have found their way into the homes of rich merchants, who in this way combined, in a fashionable manner, intellectual curiosity with more professional interests.[137]

Nor did people of higher education think it beneath themselves to engage in navigation. Leading humanists took a part in the translation of nautical works into Latin. The Amsterdam minister Petrus Platevoet (Plancius) was actively involved in the organising of expeditions to discover new routes to the Indies. He gave navigational and astronomical instruction to the officers of the expedition. Plancius also developed a theory to determine longitude at sea by means of magnetic variation.

Under the influence of these developments, practical mathematicians became persons of some standing; they were rather self-conscious and not afraid to speak out on any issue they deemed fit. However, their attitude was quite different from that of Leiden scholars. As most of them had little theoretical schooling, they restricted themselves largely to the traditional problems of their trade. The squaring of the circle was their pet subject. Those who ventured into the fields of cosmography and astronomy did not use their mathematics to speculate about the constitution of the universe; instead, they put their knowledge to use in the pursuit of more practical goals.

An author who illustrates this attitude is Aelbert Hendricksz, alias Aelbert Haeyen, a teacher of navigation. In 1585, he published a collection of nautical maps and in 1600, a book on some problems of navigation. Among other things, the book deals with methods of finding longitude and refutes Stevin's method of navigation by using the magnetic variation. Haeyen was not a scholar or a theoretician, but a former seaman. Still, he claims that his 1600 book is founded upon learned authors and historians and he extensively lists his sources. There are some references to Copernicus in his text; for example, to Copernicus' point of reference for giving longitude. At another place, he mentions Copernicus' stating the sun's parallaxis. Apparently, then, Copernicus' book was known and studied not just by the learned. One might surmise that Haeyen had had some help in mastering this learned work. But whereas the Leiden humanists were interested in Copernicus for the view of the cosmos he offered, Haeyen referred to astronomical details only.[138]

One of the most prominent mathematicians of the early Dutch Republic was Adriaen Anthonisz. He was born in 1541 in Alkmaar (northern Holland)

---

[137] Van der Krogt (1993) 250.

[138] Haeyen (1600) 3, 13, see also 4. The references are to book 2, Chapter 13 and book 4, Chapter 19 of *De revolutionibus*. On Haeyen, see Davids (1985) 313, 58-59 and *passim*.

and started his career as a surveyor. He may have attended the Latin school, but he received no higher education. Without the Dutch revolt, he probably would have remained an obscure practitioner, hardly noticed even at the local level. During the troubles, however, he was called upon to assist in the construction of Alkmaar's new fortifications, which withstood a Spanish siege in 1573. This launched his career. Other towns in Holland, which were much in need of new defences, called on him. In 1579, the States of Holland appointed him as their master of fortifications. In 1584, he was employed by the States-General as superintendent of the fortifications of Holland and Utrecht. For years, he was the chief military engineer of the Dutch Republic, active in a large number of fortification projects.[139] All this made Adriaen Anthonisz a person of note and decidedly affected his public standing. He was in touch with Holland's political leaders, among them the Prince of Orange. He even became mayor of Alkmaar. One should add that his case was rather exceptional and that most engineers remained more obscure.[140]

Adriaen Anthonisz was a practician, not a theoretician. His fortification designs represented real innovations, partly improvised because of lack of time and money, partly resulting from clever adaptations to local circumstances. But he never wrote a treatise on fortification. He published only on subjects familiar to traditional mathematical practitioners, i.e. on sundials, astronomical tables (he calculated the tables of the sun's declination in Waghenaer's famous handbook of navigation) and the calendar. A recently discovered broadsheet from 1595 on the Gregorian perpetual calendar 'reveals great originality and insight in every respect, resulting in a valuable contribution to chronometric techniques.'[141] He also devised or improved various astronomical instruments. Nor did the ever-present bone of contention – the squaring of the circle – escape his attention. At the request of stadholder William the Silent, he wrote a judgement on an earlier book by Simon van den Eycke. The manuscript was left unpublished and is now lost, but it appears that in it he introduced the 'value of Metius' (355/113) as an approximation for π. The finding of this proportion seems to have been largely a stroke of luck.[142]

All this indicates that Adriaen Anthonisz was an able and industrious man; however, little in his works indicates that he was an innovative mathematician. His fame rests on his prominent position as an engineer. Nor are there indications that he ever sought to apply his art to new fields of learning.

---

[139] See Wortel (1990) and Westra (1992) 36-44 for an enumeration of his activities.

[140] On Dutch engineers, see Westra (1992) *passim*.

[141] Breugelmans and Dekker (1989) 153. The broadsheet itself is reproduced at the end of the volume.

[142] Bierens de Haan, I, 219-253; De Waard in NNBW, I, 156-159.

However, he appears not unaware of the cosmological debate of his time. Among the instruments he devised, one deserves special mention: a model to reproduce mechanically the third motion of the earth as described by Copernicus, which should keep the axis of the earth in constant position with respect to the fixed stars, as well as cause the precession of the equinoxes. A description of the instrument is preserved in the writings of Nicolaes Mulerius, who learnt about it from Adriaen's son, Adriaen Metius.[143]

Apparently, Adriaen Anthonisz constructed the model for didactic purposes only. We are certainly not justified in deducing that he endorsed Copernicus' theories. He simply did not speak out in public on the subject. The only place which offers an indication speaks against Copernicus. In his 1595 broadsheet on the calendar, he wrote that time is measured 'after the violent daily course all round of the sun, which is occasioned by the *primum mobile*'.[144] True, such a statement may be a manner of speech rather than a deliberate theoretical stance. But regarding Adriaen Anthonisz's on the whole rather traditional stance, one may well doubt that he would have welcomed a radical cosmological innovation. He may well have built the instrument in order to fulfil the special request of some patron who had difficulty in understanding this aspect of Copernicus' work.

Adriaen Anthonisz' sons also applied themselves to mathematics. But, interestingly, the father's practical attitude persists throughout this second generation as well. His son Adriaen Metius, as noted above, studied at university and was appointed to the chair of mathematics at Franeker. But to him, too, mathematics was primarily a practical occupation. Metius did engage in astronomical calculations, but not in cosmological speculation. Inevitably, in his many works he touched upon the subject of the system of the world. But unsurprisingly, he fully endorsed the traditional position of Aristotle and Ptolemy.[145] Practicians, it appears, took a much more limited view of mathematics than humanist scholars.

The role of mathematical practitioners in the development of the new science in general, and the reception of Copernicanism in particular, has been much debated. Copernicanism was a mathematical theory and its acceptance was partly dependent on the measure of respect mathematical arguments could command. The rise of mathematics as an independent discipline, able to assert itself against tradition, philosophy or scholarship, is thought to

---

[143] Copernicus (1617) 28-31. From Mulerius' words, it is not clear whether the model was actually built.

[144] '*nae de gheweldige dagelijcksche omdrijvinghe der Sonnen, die door den* Primum mobile *is geschiedende*.' Broadsheet reproduced by Breugelmans and Dekker (1989), dedication.

[145] Metius (1614) 13-17; Metius (1633) 7.

have been a vital factor in the emerging of new mathematical theories, such as the one by Copernicus. But the status of mathematics as a discipline was closely linked to the status of its practitioners: 'the epistemological legitimation of the mathematical method that characterised the scientific revolution involved and depended on the social legitimation of mathematical practitioners.'[146]

However, without denying that such developments were important, it is evident that a new professional ethos or a new social respectability will not automatically generate a new scientific attitude. The case of Adriaen Anthonisz is telling in this respect. He certainly did not lack social standing. What he did lack was a new concept of the aims of his discipline, a programme which would enable him to use mathematics as a key to the interpretation of reality. He was, so to speak, a surveyor who more or less by accident became a VIP, but who maintained the outlook of a surveyor. The study of general tendencies or social factors may well be helpful, but what we are looking for in the first place is an intellectual programme. And intellectual programmes should be looked for in intellectual rather than social history.

Intellectual programmes which allotted a central place to mathematical learning certainly existed in Renaissance Europe. In most cases, they drew their inspiration from higher learning. The mathematician Johannes Kepler, for one, did not hesitate to enter completely new territories. It has been argued that this attitude sprang from his position as a court mathematician. But intellectual factors may have been of some importance, too; for instance, the fact that he was a theologian by training who did not need to feel abashed by the frowning of the learned.[147] In the Netherlands itself, a centre of mathematical studies had been built up in the first half of the sixteenth century at Louvain. Louvain University was the intellectual centre of the Spanish Netherlands, including the northern provinces which would later become independent. Louvain did not have an official chair of mathematics, but there was nonetheless an important mathematician present: Gemma Frisius. Gemma had taken a degree in medicine to assure himself a decent living and a public standing, but he was a mathematician by profession and calling. He designed globes and instruments and wrote a lot of works in the field, many of which became quite popular. He also gave private courses at Louvain from 1543 until his death in 1555. He was among the first admirers of Copernicus' theories, which he appears to have fully endorsed. His example and tuition inspired a lot of others. It may be inappropriate to speak of a Louvain

[146] Biagioli (1989) 41; see also ibid. 47-50; Westman (1980).
[147] Bauer (1989).

'school', but there can be no doubt that in his wake, a new group of mathematicians (cartographers etc.) arose which considered mathematics a study in its own right.[148] The most outstanding of these was Gerard Mercator, presently mainly known as a cartographer, but whose pretensions as a mathematician were much higher. Starting from a description of creation, he planned a new description of the whole universe, which was intended to replace established philosophical learning.[149]

Gemma was an influential figure and his activities at Louvain left some marks on the mathematicians in the later Dutch Republic. But that his influence stretched to the status of the discipline appears doubtful. There are no clear lines of influence from Gemma to important figures in the northern Netherlands. However, ambitious mathematicians in Holland had an alternative. They could draw inspiration from the mathematics of the Leiden humanists.

Simon Stevin, a mathematician with pretensions

The mathematician Simon Stevin is normally regarded as a representative of the rising discipline of early modern 'mathematical practitioners' – and not without reason: he started his career as a practitioner.[150] Stevin originated from Bruges, where he was born in around 1548. He arrived in Leiden in 1581. Initially, he was not taken very seriously by the Leiden scholars. Lipsius described him as 'a mere mathematician, knowing no other art, hardly even the language; moreover of the mechanical rather than the theoretical kind.'[151] As it seems, Stevin made his living at that time as a kind of civil engineer, mainly by designing and constructing windmills. From 1584 onwards, the States-General granted him letters patent for several inventions. Moreover, by 1584 Stevin was teaching mathematics to Maurice of Nassau, the son of William the Silent, with whom he would remain in close contact during the remainder of his career.[152]

---

[148] On Gemma's career: van Ortroy (1920). On his intellectual programme: Vanpaemel (1995)a. His Copernicanism has been discussed by McColley (1937), de Smet (1973), Waterbolk (1974), Hooykaas (1976) 34-35, Vanpaemel (1995)b and most recently Hallyn (1998).

[149] Blotevogel and Vermij (1995). For an overview of Mercator's life and work, Averdunk and Müller-Reinhard (1914) remains the best introduction.

[150] On Stevin, see the biography by Dijksterhuis (1943). Dijksterhuis (1970) is an abbreviated English translation of this work.

[151] Lipsius to Dudith, [April 1584]. Lipsius, *Epistolae*, II, 83 (no. 328), cf. 76 (no. 325) and I, 356 (no. 214).

[152] Maurice to Lipsius, 29 August 1584. Lipsius, *Epistolae*, II, 146 (nr. 368). In 1598, Maurice referred to Stevin as 'my mathematician': Westra (1992) 66.

During his first years at Leiden, Stevin published an amazing range of books. In general, these were didactic works explaining mathematical theory. Some of them were of a purely practical-mathematical nature, such as his interest tables from 1582, which are commonly taken to be his first publication, and the famous work propagating the decimal system (*De thiende*, 1585). However, *pace* Lipsius, he also published important theoretical works. The most famous of these are his works on statics and hydrostatics, which are counted among the most important scientific works of the century. Although Stevin legitimised his theoretical work because of its eventual use, he maintained that theory should be developed in its own right, and not with an eye to practice.

However, Stevin did not write only about mathematics. He was interested in language and logic from very early on. In 1585, he published a manual of dialectics, 'teaching to judge in a correct and artful way of all things, and opening the way to Nature's deepest secrets'. In 1586, he published with his work on statics a 'Discourse on the worth of the Dutch language'.[153] Dutch, he explained, was the most perfect and probably the oldest language in the world. It was particularly apt for expressing scientific ideas. This view returns in the works on hydrostatics, also. The fact that Stevin discusses language in a work on mathematics may appear strange to us, but in his view there was a close connection between language and correct reasoning, and hence science.

In the course of time, it seems, this idea of a pristine language, perfectly apt for doing science, turned into the idea of a pristine science as such. The Dutch language had originally been developed by an age-old civilisation, which must have possessed a perfect knowledge of nature. Stevin calls this period the *wijsentijt* ('age of the sages').[154] The knowledge of the *wijsentijt* had been lost already before the time of classical antiquity. Only some fragments had survived, known in a distorted form by classical authorities. Now, the task of science in Stevin's own age was to reconstruct this lost pristine knowledge. This could be done by the coordinated effort of many people, by using the right language in practising science (Dutch, as stated), and by collecting many observations.

Science as the recovery of pristine knowledge was a common humanist ideal. At Leiden, as noted, it was upheld by Willebrord Snellius, whose ideas appear rather similar to Stevin's. Most probably, Stevin picked up the concept of the *wijsentijt* from his learned friends. Anyhow, it turned up rather late in his career. In 1586, in the dedication of his work on hydrostatics, he had proudly announced that many of the propositions in the book 'were not

---

153 PW, I, 58-93.

154 Cf. Dijksterhuis (1943) 317-320.

known to any mortals before us'.[155] The *wijsentijt* was first discussed by Stevin in his geography, published in 1608 in the first part of his 'Mathematical memoirs'.[156] It thereafter increasingly came to dominate his work. In his work on cosmography and his unpublished work on music, he no longer aims at discovering things hitherto unknown. The theme is rather that of recovering the lost knowledge of the *wijsentijt*.

Stevin's ideas in this respect testify to his willingness to integrate scholarly concepts into his thought. Because Leiden was small, this influence must have been strong. Nicolas Stochius, rector of the Latin school, a respected scholar and a friend of Justus Lipsius, was Stevin's landlord for many years. Many years later, Stevin would intercede on behalf of Stochius' son Paul, when the latter was arrested after the Arminian troubles.[157] The famous scholar Hugo Grotius was the son of a Delft mayor, with whom Stevin collaborated in his projects on windmills. Indeed, Hugo Grotius provided him with arguments for the existence of a *wijsentijt*, which Stevin included in his work with due reference.[158] In his discussion of the prehistory of astronomy, Stevin even referred to an interview with Joseph Scaliger himself.[159] As a result, the current humanist ideal of a recovery of lost knowledge became combined with Stevin's ideal of a mathematical science.

As Stevin came to see his role less as that of a mathematician and engineer, whose task was to offer practical solutions for the common good, and more that of a scientist whose primary task consisted of the recovery of lost wisdom, the range of topics he dealt with became wider. He still engaged in traditional mathematical subjects, such as the theory of music or architecture. But his work came to include other topics as well: the explanation of the tides, ideas on what we would nowadays call physical geography, and even politics. Some of this work was published in the 'Mathematical memoirs'. Other parts remained unpublished and probably many were lost.

Stevin's astronomy

One of the areas where the quest for lost wisdom was to be conducted was astronomy. According to Stevin, astronomy was a science which had blos-

---

155 PW, I, 390-381.

156 Stevin (1605-1608), vol. 1, part 2, pp. 9-17. PW, III, 591-615.

157 Dijksterhuis (1943) 17, with reference to Brandt, IV (1704) 173-175. Brandt's principal source for the episode appears to have been a manuscript relation, now lost, by Paulus Stochius.

158 Stevin (1605-1608), vol 1, part 2, pp. 16-17. For Stevin's relations to Grotius, cf. van Dam (1996) 73-74.

159 PW, III, 598, 600-601. Cf. Grafton (1993) 458.

somed during the *wijsentijt*, but which had fallen into decay thereafter. In order to restore the discipline to its pristine glory, many accurate observations, made at many places and over a long period of time, would be needed. However, the number of available observers was much too small. Stevin was therefore pessimistic about the likelihood of arriving at a better astronomy within a reasonable time. This may have caused him to restrict himself to basic theory. As far as is known, Stevin never engaged in the practice of astronomy. He calculated no ephemerides, made no tables, and wrote nothing on the calendar.

His ideas on astronomy and the system of the world, then, appeared in his *Wisconstige gedachtenissen* ('Mathematical memoirs'), which were published in the period 1605-1608 and contain 'the things wherein his Excellency... Maurice, Prince of Orange, has exercised himself'. They were the result of the private lessons Stevin gave the stadholder. It was not really an elementary course; few universities at the time offered a mathematics course at this level. On the other hand, as it is primarily a didactic work, we should not expect too much originality. It is mainly an exposé of existing knowledge. Still, at several places Stevin manages to bring in his own views, to find new applications, etc.

The first part (of five) of the *Wisconstige gedachtenissen* is entirely devoted to what Stevin calls cosmography (*weereltschrift*). It comprehends about half of the work. One should note that Stevin's cosmography comprehends the whole science of the heavens. It is subdivided in its turn into three parts, on spherics (*driehouckhandel*), geography (*eertclootschrift*) and astronomy (*hemelloop*, literally 'course of the heavens'), which will interest us here. This astronomical part [160] is mainly concerned with mathematical calculations of planetary positions. However, it is not really a traditional textbook. Normally, one would expect a summary of established theory, followed by an elaborate explanation of how to put existing astronomical tables to use. Stevin, however, preferred to proceed in the same way that astronomical theory had been developed. In his view, the astronomers of the age of the sages compiled 'empirical ephemerides', large collections of day-to day observations of the heavenly bodies. From the study of these, they derived their knowledge of the motions of the heavens.

Of course, he does not have any 'empirical ephemerides'. As he explains, the original ones had been lost, already before Ptolemy's time, and one would need many years and the cooperation of many able observers to compile another set. Stevin therefore took recourse to a set of calculated ephemerides,

---

[160] Discussed by Dijksterhuis (1943) 146-162; idem (1970) 70-80. Partly printed in PW, III, 26-319.

those of Stadius. For his didactic purpose, these will do, but of course, only in a limited sense. One cannot derive the motions of the heavens in all their detail from them, but only the most basic principles. Stevin limits himself to the theory of 'first inequalities'. He demonstrates from his ephemerides that the planets move in eccentric circles, and calculates their periodicity, their eccentricity and the motion of their apogee. Moreover, he shows how it follows that – assuming the earth is immobile – Mercury, Venus, Mars, Jupiter and Saturn move in eccentric epicycles, and that the centre of each epicycle moves in an eccentric circle around the earth. However, he does not mention at this point Ptolemy's theory of the '*punctum equans*', or Copernicus' construction with a small epicycle meant to replace it. Still less does he mention the even more complicated constructions of both which should account for the apparent motions of the moon and Mercury. All these theories are relegated to an appendix at the end of the work.

This restriction to first inequalities may have been inspired by Stevin's didactic ideas, especially as the pupil at whom these lessons were aimed in the first place – Maurice of Nassau – of course did not really need to master all technical details. The stadholder wished to be taught astronomy, and was not interested in earning a living calculating almanacs. However, a deeper motive for Stevin's design probably was his conviction that the mathematical theories of both Ptolemy and Copernicus were wrong in the first place. As he saw it, only the theory of eccentric circles is reliable. This theory had been found in the *wijsentijt* and transmitted to posterity. This was not the case with Ptolemy's equant theory or Copernicus' alternative theory. These hypotheses had been framed by themselves, and Stevin clearly does not regard them highly. He calls them 'unnatural, obscure and erroneous'. [161] Small wonder, then, that he did not feel it necessary to include them.

These theories are rejected not because they would not save the appearances, but because they do not conform to reality. As Stevin states, they are 'unnatural'; that is to say, they do not occur in nature. The rejection of Copernicus' theory of second inequalities is remarkable. It was by means of this very construction that Copernicus had maintained the regularity of circular motion in the heavens (which had been violated by Ptolemy's equant construction). As we have seen, most scholars regarded this as one of Copernicus' main innovations. Stevin does not clearly state his objections on this point. There is no indication that he rejects the principle of uniform circular motion as such. A clue to his feelings can be found in his discussion of the third motion which Copernicus attributed to the earth. Here, Stevin acknowledges

[161] *PW*, III, 52-53. (I corrected the translation.)

that 'any notion of the Heavenly motions as imagined equally fast and made to fit into one another, as the wheels of a timepiece are made to fit together, does not satisfy me, as not seeming to happen in nature.' [162] Stevin is speaking here of Copernicus' third motion (that of the earth's axis) only. But the argument is equally valid against Copernicus' theory of the second inequalities. Copernicus had replaced Ptolemy's equant by an excentre and a small epicycle, which turned with the same periodicity (and in the same direction) as the larger circle. So, the motion of the small circle indeed had to be adjusted with watchmaker's exactness to that of the larger circle.

The deduction of the basic principles of astronomy from 'observations' and the construction of astronomical tables make up the first book of the part on *hemelloop* (astronomy). In the second book, Stevin continues with more common astronomical problems. Among other things, he teaches how to calculate a planet's position from the tables if its position on a certain date is given. All this uses the basic, simplified theory, based on eccentric circles only, and presupposes a geocentric universe, with planetary epicycles replacing the earth's motion. Stevin acknowledges that the geocentric theory is erroneous. Still, he uses it at this point, firstly because it seems to be older than the heliocentric theory, and secondly because it is simpler to use. [163]

In the third book, Stevin moves on to the 'true' theory of a moving earth. It is not clear how far this heliocentric theory was taught to the stadholder as well. In the first two books, there are some instances (as there are throughout the 'Mathematical memoirs') of Maurice further elaborating upon the matter proposed, his contributions respectfully inserted by Stevin. [164] There are no such instances, however, in the third book. One might doubt, therefore, whether this book was part of the original course. Indeed, the 'higher' theories, which Stevin later decided to relegate to an appendix, were originally intended to be explained in the second book, after the basic theories with eccentric circles. So, it may well be that originally the third book had not been intended at all. Only when Stevin was about to publish would he have added the heliocentric theories he had developed by this time, and amended the work accordingly. This is not to say that the stadholder was innocent

---

[162] *PW*, III, 128-129.

[163] Stevin follows a similar procedure in his book on spherics: 'Although it is apparent to some astronomers [*Hemelmeters*] of this time, as it was to many astronomers of the Age of the Sages, that the earth makes a yearly turn around the sun, as well as a daily turn in its place, it seems nevertheless apt to start with the feigned proposition, that is, that the earth is fixed as central point to the fixed stars...' Stevin (1605-1608) vol. I part I (*Driehoekhandel*), 347.

[164] E.g. Stevin (1605-1608) vol. I part III (astronomy) pp. 64, 73 in book I and p. 168 in book II.

about heliocentrism. Although Maurice must have known about the activities and ideas of 'his' mathematician, how far he agreed or participated is obscure.

Stevin's third book has three main subjects. In the first place, he argues for the reality of the heliocentric theory. He acknowledges that the question of the motion or rest of the earth is irrelevant for astronomical calculations – if anything, the geocentric view is the easier one – but he prefers a 'realistic' theory. In the second place, he demonstrates the mathematical equivalence of the heliocentric and the geocentric theory. Copernicus had assumed such an equivalence, but had not proved it in a strict way. Stevin thought the matter needed further clarification. As he explains, for some time he was of the opinion '(as more among those who study the theory of a moving earth) that by imagining the moving sun to be in the place of the fixed earth, and on the other hand the earth as moving in the circle of the sun, one would have the beginning of the theory of the moving earth'. But 'the propositions based on such a foundation turned out defective', and Stevin realised he had to draw a different circle for the earth.[165] Of course, the heliocentric theories he exhibits are as simplified as the earlier geocentric ones: all constructions are from eccentric circles only.

In the third place, Stevin gives a theory of the motion of the planets in latitude; that is, the distance they appear north or south of the ecliptic. While the earlier parts of the third book had been written well before printing, the part on motion in latitude was added at the last moment. Originally, he had intended to explain these theories in the second book, along with the other astronomical details. But when preparing his work for the press, he had decided that they were better explicated by a heliocentric theory, and therefore wrote a new explanation. Stevin realised that the apparent motion in latitude was due to the inclination of the planets' orbits towards the plane of the ecliptic. This had been upheld already by Ptolemy. But as Ptolemy took the earth as the point of reference, he had to allow a certain swinging of the orbits, which made his theory rather complicated. Copernicus had simply adapted Ptolemy's theories to his own system, thereby tacitly implying that the orbits intersected at the centre of the earth's orbit. His theory was therefore as complicated as Ptolemy's. Stevin tried to find the point of intersection from the astronomical data. For the outer planets, he came up with results which agreed well with Ptolemy's (Stevin compares his results only with those of Ptolemy and does not even mention Copernicus), but the orbits of Mercury and Venus intersected with the earth's orbit at quite a distance from its centre. Stevin posed the right question, but of course his solution is marred by the

---

[165] *PW*, III, 202-203.

fact that his calculations are based on his simplified planetary theory, which does not account for higher inequalities. The matter would be solved by Kepler, who postulated that all orbital planes intersect in the sun.

Stevin's Copernicanism

Stevin's reasons for accepting a heliocentric world were mainly mathematical, as one might surmise. 'Mathematical' not in the sense that such a configuration was most practical for calculating stellar positions, but in that a heliocentric order was in agreement with the mathematical way the real world was thought to be constructed. Interestingly, Stevin not only gives several arguments to support the validity of the heliocentric theory, but also speaks of various doubts and misgivings which initially made him hesitate to accept it, and the ways he solved them.

At the beginning of the third book, Stevin offers three main arguments why the Copernican system is 'in accordance with things as they exist in nature' [166]. Later, in the section on motion in latitude, he added a fourth argument. Two of them invoke the greater simplicity, or economy, of the system. In the first place, one does not need to assume the existence of epicycles to explain the way the planets follow the course of the sun. [167] Moreover (this is the fourth argument), the fact that the planet's epicycles are always parallel to the plane of the ecliptic is a necessary consequence of heliocentricity.

The other two arguments points to the order displayed by the heavenly spheres. At first sight, this sounds very similar to the argument which gave rise to the Capellan system. As Stevin explains, the larger spheres move slower than the smaller ones, which are comprehended by the former. Jupiter moves slower than Mars, Saturn slower than Jupiter, etc. According to Ptolemy, however, the largest, outermost sphere, instead of being slowest, would turn around in 24 hours, and thus have the greatest velocity of all. This contradicts the 'general arrangement' of the heavens. 'It is more in accordance with natural reason to believe and to assume that this fastest motion of all is to be assigned to the smallest circle, to wit, the circle of the earth in its place.' [168] (Mulerius also asserted this.) A second argument is akin to the former. All planets move from west to east, that is, with respect to the fixed stars. According to Ptolemy, however, the fixed stars should move from east to west. This, again, appears contrary to the general arrangement of the uni-

---

[166] *'lijckformich ... mettet ghene in de natuere bestaet.'* PW, III, 122-123.

[167] PW, III, 122-125.

[168] PW, III, 124-125 (translation Dikshoorn).

verse. According to natural reason, one should ascribe the earth a motion from west to east instead.[169]

To Stevin, this was not just a convenient argument with which to defend heliocentrism, but a principle which governed many other aspects of the universe as well. Stevin took it for a general principle that 'all bodies contained by other bodies must take the course in which their containing bodies carry them.' That it to say, the spheres of the planets should be carried on by the surrounding spheres. For instance, the apogee of Mars should be carried along by the sphere of Jupiter. Now, in practice, the planets did not obey this rule. Their apogees remain nearly fixed with respect to the fixed stars. Stevin says that he was puzzled about this for some time. He wondered whether it could be 'that the Planets were not attached to Heavens, but were flying through the air like birds about a tower, without the motion of the one causing the motion of the other; but other reasons again made me think differently.'[170] In a like way, Stevin was puzzled by the fixed direction of the earth's axis with respect to the fixed stars, whereas it should, according to him, participate in the earth's motion around the sun. Copernicus had made up for that by giving it an annual motion which compensated for the motion of the earth as a whole. Stevin was not satisfied by this construction, as it supposed that the various motions were adjusted to each other, which he thought unnatural. 'Nevertheless this motion had to be admitted in order to give a sure basis for all other natural correspondences that follow the theory of a moving earth.' So, 'this supposition long troubled me in my mind...'[171]

The solution was provided by William Gilbert's book on the magnet, which was first published in 1600.[172] Gilbert demonstrated that the earth was one big magnet. Stevin realised that this would solve his problems. The keeping in position of the earth's axis can be regarded in the same way as the constant position of a magnetic needle despite the movement of the ship: both are kept in constant position by magnetic forces. The further motions of the earth simply does not affect its axis, just as the movement of a ship does not affect the magnetic needle of its compass. In the same way, the spheres, and hence the apogees, of the planets were kept in position. They did not

---

169 PW, III, 124-125.

170 PW, III, 132-133.

171 PW, III, 128-129.

172 PW, III, 128-129. William Gilbert, *De magnete* (Londini 1600). Gilbert's ideas had quite some influence on the debate on the Copernican system; they were used by both adversaries and supporters. See Bennett (1981); Baldwin (1985); Pumphrey (1990). (S. Günther, *J. Kepler und der tellurisch-kosmosche Magnetismus* (Wien-Olmütz 1889) offers little help in this respect). None of these authors mentions Stevin's work.

affect the motions of the comprehensive orbs because they were kept in place by magnetic forces.

In order to keep the orbs, or the earth's axis, immobile, these magnetic force had to originate in the region of the fixed stars. But the mechanism proved perfectly apt for accounting for other seeming anomalies in the system of the world. At first, Stevin was amazed that according to the Copernican hypothesis, he found a motion of the moon's apogee which seemed 'against the order of the degrees' (that is, moving from east to west, instead of from west to east, as all other heavenly motions). 'It seemed strange to me that the true theory of the moving earth should involve that something was found against the order of the degrees.' On closer inspection, however, the problem could easily be solved by assuming that the motion of the moon's apogee was caused by an attractive force in another heaven, somewhere between Mars and Jupiter, which would perform one revolution of about nine years, according to the order of the degrees. A similar construction, according to Stevin, would explain the motion of the line of nodes. [173]

Now, the arguments by which Stevin defends heliocentrism are akin to the ideas of the Leiden humanists. The Leiden scholars had adopted Capella's system because it maintained the harmonious order of the spheres: the larger spheres should have longer periods of revolution. Stevin carries the idea further, but the argument appears as principally the same. And yet, there is a difference. To the Leiden humanists, the order in the universe was a reflection of divine harmony. As commonly conceived, it was not in the last place an aesthetic order. Copernicus had linked 'the law of order' (*ratio ordinis*) with 'the harmony of the whole world'. In the dedication of his book to the Pope, he dismissed the Ptolemaic system by drawing an analogy with the fine arts. He compared it to a painting wherein the various elements did not fit each other. In a similar way, the Ptolemaic system failed to demonstrate the form (*forma*) of the world and the certain symmetry (*certa symmetria*) of its parts. [174]

Such aesthetic considerations are lacking in Stevin. Here, it seems, the very terms in which Copernicus spoke are meaningless to Stevin. This does not appear in his astronomy, but in his work on architecture, where it was impossible to evade aesthetics altogether. Here, he rejects the common idea that harmony would follow from applying the 'right' proportions. He does so with linguistic arguments. In architecture, the term 'symmetry' was commonly used to mean 'right proportion of the parts to the whole'. In this sense Copernicus, in the passage referred to above, had spoken of the symmetry of

---

[173] *PW*, III, 134-135.

[174] See Stumpel (1990) for a detailed analysis of this passage by an art historian.

the parts of the universe. However, according to Stevin, 'symmetry' only meant that the two halves were similar. Any other use was mistaken. In his view, there was no higher harmony here.[175]

Now, it seems unlikely that a notion which he rejected in architecture would still apply in astronomy. Indeed, Stevin's 'general arrangement' is rather a mechanical construction. As indicated, in Stevin's view all bodies contained by other bodies must take the course in which their containing bodies carry them. Hence, 'it would follow that every planet must have a motion consisting of a combination of all the motions of the planets that are above it.'[176] So, the order of the heavenly motions, whereby the inner spheres move faster than those at the outside, is not the expression of divine harmony, but a consequence of rather down-to-earth mechanics.

Stevin did not regard the heavens as a divine region, elevated above the world of man. Indeed, he does not appear to have attached much meaning to the common division into a supralunar and a sublunar world. In his geography, he defines the earth as 'the moving celestial body [*weereltlicht*] which we inhabit, and the eighth planet besides the other seven.' To show that the earth really is a planet, he invites his readers to imagine themselves placed on the moon. From that viewpoint, the earth will show all the characteristics (phases, luminosity, etc.) we observe when we look at the moon from the earth.[177]

So, whereas at first sight Stevin's cosmological ideas may appear similar to those of the Leiden humanists, the intellectual background is radically different. What seems to have happened is that Stevin had picked up the argument of the harmonious arrangement of the heavenly spheres from his learned friends, but then gave it an interpretation of his own. He integrated elements of humanist learning into his own thought, but in the end he remained an engineer. As it seems, this stance resulted not from mere lack of understanding, caused by defective education. Rather, Stevin demonstrates the self-assertive attitude of someone who develops his mathematical science as a new gateway to knowledge. He simply thought he could do better than others.

Willem Jansz Blaeu

Another person who came from a background of practical mathematics which he apparently combined with more scholarly studies, is Willem Blaeu (sometimes, he Latinised his name as Caesius). We shall deal with him rather briefly. Blaeu came from a family of well-to-do fishmongers. From an early

---

175 Van den Heuvel (1995) 56-57.
176 *PW*, III, 132-133.
177 Stevin (1605–1608) I, *Eertclootschrift*, 5.

date, he applied himself to mathematical studies. In 1595-1596, he spent some time at Tycho Brahe's observatory in Hven. It is not known whence he obtained a recommendation. As he seems to have spent his youth in Alkmaar, Adriaen Anthonisz might have been instrumental, but this is mere speculation. Having returned to the Netherlands, Blaeu settled in Amsterdam and applied himself to geography. He dealt in maps and globes and eventually entered the book trade. The Blaeu firm became one of the leading publishing houses in the Netherlands. It became particularly famous for its very fine maps, globes and atlases, as well as works on navigation. Another speciality of the house was less conspicuously advertised, but probably much more remunerative. Blaeu published large quantities of Catholic devotional works, which were formally forbidden in the Dutch Republic. Blaeu himself was born a Mennonite, but does not seem to have been very committed to this creed. He appears rather as a man of rather libertine opinions. He even published Socinian works, which were considered downright heretical. [178]

In the meantime, Blaeu remained active in mathematical practice. The globes and some of his maps were his own design. In 1623, he published a large work on navigation. This work also contained astronomical tables, which were later (1625) published separately (in Dutch): 'Tables of the declination of the sun and the principal fixed stars, as well as different ways to use the polar star'. Blaeu had calculated these tables from the observations of Tycho Brahe. As the 1625 title page stated, he had done so 'for the use of all navigators'. In 1625, he also published a ten-year almanac, 'accommodated to the meridian of Amsterdam'. In 1633, he was appointed the official cartographer of the Dutch East India Company. He also continued Snellius' work in geodesy. On the whole, Blaeu was a *mercator sapiens* – a learned merchant who combined material and scholarly pursuits.

As a cartographer and globe maker, Blaeu propagated the Copernican system. In 1619, his firm published a large new map of the world. [179] The world was depicted in the form of two hemispheres, which left considerable space for allegorical or instructive decoration at the borders. Among these decorations is a representation of the system of Copernicus. It was not uncommon for world maps to show cosmological schemes, although Blaeu's is the first example from the United Provinces (in a later chapter, I shall say something more on images on maps). During this period, however, Blaeu was unique in depicting the universe as a heliocentric system only.

More famous were Blaeu's efforts to produce a heliocentric sphere. The com-

[178] On Blaeu: Keuning & Donkersloot-de Vrij (1973); De la Fontaine Verwey (1979) 9-34.
[179] Shirley (1983) no. 300. Schilder (1990).

mon spheres or celestial globes showed the heavens from a geocentric point of view. This apparently did not satisfy Blaeu. He tried to construct a sphere which would reproduce the heavenly motions according to the system of Copernicus, thus being, according to him, more in agreement with reality. The project occupied him for several years. After some earlier designs, this led to his famous *tellurium*, a construction which showed the earth in its daily and annual motion around the sun. On 22 September 1634, the States-General granted him a patent for it. It was widely acclaimed as a piece of excellent instrument-maker's craft. However, it was not widely used, probably because of its price, nor does it appear to have been copied by other instrument makers. [180]

One result of his activities as a globe maker was his book 'Double instruction of the celestial and terrestrial spheres; the one according to the opinion of Ptolemy with a fixed earth; the other after the natural proposition of N. Copernicus with a moving earth'. [181] It was a kind of instruction manual which taught the buyers of his globes how a globe was to be used as a mathematical instrument. The book was first published in 1634, the year he was granted the patent for his tellurium. It was reprinted in 1638, 1647 and 1666, and its success was not restricted to the Dutch Republic. Martinus Hortensius made a Latin translation, which appeared in 1634. [182] This Latin translation was then translated into English and published in London by Joseph Moxon in 1654. [183] A French translation appeared in 1642 and 1669. In the second half of the eighteenth century, there was even a partial translation into Japanese. [184]

The title leaves no doubt as to Blaeu's Copernican conviction. Still, Blaeu announces that he will say only little about it, in particular because such has been done already by others, even in the Dutch language. Blaeu refers to Stevin and Lansbergen. 'Someone who is not prejudiced, and who likes to give place to reason, will find there sufficient proof, and abundant material to refute anything which may be brought against it.' [185] Indeed, Blaeu does not explain the background of his ideas, so little more can be said of it.

---

[180] A fine specimen is in the Nederlands Scheepvaart Museum in Amsterdam. Information on this topic derives from the papers of Hooykaas, who collected a lot of documentation on the *tellurium*. Hooykaas presumes that the tellurium was ready and on the market only by 1633. Earlier datings of existing specimens by Zinner he regards as unreliable. Cf. Zinner (1943) 390-391. See also E. Zinner, *Astronomische Instrumente des 11.-18. Jahrhunderts* (München 1961) 250, 252.

[181] Blaeu (1634).

[182] Blaeu (1640). Reprints are reported from 1650 and 1655; I have not seem them myself. Moreover, the title page of a work from 1646 with the same title, but under the name of Jacobus Colom, is reproduced in *JHA* 17 (1986) 59.

[183] Russell (1972) 224.

[184] Nakayama (1972) 163-164.

[185] Blaeu (1666) preface.

However, it may be worth dwelling a moment on another aspect. As seen above, there cannot be any doubt that Blaeu was a convinced Copernican from early on. Yet, at times he made remarks which suggested the opposite. The world map of 1619, with the system of Copernicus on it, also hinted at geocentrism. The map was reissued in 1645-1646 by Willem's son, Joan. There are some modifications, but these do not affect the world system. The map still shows the system of Copernicus.[186] Now, this map has a text part, in Latin, 'Short and clear explanation of the division, constitution and proprieties of the countries of the whole world'. This text starts with the story of Creation and therein clearly follows a geocentric picture. God put the earth and the heavy elements in the centre of the world, being the lowest place (*in centro ut loco inferiori*). The heavens with the heavenly bodies were put over it as a tent.[187] It is striking that, even if people agree that the world should be seen as heliocentric, in a description of the Creation they take the traditional order of the world for granted, although in a case like this, one can understand why Blaeu preferred such a version: meddling on one's own responsibility with the traditional version of biblical episodes was asking for trouble.

The original issue of 1619 was published with a text, too. Regrettably, the text part is missing from the only copy of this issue to have been preserved. Therefore it cannot be proven that the same text was used in both 1619 and 1645-1646, although it seems probable. Joan Blaeu, who did the reissue, was also a supporter of Copernicus, as is clear from his wall map of 1648, to be discussed later. He also appears to have given private lessons in astronomy. From a letter written by Pieter Cornelisz Hooft, who was sheriff of the town of Naarden and related by marriage to Blaeu, we learn that he taught the Copernican theory to Hooft's son.[188] The phenomenon was not restricted to wall maps. We meet more or less the same phenomenon in Blaeu's atlases. At the head of an atlas is generally placed a short introduction to cosmography and geography, taken from some well-known textbook. As all textbooks at this time still presupposed a geocentric order of the world, these introductions do so as well.

Here, it seems, is a fundamental problem. Heliocentricity might appear probable from an astronomical point of view, but how did one accommodate it in one's general world-view? The 'order of the universe' was supposed to be not only a mathematical but also a religious and moral order. The problem does not surface as long as one restricts oneself to mathematical works. But

---

[186] Shirley (1983) no. 366. Schilder (1990); see in particular 179, 184.

[187] Schilder (1990) 315-329, an appendix which renders all texts on the map, in the original as well as in Dutch and English translation. The passage mentioned is on page 317.

[188] P.C. Hooft to Arnout H. Hooft, 7 June 1646. Briefwisseling Hooft, III, 735.

the large wall maps were showpieces, not so much meant to demonstrate technical details but to instruct in a more general sense. They showed the moral order of the world as well. Hence the inclusion of a text on the Creation, which remained quite traditional. Quite understandably, people were hesitant to accommodate general religious notions because of mere astronomical considerations. As we saw in the case of Mulerius, such wider implications, once realised, could be a serious obstacle for the reception of the Copernican theory. In the case of Blaeu, it seems probable that he put the text on the map simply because his customers expected something of the kind, not because he endorsed it himself. As noted, Blaeu did not bother very much about the religious tenor of the works he published. How far his own Copernicanism was part of a wider world-view is a question which cannot be answered.

5. Philips Lansbergen's Christian cosmology

Lansbergen's life and work

The Reformed minister Philips Lansbergen was one author who did integrate heliocentric theory into an all-embracing Christian world-view. Lansbergen is one of the best known and most influential Copernican authors of the Netherlands. Because he published a lot, it is possible to get a close look at his ideas. Like Stevin, he was not a humanist, but still deeply influenced by the views of the Leiden scholars. That is, humanist ideas were used but did not form his world-view as such. In these circumstances, they could lead to unforeseen conclusions. Whereas Stevin's world-view was largely constituted by his mathematical background, Lansbergen's work was dominated by his interest in religion and Hermeticism.[189]

Much is uncertain about Lansbergen's education as he grew up during a chaotic period and wandered a lot in his youth. He was born in Ghent in 1561, to a well-to-do family. His parents were Calvinists. After the troubles of 1566 they had to flee the country, and Philips spent his youth in France and England. Later on in life, he maintained his contacts with English scholars. In 1635, the London mathematician Henry Gellibrand reported that he had had a conference with Lansbergen, but the date of this conference is not clear.[190] Lansbergen grew up a convinced Calvinist and applied himself to theological studies. Mathematics attracted his interest at an early date, but for the time being, theology would remain the dominating force in his life. When after the Peace of Ghent in 1576 it became possible to return to his

---

[189] The main biographical data are presented by De Waard in *NNBW*, II, 775-782. See on his work also Delambre (1821) 40-47; and Vermij (1998). I also used the Hooykaas collection as well as an unpublished lecture by H. Braam, 'Reasoning, religion and experience: the theological Copernicanism of Philippus Lansbergen', given at the conference 'Philosophy in the Netherlands in the 17th and 18th centuries', Rotterdam 1988. I thank Dr Braam for giving me a copy of this lecture.

[190] Feingold (1984) 11. (On Gellibrand, ibid. 113.)

fatherland, he showed his zeal for Protestantism by dedicating himself to the propagation of his faith.

He returned from London to Ghent in March 1579 and preached in several places in Flanders.[191] In 1580, despite his youth, he was appointed one of the ministers of the Calvinist community in Antwerp. He refused a call to Malines[192] and remained in Antwerp until the conquest of the town by the Prince of Parma, the Spanish governor, in 1585. As Parma allowed the Protestants to leave in peace, Lansbergen went to Holland, the centre of resistance to Spanish rule. He matriculated at Leiden where he continued his theological studies. He remained at Leiden for about a year and apparently made friends with the Leiden humanists. His first book – which was published in 1591 by Plantin (Leiden) – was accompanied by laudatory poems by Janus Dousa Jr. and Jacob Gruterus. In some of his later works, he printed letters written to him by Willebrord Snellius.[193] It may well be that these contacts kindled his interest in astronomy. His ideas seem to have been fully developed by this time and, as I will show later on, Lansbergen's earliest recorded observations date from 1588, that is, shortly after he had left Leiden University. Still, one has to admit that his stay in Leiden was only short and, although young, he was not really an inexperienced boy at that time. Lansbergen remained a theologian, not a humanist thinker. In 1604, he wrote a Latin poem to commemorate the new star of that year; this poem refers almost exclusively to biblical verses, not to classical authors.[194] Moreover, it seems probable (although evidence is lacking) that his Hermetic and alchemical interests were mainly formed in the southern Netherlands.

In 1586 Lansbergen was appointed minister of Goes (province of Zealand). In 1597, he was called to Amsterdam, but his ministerial office there did not materialise. It is not clear what happened. It seems that the church council of Amsterdam had tried to circumvent the magistracy, which nevertheless in the end stood in their way. Some 20 years later, the Amsterdam burgomaster Hooft still recalled the procedure as a flagrant feat of ecclesiastical presumption.[195] Lansbergen himself was not averse to such 'presumption'. Goes was near his homeland Flanders and, perhaps more to the point (as he cannot have

---

[191] Decavele (1984) 81-83. I owe this reference to Dr G. Marnef. Lansbergen's name is not found among those sent as missionaries by the London Reformed congregation to the Netherlands (personal communication by Dr O. Boersma). Regarding his youth, this is not very surprising. Probably he developed his talents on the spot.

[192] Marnef (1987) 261.

[193] Lansbergen (1619) (Snellius to Lansbergen, 5 Dec. 1606), Lansbergen (1616) (Snellius to Lansbergen, 9 Oct. 1607).

[194] Hallyn (1997).

[195] C.P. Hooft (1871-1925) I, 43, 137; II 226-229.

developed very strong bonds with Flanders in his youth), quite close to the frontier with the Spanish. It was a position which called for a militant attitude, and Lansbergen certainly did not compromise. His implacability brought him into conflict with the local magistrate, who, especially as the Spanish menace had diminished since the truce of 1609, saw less need for such an attitude. Lansbergen thought himself entitled to tell the magistrate how to act. A vigorous attack on the leniency of the town government of Goes, which allowed people suspected of Catholic sympathies to stand for the magistracy, eventually led to Lansbergen's downfall. In 1613, he was relieved of his ministry. He thereupon went to Middelburg, the capital of Zealand, where he probably had some relations [196] and where he lived as a private citizen until his death in 1632.

Lansbergen wrote a number of books. Surprisingly for a man for whom religion was a driving force in his life, he did not write books on theology proper. He did publish, however, a collection of 52 sermons on the Calvinist catechism (the Heidelberg catechism). Originally published in Latin, Lansbergen's book was later translated into Dutch; this version was republished several times in the seventeenth century. [197]

Lansbergen wrote this book early in his career. Later, he only published on mathematics, a field of study he had been engaged in from an early age and which appears to have increasingly engross him. His first book, from 1591, was a Latin work on trigonometry which contained some original contributions. Moreover, Lansbergen wrote manuals on astronomical instruments, such as the quadrant and the astrolabe. These [two] books were written in Dutch and appear to have been rather popular. Several new editions were published during the seventeenth century. Both were translated into Latin as well. In his *Opera omnia*, a book on sundials was posthumously published, and was soon translated into Dutch. Lansbergen also wrote on biblical chronology and, inevitably, on the quadrature of the circle. His predilection, however, was astronomy. By 1597, he was so much addicted to it that it was used as an argument against calling him to Amsterdam: 'You have Plancius as a geometer, perhaps you will receive an astronomer: think over it whether such a ministry will be useful for your church. For the nature of astronomy is such

[196] Lansbergen's relation to Middelburg is somewhat unclear. He had dedicated his book on trigonometry (1591) to the government of Middelburg. In the letter of dedication, he writes '*libros... aliquot ante annis in vestra hac insula scripsi, & perscipsi*' (*vestra insula* should not be taken too literaly, as he heads the letter *Goesae in insulae vestrae urbe*; in fact, Goes lies on the island of Zuid-Beveland, and Middelburg on the island of Walcheren.) In the second edition (1631), this has been changed to: '*primum in urbe vestri concepi, post Goesae scripsi, & perscripsi...*'

[197] Lansbergen (1595), (1645); Ursinus (1606).

that it induces the whole man to love it and indulge in deep speculation. Who takes care of many things, can give less attention to each thing apart.' [198]

As an astronomer, Lansbergen was not satisfied with the existing theories and tables, either those of Ptolemy or those of Copernicus. The nature of his objections is not clear, however. The main task he set himself was to procure new ones which should be perfect. This was regarded as a quest for lost knowledge, as the dominating humanistic ideal had it. As Lansbergen explained, God had put the sun and the moon in the heavens to indicate the time. Pristine man had had a perfect knowledge of their motions, as attested by the Hebrew calendar. But after the introduction of the Julian calendar, this knowledge had sunk into oblivion. It was the task of scholarship to recover it. [199]

This recovery of astronomy should be based on observations. Lansbergen followed a double course. In the first place, he unconditionally accepted all observations reported by ancient authorities. This was a rather traditional way. But then he recognised that those data did not suffice. As Lansbergen wrote, many people had already been urged by God to reinstate astronomy in its pristine glory, but had failed largely because of lack of observations. 'When this dawned on me, now 30 years ago,' Lansbergen wrote in 1619, 'I started to entertain some hope to repair the Art.' To obtain the observations his predecessors had lacked, he procured himself suitable instruments and observed for several years the motion of the sun, on which the other planets depended, and of the other planets, 'each time that I found occasion'. [200] As we have seen, Simon Stevin had developed similar ideas on the inadequacy of existing astronomical theories and the need for systematic observations to restore them. How far they were influenced by each other, or by Leiden scholarship, is impossible to say.

The self-imposed task of restoring astronomy proved harder than Lansbergen had imaged. His own observations could readily be accounted for. But when he compared his findings with the records of observations made by classical or Arabic astronomers, like Ptolemy or Albategnus, they deviated widely. Lansbergen then tells how for many years he continually tried ever new calculations, 'with numbers which my spirit shrinks back from recalling'.

---

198 From a letter by the Leiden theologian Joh. Kuchlinus to the Amsterdam burgomaster Claes Fransz, 1 Oct. 1597, published in Hooft (1925) 228. '*Plancium Geometram habetis, Astrologum accipietis forte, cogitabis pro tua prudentia an sit vestrae ecclesiae futurum utile ipsius ministerium. Nam ea est astronomiae ratio ut totum hominem ad sui amorem et profundam alliciat speculationem. Pluribus intentus minor est ad singula sensus.*' In a following letter, Kuchlinus wrote on Lansbergen: '*addictissmus est astronomiae studio,*' ibid., 229.

199 Lansbergen (1619), dedication.

200 Ibid.

Finally, however, God showed him the way out of the labyrinth. And so, 'I obtained the [theory of the] motions of sun and moon, as of the fixed stars, restored to their original state, so that they accord with the heavens, not only for this, but also for former centuries, up to 2,000 years back.[201]

By 1619, the project had advanced so far that Lansbergen decided to publish. This resulted in his *Progymnasmatum astronomiae restitutae liber I. De motu Solis* ('Preparatory exercise of the restored astronomy, book 1: On the motion of the sun'). The apparent motion of the sun in the heavens is explained from theory and with the help of a large number of tables, by means of which the sun's position can be calculated for any given moment for any place on earth. In the dedication, Lansbergen announces his intention to publish a second book on the motion of the moon. He also announces a third book on the motion of the fixed stars (that is to say, the precession of the equinoxes).[202] None of these books appeared, however. Lansbergen remained active in other fields, but his project to restore astronomy seems to have reached a deadlock after 1619. He may have been disappointed by the lack of response to it. The *Progymnasmatum* appears to have drawn but little attention. Today, this edition is extremely rare.

This situation lasted until Maarten van den Hove (Martinus Hortensius) appeared on the scene. Hortensius, born in Delft in 1605, had been a student of Isaac Beeckman at Rotterdam and of Willebrord Snellius at Leiden. Beeckman had been born in Middelburg and he knew Lansbergen personally. In 1628, he introduced his young pupil to him.[203] Hortensius quickly became enthusiastic supporter of the aged astronomer. He seems to have been mainly responsible for Lansbergen's project finally reaching completion. Lansbergen later thanked him publicly for the work he had done, and said that he was lucky 'that, by divine providence, in my old age, pressed by sickness, such a strong helper came to my aid, as formerly the learned Rheticus to the great Copernicus.'[204]

From 1628 onwards, Lansbergen published in quick succession quite a number of astronomical books. The first was a new edition of the *Progymnasmatum*. In fact, this was only the old edition in a new cover. The original publisher, Schilder (in Middelburg), apparently still had a fair number of copies in stock and must have been glad to get rid of them. They were provided with a new title page and brought out by a new publisher, Zacharias Roman,

---

[201] Ibid.

[202] Lansbergen (1619) dedication; 6, 115.

[203] Van Berkel (1983) 143.

[204] Lansbergen (1632), preface.

also in Middelburg.[205] Lansbergen's 'Reflections upon the daily and annual course of the earth. The same on the true image of the visible heaven; wherein the wonderful works of God are discovered' appeared the following year (1629), in Dutch. This book is a defence of Copernican cosmology. It is in fact a work of popular education, and is rather unconnected with Lansbergen's learned project of the restoration of astronomy. It will be discussed more fully further on. In 1630, Lansbergen's *Uranometria libri III....* appeared. The contents are clearly indicated by the full title: 'Three books on the measurement of the heavens, wherein the distances to the earth and the sizes of the moon, the sun, the remaining planets and the fixed stars, hitherto unknown, are clearly demonstrated'.

The fruit of Lansbergen's long-cherished project of a restored astronomy finally appeared in 1632. Its original form had been somewhat adapted. Instead of supplementing his work on the motion of the sun with similar works on the motions of the moon, the fixed stars and, probably, the planets, as originally foreseen, Lansbergen decided to put all theories, irrespective of whether they concerned the sun, the moon or any other planet, in one part, and all astronomical tables in another. (The theories and tables on the sun, which had been published before in the *Progymnasmatum*, were included.) The part with tables was entitled *Tabulae motuum coelestium perpetuae...*: 'Perpetual tables of the heavenly motions, made up from, and in accordance with, the observations of all times'. The other part was entitled: 'New and genuine theories of the heavenly motions'. This explained the mathematical models Lansbergen had used to calculate his tables. Finally, Lansbergen added a 'Treasure of astronomical observations', with observations of eclipses etc. over the ages.[206] Probably, Lansbergen was following here the example set by Nicolaus Mulerius, who had appended a similar thesaurus to his 1617 edition of Copernicus' *De revolutionibus*. In it, Mulerius brought together all observations to which Copernicus had referred.[207] Lansbergen's overview is far more exhaustive; he had obviously taken recourse to many more observations than Copernicus had. In each instance, he shows how nicely the respective values calculated from the tables agreed. Thus, he tried to prove his point that his tables agreed with the observations of all periods. Without any doubt, these

---

[205] A year earlier, 1628, Roman had already reissued another book by Lansbergen, formerly published by Schilder: his *Cyclometriae novae libri duo*. In this case, he simple pasted on the old title page a piece of paper bearing his own name over Schilder's. On Roman: Meertens (1943) 420. On Schilder: Aarssen (1969).

[206] Lansbergen (1632). The book starts with a number of *Praecepta calculi*. Then follow the *Tabulae*, with separate pagination. Pagination starts afresh with the *Theoricae* and continues in the *Thesaurus*.

[207] Copernicus (1617) 471-487: *Astronomicarum observationum thesaurus, e scriptis Nic. Copernici collectus: servata serie qua usus fuit Copericus.*

Figure 3: Title-page of Philips Lansbergen's *Tabulae motuum caelestuum perpetuae*, 1632. The figures seated above are the ancient astronomers Aristarchos of Samos, holding a heliocentric sphere, and Hipparchos of Rhodeś, holding a geocentric one. While Hipparchos apparently is demonstrating something from a book, Aristarchos points to the heavens. Bottom right a portrait of Lansbergen himself. (Utrecht university library.)

three volumes were Lansbergen's *magnum opus*. Having seen its fulfilment, he died the same year (1632).

Lansbergen's defence of the Copernican system was part of his restoration of ancient astronomy in general. It is therefore worth looking a little closer at this larger project. As stated, Lansbergen wanted to found his theories on long series of observations. Sixteenth-century astronomy was still largely based on observations from classical antiquity. Copernicus himself largely used Ptolemy's data. Moreover, this concerned only special events (eclipses, conjunctions, etc.). Tycho Brahe was the first to start a programme of systematic observations. At his observatory in Uraniborg (Denmark), every day for many years he recorded the positions of the heavenly lights, so far as the weather allowed. [208]

Lansbergen's project was similar, but on a smaller scale. Tycho had a fully equipped observatory and a staff of assistants, while Lansbergen had to steal his observing time from his pastoral duties (this, by the way, does not seem to have given him any pangs of conscience). In *De motu solis*, he gave a record of the sun's altitude at noon in Goes on 27 days between March and September 1589, apparently the only days on which he was able to make the observation. [209] As he alleges, these printed data were taken from a longer series of observations, which extended from the end of 1588 to the end of 1590. Lansbergen undertook a new series of observations of the sun's altitude in the period 1599-1605, of which he offers only an extract. [210] The new series of observations was undertaken after he had acquired a new instrument. The first series had been made using a wooden quadrant with a radius of four feet, the second with a larger quadrant of bronze. [211] It may be that making these observations indicates a deadlock in Lansbergen's calculations, but this remains just a guess. Lansbergen does not regard his observations with the wooden quadrant as less reliable than with the bronze one.

Lansbergen gave such a record of observations only in *De motu solis*, not in any of his later works (the observations in the *Thesaurus* concern special events such as eclipses etc.). These observations, of course, concern only the sun. It is more difficult to say something about his observations of the moon and the planets. In his *Uranometria*, Lansbergen records three observations of the position of the full moon, from 1600, 1601 and 1602. His words seem to indicate, however, that these too are a selection from a longer series and that in fact for

---

[208] Thoren (1990).

[209] Lansbergen (1619) 15-16.

[210] Lansbergen (1619) 16-17.

[211] Cf. Lansbergen (1631) 2. The later quadrant was also used by Isaac Beeckman for an observation in 1616: *IBJ*, I, 106.

some years he had tried to observe the position of the moon at its fullness as often as possible.[212] Lansbergen nowhere mentions observations of the planets, although this does not necessarily mean that he never made them.

One does find, however, both in *Uranometria* and in the *Thesaurus*, a number of observations by Lansbergen of eclipses of the sun and the moon, as well as of a conjunction of the moon with the star Aldebaran. For the most part, these date from the period 1598-1612. That is, observing abruptly stops after Lansbergen's dismissal as a minister and his move to Middelburg. During this later period, there are only two observations (in 1621 and 1630) and two determinations of the sun's altitude (on 23 December 1616 and 22 June 1617[213]). Perhaps after 1614 he concentrated on the theoretical elaboration. Once his theories had been completed, he would have felt less need for new data.

All the efforts, however, could not guarantee the agreement with future observations. In particular, Lansbergen's tables proved inferior to the 'Rudolphine tables' published by Kepler five years earlier (in 1627). This was not immediately clear. For a decade or so, Lansbergen's tables seriously rivalled those of Kepler.[214] In 1635, Noël Durret published a set of astronomical tables in Paris which were almost exact copies of the tables of Lansbergen; he had simply adapted them to the meridian of Paris. A few years later, however, he appears to have switched to Kepler's side.[215] Father Octoul was still using Lansbergen's tables in the 1640s, as is attested by his *Inventa astronomica* (Avignon 1643).[216] In England, the almanac maker Vincent Wing also used Lansbergen's tables until the middle of the 1640s. John Pell started translating them into English in 1634, but never accomplished the work.[217] Henry Gellibrand and John Bainbridge discussed Lansbergen's tables in the 1630s.[218] Gellibrand also brought the tables to the attention of Jeremiah Horrox, who at first was very enthusiastic about them. But as he discovered their defects, enthusiasm turned into violent dislike. He, too, switched over to Kepler's tables and he wrote several very hostile tracts against Lansbergen,[219] for example: 'Nobody ever treated astronomy in a less fortunate way than Lansbergen; nobody ever sinned more.'[220] In Spain, Lansbergen's tables seem to have re-

---

212 Lansbergen (1631) 2-5.

213 The latter two are recorded in his book on the quadrant. Cf. IBJ, I, 106.

214 Russel (1964) 9.

215 Wilson (1989) 165. Russel (1964) 12. Delambre (1821) II, 236.

216 Thorndike, VII, 105.

217 Pell's manuscripts are in London, BL, Add. Ms. 4416 f 1-3; idem 4418 ff. 4-9; idem 4459 ff. 123-137.

218 Feingold (1984) 145.

219 Horrox (1673). See especially the preface by Wallis. Horrocks describes his astronomical *Werdegang* in the Prolegomena, 1-5. See also Russel (1964) 12.

220 Horrox (1673), prolegomena, 21.

mained in use for somewhat longer. Malvasia based his ephemerides for the year 1662 on them,[221] while in 1670, the Spanish Jesuit José Zaragoza wrote an astronomy according to Lansbergen's hypotheses, adapted to the meridian of Madrid.[222] But generally, Lansbergen's tables quickly lost favour and his fame as a restorer of astronomy was short-lived.

The critique on Lansbergen's tables went so far that several seventeenth-century authors accused him of falsifying his observational results in order to bring them into agreement with his tables.[223] Even though Lansbergen's observations are open to some criticism, this seems a bit far-fetched. Pingré remarked that Lansbergen's (as well as Hortensius') observations of solar eclipses were inaccurate because the diameter they assumed the sun to have was too large, probably because they were misled by the sun's radiation.[224] But deliberately falsifying the results would have made the whole project worthless, as Lansbergen could not have failed to realise. He seems quite honest in stressing the extreme efforts he made in continuously adjusting his theory to observation. In fact, it seems that it was his strict adherence to given data which spoiled his results, as he attached too much value to rather dubious ancient observations. This was certainly counterproductive in the long run.

Lansbergen's Copernicanism

Fame, however, came from another quarter. As a kind of a by-product, Lansbergen had written for a more general public his 'Considerations on the daily and annual rotation of the earth', a defence of the system of Copernicus. The fame of this booklet far outlived that of his more serious scholarly undertakings. The *Bedenckingen* (as the title reads in Dutch) was not the first defence of the Copernican system in the vernacular (Stevin had already produced one), but it was the first which aimed at a popular, non-mathematical audience (it was published well before Galileo's *Dialogo* or Wilkins' *Discovery of a world in the Moon*). The book was published several times in Dutch and had quite an impact on the general public.

Moreover, within a year after its publication, Hortensius produced a Latin

---

[221] Russel (1964) 9.

[222] Vernet (1972) 283-284.

[223] Pingré (1901) 14: '*Riccioli remarque quelque part qu'il se rencontre presque toujours quelque différence notable entre les observations de Lansberg et celles des autres astronomes, soit que, comme le pense Wendelin, Lansberg ne distinguât pas assez l' ombre et le pénombre, soit parce qu'il altéroit ses observations pour les faire cadrer avec ses tables, ce qu'Horrox et d'autres lui ont reproché.*' See also ibid., 91. At other places, Pingré defends Lansbergen against critique by his fellow astronomers, ibid. 20, 21, 25. Lansbergen was also criticised by Holwarda and Wendelin; see for the latter, Monchamp (1892) 111-112.

[224] Russel (1964) 9. Pingré (1901) 113, see also 3-4.

translation of the book, with an elaborate introduction. The Latin version drew attention all over Europe. It was the object of several anti-Copernican attacks by Alexander Ross, a conservative Aristotelian from Scotland; [225] by Jean-Baptiste Morin, a professor of astronomy at Paris, who was famous for his work on astrology; and by Libert Froidmont (Fromondus), a Jesuit from the Spanish Netherlands. [226] These attacks were answered by Hortensius and by Lansbergen's son, Jacob. [227] As these polemics would carry us outside the Netherlands, I will not discuss them here, but it is mainly to these controversies that Lansbergen's name is attached. When Lansbergen is remembered nowadays, it is not because of his great astronomical enterprise, but as an early adherent of Copernicus. [228]

Quite at what stage Lansbergen decided in favour of Copernicus' cosmology is not clear. It is clear, however, that he was critical regarding Copernicus as an astronomer – undoubtedly, he reckoned him among those astronomers who had failed to restore astronomy because of lack of observations. It seems typical that even in the dedication of his tables, he calls Copernicus *Ptolemaeo suppar* ('almost equal to Ptolemy') while the latter is denoted as *Astronomiae Princeps et Pater* ('prince and father of astronomy'). [229] His decision to accept the Copernican cosmology was not taken on Copernicus' authority. In 1598 Herwart von Hohenburg reported to Kepler that Lansbergen was considering a hypothesis, 'wherein he puts the earth's centre in a fixed place, but attributes the other motions of the primum mobile to its [the earth's] circumference;' [230] that is, a system wherein the earth had a daily but not an annual rotation. Some caution is required as it is not known where Herwart got his information from, but that Lansbergen considered several possibilities before definitely opting for heliocentrism seems quite believable.

The first indication that Lansbergen had decided in favour of the Copernican system is provided by his *Cyclometria* of 1616, the title page of which bears a vignette representing the system of Copernicus. It is in his *Progymnasmatum*, however, that he really comes out into the open. After the main, technical-astronomical part of the book, he concludes with a few chapters on the system of the world. He argues successively that the earth has a daily rotation on its

[225] Russell (1972) 230. See also London, BL Sloane Ms 955 ff. 64-68v, containing notes by Pell of Lansbergen's arguments, with refutations.
[226] On him, see Monchamp (1892) 34-44, 50-53, 72-112, 144-148; Redondi (1988); Nouhuys (1998) 240-250, 276-303, 522-528.
[227] The discussion between Froidmont and Jacob is discussed by Monchamp (1892) 100-112.
[228] Cf. Bailly (1779) 154: '*Il n'a de vraiment recommendable que d'avoir été le partisan de Copernic & d'avoir écrit pour démontrer le mouvement de la terre.*'
[229] Lansbergen (1632), dedication.
[230] Kepler, *Werke*, XIII, 178.

axis, that it has an annual rotation round the sun, and that such a double motion is not absurd.[231] Finally, in 1629 his *Bedenckingen* appeared, which was entirely devoted to the system of the world. The book consists of two parts. In the first, Lansbergen argues for the daily and annual motion of the earth. In the second, he considers the structure of the universe; he rejects the systems of Ptolemy and Tycho Brahe and accepts the Copernican one.[232]

In the *Progymnasmatum*, the arguments Lansbergen uses for the motion of the earth are rather diverse. Most of them are philosophical in character. Nature does nothing without purpose; it is impossible for a single part to be at rest while the whole is moving; nature works in the simplest way. Lansbergen rejects Ptolemy's ninth sphere; above the fixed stars, there is nothing which would be able to move or put the universe in motion. One wonders whether some of these arguments were perhaps rather ad hoc, for instance when he turns the argument of the earth's heaviness against those who claim that this argues for the earth's immobility. How could such a heavy earth remain motionless in the air? A stone which is thrown moves naturally through the air. But nothing but a miracle can suspend a heavy object in the air. Still, the same argument appears later in the 'Considerations'.[233]

In the 'Considerations', the same or similar arguments reappear. By now, Lansbergen admits that they are only probable. However, he claims that the motion of the earth can also be proved in a strictly mathematical way, and this is what he sets out to do in the present work.[234] Correspondences between the motions of the heavenly bodies, which appear completely fortuitous in the system of Ptolemy, were explained by Copernicus in a logical way from the very properties of his system. As Lansbergen states, whereas Ptolemy's aim was simply to show the motions which are seen in the heavens, Copernicus also tried to explain their causes.[235]

The daily rotation of the earth is now argued from a calculation of the velocity which the outer sphere should have in order to turn round every 24 hours. To that end, he first calculates (referring to his still unpublished *Uranometria*) the diameter and the circumference of the sphere of the fixed stars. The velocity he arrives at is of such magnitude that according to Lansbergen such motion is simply impossible. In his view, this proof of the earth's motion is evident and infallible.[236] In order to argue the annual motion of the

---

[231] Lansbergen (1619) 102-115.
[232] Lansbergen (1629) 3-32, 32-103, respectively.
[233] Lansbergen (1619) 102-105. Lansbergen (1629) 16-17.
[234] Lansbegen (1629), dedication.
[235] Lansbergen (1629) 3.
[236] Lansbergen (1629) 8-12.

earth, Lansbergen has recourse to the motion of the moon. In the Ptolemaic system, the sun and the moon move independently in the heavens: each has its own orbit. In the Copernican system, however, the moon and the earth turn together around the sun. This means that the motion of the moon is composed of its own motion and that of the earth; or, the apparent motion of the moon in the sky is composed of its proper motion around the earth and the apparent motion of the sun in the Zodiac. That the latter case actually holds is proven by the authority of Ptolemy (Almagest, book IV Ch. 3); however, the quote is taken rather out of context. [237]

From the point of view of astronomy, the latter argument is nonsense. The period of the synodic month is determined by the moon's position with regard to the sun, be it in the Ptolemaic or the Copernican system. The argument appears to have failed to convince anyone. One suspects that it was too technical for the common man; moreover, Lansbergen's somewhat verbose explanation obscures rather than elucidates the matter. But even astronomers and fellow Copernicans were very sceptical. [238] Lansbergen, however, regarded this argument as decisive: 'so firm and infallible... that it cannot be disproved by any argument.' [239] For 'it has been taken from the nearest cause of the issue to be demonstrated. For such a proof is mathematical, and cannot be contradicted by argument.' Lansbergen trusted, therefore, that the argument would settle the dispute 'which has divided for so many years the students of astronomy', viz. whether the sun or the earth is moving, 'at least, among such people who are ready to allow room for reason.' [240] The episode at least illustrates the power Lansbergen attributed to mathematical arguments.

---

[237] Lansbergen (1629) 26-27; the same argument in Lansbergen (1619) 108-109. Cf. Ptolemy (1984) 179.

[238] For instance [Ismaël Boulliau], *Philolai, sive dissertationis de vero systemate mundi, libri IV* (Amsterdam 1639) 130-131. Only Daniel Lipstorp was to comment favourably upon it, in 1653, whereas Lansbergen's son Jacob endorsed it fully and tried to explain it to Morin and Froidmont (Jacob Lansbergen (1633) 67-70; see also the dedication). Modern commentators are as sceptical. Delambre (1821) 44 remarks dryly: *l'argument qu'il tire de la Lune, et qui lui paraît victorieux, paraîtra sans doute moins fort qu'il ne l'a cru.* Hooykaas thought the argument to be grounded in Lansbergen's belief in solid orbs: 'Lansbergen's triumphantly announced (great) argument for the earth's annual revolution, however, is disappointing. He pretends that only this argument can account for the length of the synodic month (...) It turns out to be based, however, on the gratuitous assumption that the orbit of the moon is immovably fixed to that of the earth (p 27-30). If this be true, the synodic and not the siderial month is the duration of a complete turn of the moon along its orbit.' (Hooykaas collection.) Strictly speaking, there is no mechanism implied, but it is certainly easier to invent such an argument when adhering to solid orbs.

[239] Lansbergen (1629), preface.

[240] Lansbergen (1629) 26.

As it seems, the above-mentioned arguments were justifications of the heliocentric theory, meant to convince the general public. So why had Lansbergen accepted it in the first place? Most probably, his view of the world originated from the cosmological principles as they had been developed by the Leiden humanists.

Lansbergen's Copernicanism was, in a sense, a consequence of his general pretensions. Lansbergen aimed at writing 'New and *true* [*genuinae*] theories' – where 'true' meant in accordance with physical reality. Therefore, he had to take a stance on Copernicus' views. In the preface to his tables, Lansbergen discerns two different types of knowledge of the motions of the heavenly bodies. The first is theoretical astronomy, which explains the motion of the heavenly bodies from fixed principles. The second is called *ποιητικη* or practical astronomy and offers the numerical values. [241] The latter does not offer theoretical insight, but just serves practical purposes. Practical astronomy is offered in his tables, theoretical astronomy in his 'New and true theories'. It is clear that Lansbergen esteems theoretical astronomy highest. Practical astronomy should be based upon these true principles.

Of course, these true principles do not just concern heliocentrism. Like Mulerius, Lansbergen states that the main principle of heavenly motions is that all bodies move regularly in circles. In his tables, he referred to the 'axiom of the astronomers', '*that the heavenly motions are circular, or constructed from circular motions*, which are eternal and invariable.' [242] At other places, too, he was adamant that the heavenly motions were circular. '... as every planet has but one circle wherein he makes his course in the heaven; therefore no theory can be in conformity with the heavens unless it exhibits the course of every planet in a circle.' [243] The idea of circular motion appears linked to the idea that the planets are moved by solid orbs. However, constant circular motion was not so much a mechanical as an aesthetic and philosophical principle. Lansbergen hinted at its religious dimensions on the title page of his book on the measuring of the circle. Accompanying an emblem of the Copernican world system, he put as a device *ὁ Θεός ἀεί κυκλομετρει* ('God always cyclometrises'), which was his version of the famous Platonic dictum 'God always geometrises'. [244]

Like other astronomers originating from Leiden, Lansbergen made much of the harmonious order of the cosmos. He not only defended heliocentric

---

[241] Lansbergen (1632), *Tabulae,* 1. The same distinction is made in Lansbergen (1619) 61-62.
[242] Lansbergen (1632), *Tabulae*, 3.
[243] Lansbergen (1650) 37. (I translate from the original Dutch version. Hortensius' translation, however, is faithful and can be used without restriction.)
[244] Lansbergen (1616/1628).

orbits for Venus and Mercury on this ground, as was common among the Leiden humanists, but – like Stevin and Copernicus himself – he also defended the heliocentric motion of the earth. Against Tycho he argued that, since the periods of revolution of the earth, Mars and Venus are 365, 687 and 225 days, respectively, the orbit of the earth should be situated between the other two. In the same way, Venus should move between the earth and Mercury.[245] Lansbergen stretches the argument even further: there is a fixed ratio between the distance of the spheres to the centre, and their period of revolution. Lansbergen is confident that this ratio is valid up to the sphere of the fixed stars. In this way, Lansbergen argues that the sphere of the fixed stars should be 28,000 times as far from the sun as it is from the earth, as the earth revolves around the sun in one year, and the fixed stars, in the motion called precession, in 28,000 years.[246] (Note that Lansbergen tacitly rejects Copernicus' explanation of the precession from a motion of the earth's axis.)

The fact that Lansbergen is one with Stevin in interpreting the order of the universe as implying heliocentricity, should not be taken as proof of a common approach. On the contrary, while both are stretching the common argument in the same direction, they do so for quite different reasons. Stevin, as an engineer, had interpreted the concepts of the humanists in a rather down-to-earth, mechanical way, ignoring all aspects of beauty or divine harmony. Lansbergen tends towards the other extreme. To him, the divine order of the universe was expressed by beauty and elegance. The proportions of the heavenly spheres and their eccentricities are so excellent, 'that they cannot be altered, without making the whole fabric of the heavens unorderly and ugly.' From the harmonious proportions following from his own theories, one can 'conclude with certainty and prove with clarity, that it is a true image of the heavens.' In other words, Lanbergen's model is true because it is beautiful. However, Stevin's world picture is, according to Lansbergen, ugly and therefore untrue: 'who only has a look at it, remarks immediately, that this cannot be an image of that beautiful heavenly building, that God has made in the beginning: but an invented image, full of confusion and disorder, not corresponding to the heavens.'[247]

Neither Stevin nor Lansbergen were really part of the learned, humanist culture in the Netherlands. Both, however, were close enough to scholarly circles to borrow their ideas from them: a search for pristine knowledge, the investigation of the cosmos by mathematics, with an emphasis on the order of the spheres. Both interpret them in their own way, in the process carrying

---

[245] Lansbergen (1619) 111.
[246] Lansbergen (1631) 70.
[247] Lansbergen (1629) 55-56.

them further than in their original context. In the case of Stevin, his interpretation can be linked to his upbringing as an engineer and his work in practical mathematics. In the case of Lansbergen, we have to look elsewhere. His background as a theologian is certainly of importance. Lansbergen's ideas on the cosmos were much more religiously coloured than was usual. However, his Christian inspiration was not restricted to the common Reformed creed. Perhaps as a result of his prolonged stay in the southern Netherlands, his worldview appears to have undergone strong influences of neo-Platonist philosophy and alchemical speculation. It is because of this background that in Lansbergen the ideals of cosmic beauty and harmony, which were common among his contemporaries, took on such force that they overturned the established world-view.

Lansbergen's Christian cosmology

The main source for Lansbergen's private feelings are his *Bedenckingen* or 'Considerations', his main defence of the Copernican system. As this book was not so much a scholarly work as a piece of popularisation and education, it allowed Lansbergen scope for reflections which, perhaps, did not belong in an astronomical work proper and indeed are lacking in his other works. He rather elaborately explicated his general view of the cosmos, a view with clear religious overtones. Indeed, Lansbergen himself testified that the book 'is built not only upon the foundations of geometry, but also upon the testimonies of the Word of God. Both of which are so infallible, that one cannot doubt its certainty and truth.' [248] Whereas in the publications of Mulerius or Blaeu, heliocentrism appears at odds with their notions on the Creation and the moral order in the universe, Lansbergen succeeds in drawing an overall picture which integrates the physical, moral and religious elements. Although by the time he published them his views had been overtaken by newer developments in cosmology, not everyone was immediately aware of the newest developments. Lansbergen's Christian interpretation of the cosmos may partly account for the success of the 'Considerations' and may indeed have contributed to making heliocentrism acceptable to a wider public.

A rather prominent element in Lansbergen's cosmology as put forward in the 'Considerations', is the threefold division of the heavens. The idea in itself was rather common. One comes across it in traditional Leiden disputations as well as in more daring cosmological works of the period. The division may partly have been inspired by Aristotle's' threefold definition at the beginning

[248] Lansbergen (1629), dedication.

of his book on the heavens, partly by the 'third heaven' mentioned by St Paul in his second letter to the Corinthians, 12:2 (to which Lansbergen refers). Most of these divisions – for instance, into the earthly heaven (the air), the sidereal heaven (planets and stars) and the empyreal heaven [249] – implied a geocentric world picture. As a Copernican, Lansbergen applies the idea in a somewhat different way. According to him, the first heaven is that of the planets; it stretches from the sun in the centre up to the orb of Saturn. The second heaven is that of the fixed stars; it stretches from the orb of Saturn up to the eighth sphere, that of the fixed stars. The third heaven is the empyreal heaven, the throne of God and the place reserved for the Elect. This heaven is invisible to us, but we know about it from scriptural revelation. [250]

The three heavens correspond with the three parts of the tabernacle: the first heaven corresponds with the court, the second with the holy place and the third with the most holy. [251] It is not surprising to find that Lansbergen also finds images of the Trinity in the world (although, amazingly, not in the heavens). Sun, moon and air correspond with Father, Son and Holy Ghost; equally sun, moon and earth. Such considerations of the creation as an image of the Trinity were not uncommon among neo-Platonists, particularly Paracelsians. [252]

The immense space of the second heaven is not empty. It is full of 'a host of invisible creatures'. Angels descend from the third to the first heaven, or return in inverse direction. Good and evil spirits engage in combat. [253] Moreover, the two inner heavens can only function because of the force they receive from the third heaven. According to Lansbergen, the sphere of the fixed stars is a sort of eggshell around the two inner heavens, which preserves the forces poured within. The analogy with an egg is even used to prove the existence of the third heaven. An egg contains everything necessary to produce a chicken, but nothing will come forth without the warmth of the brood-hen, which is applied from without. In the same way, nothing can be generated in the two inner heavens without the Spirit of God, who applies his force from the third heaven. [254]

The threefold division obtains clear neo-Platonic overtones. There is a gradual ascension from the first heaven. The second heaven, which reaches from the orb of Saturn up to the stars, is much more spacious than the first,

---

[249] See for instance the Leiden disputations by Murdison, 2 June 1601, th. 11; and Burgersdijk, 1627; or Kort Aslaksen (Aslacus) in his book *De natura coeli triplicis* (1597), see Moesgaard (1977) 301-302.

[250] For contemporary views on the Empyrean Heaven, see Donahue (1981) 223-246.

[251] Lansbergen (1629) 51-52. A similar analogy was made by Aslaksen: Moesgaard (1977) 301.

[252] Lansbergen (1629) 58-59. Cf. Rudolph (1980); Walker (1972) 110-122; Kepler, *Werke*, VII, 51.

[253] Lansbergen (1629) 83-87; quote on p. 83.

[254] Lansbergen (1629) 87-88, 76-77.

planetary heaven. It is also much lighter, as the first heaven is illuminated by the sun only, and the second heaven is illuminated by the innumerable stars (their splendour does not reach further down than the orb of Saturn). The empyreal heaven is even more spacious and illuminated than the second heaven. As the Bible says, God is 'dwelling in light unapproachable' (1 Timothy 6:16).[255] This third, empyreal heaven is in its turn a reflection of the splendour of God himself: 'From the beauty of the third heaven, I am finally brought unto God himself, who is the beauty of all beauty, that is, that excellent Pulchrum, or that most beautiful beauty, whereover the philosopher Plato has been so full of wonder, that he has fallen in love with it, as he testifies in his *Phaedro* [Phaedoon] and other places.'[256]

Indeed, Lansbergen appears to have been rather sympathetic towards the neo-Platonist and Paracelsian movements of his time. This is indicated by an episode from an early date. As a minister in Goes, Lansbergen also provided medical aid. Regular physicians did not like this competition. When Lansbergen's position was weakened by his quarrel with the town government, they used the unlucky outcome of a case treated by Lansbergen to attack him as a dangerous quack. The period 1612-1614 saw a bitter pamphlet war on the issue, which was not confined to Goes, but also drew participants from the two main cities in Zealand, Middelburg and Zierikzee.

In fact, the pamphlets do not deal with the philosophical background, or with medical ideas generally. They simply discuss the specific case, viz. Lansbergen's prescribing of a musk-containing medicine to a lying-in woman (his opponents generally failed to point out that he had done so only after all other remedies had failed). This is not very instructive. Some indications, however, may be found in an exchange of letters on the case between Lansbergen and the Middelburg physician Carolus Francius, which were later published by Lansbergen's son Jacob.[257]

As Lansbergen explained in a letter of 29 January (4 Kal. Feb.) 1613, the medicine objected to was his *pulverum panchraestum*, 'which we possess as a secret.' Here, 'secret' should not be understood in the modern sense. 'Secret' remedies were those which were supposed to be effective because of their occult qualities. They were rather controversial among traditional physicians, as Francus' reaction shows. In his opinion, 'secret' remedies were harmful and poisonous. 'An example can be shown by the powders of the empirics, and by the followers of Paracelsus, called *flores anthos*...' Lansbergen replied that Her-

---

[255] Lansbergen (1629) 89.
[256] Lansbergen (1629), dedication.
[257] See Lansbergen (1613). Other contributions to the debate are Ultralaeus (1613); Herls (1613); Liens (1614) a and b; N.N., *Epistola apologetica*.

metic secret remedies could have their use, too. It all depended upon their application: all medicines were harmful when applied without judgement. And then, 'not all things ascribed to Paracelsus have been approved by Paracelsus.'

In a later letter, Lansbergen returns to this in an attempt to allay the suspicion that he should be willing to propagate all kinds of unorthodox methods. 'What I wrote has not been written in the defence of the Hermetics or of Libavius, but in my own; to say that I have not diverted from dogmatic medicine, nor from Christianity.' Andreas Libavius was a German schoolmaster and an active chemist, who tried to save the chemical profession from the esoteric humdrum of the Paracelsians. He made himself a name both by his systematic description of chemical practice and by his violent polemics with the Paracelsians. As an opponent of Paracelsianism he must have been well known in Zealand, as one of his polemics had been with the physician Joseph Michelius, who lived at Middelburg. [258]

Many things remain obscure, but it would seem that Lansbergen accepted cures which were not admitted by more dogmatic physicians. The bitterness of the debate may partly have been caused by such conflicting views. This suggestion is rather strengthened when we look at his supporters. He was defended zealously by Cornelis Liens, town physician in Zierikzee. Liens' contributions to the polemics in 1614 are remarkable mainly for their extreme verbosity and literary pretensions; his style earned him the (undoubtedly somewhat ironically meant) compliments of his adversaries. [259] His views become clearer, however, from a long poem of neo-Platonic tenor he wrote later in life: 'The little world, containing the hidden cause of love, the negotiation of the soul, and the true proof of its immortality'. Herein, he also included a description of the macrocosm wherein he closely followed Lansbergen's views. He describes the three heavens, calls the sun 'Mercury-Trismegiste's god', compares the universe with an egg, and even repeats Lansbergen's analogy of the third heaven with a brood-hen. [260]

In Lansbergen's astronomical work, too, there are occasional references to Hermetic ideas. Lansbergen quotes Copernicus' remark that Hermes called the sun a visible God, and explains that one should not take this too literally, as Hermes 'understood very well that God is invisible, one in essence and

---

[258] Hannaway (1975) 75-151. Trevor-Roper (1985) 175. Some attestations of Michelius' activities in Middelburg in Smetius (1611) 721-725. Lansbergen's remark that all medicines are dangerous when applied by ignorants could be a reference to Libavius' *Alchymia triumphans* from 1607; see Wightman (1962) 1, 260-261.

[259] Liens (1614)a 48; (1614)b 37.

[260] Liens (1655) 177-180. On Liens and his poem: Meertens (1943) 354-360.

three in persons, as he clearly demonstrates in his writings.' [261] One should not attach too much value to such rather casual references to Hermes, although one should remember that Lansbergen does not often quote authorities. [262] However, there are other references. At another place, he referred to a 'very true' axiom of Hermetic philosophy, 'that everything which is present within the circumference of a circle in a diffused state, is also contained potentially collected in the centre of this very circle.' [263] This sounds very hermetic indeed. What it implies is that, as the light is in a diffused way present in the sphere of the fixed stars, it must also be present in the centre of this sphere. Hence the earth, which is not a luminous body, cannot be there; at the centre of the world should be the sun. [264]

Alchemy and the motion of the earth

Quite interestingly, Lansbergen also referred to alchemy: he called it 'lower astronomy', a name sometimes used at the time. [265] He appears to have been well versed in alchemical literature. In the *Progymnasmatum*, he expresses amazement that such a diligent alchemist as Tycho Brahe thought the motion of the earth unacceptable: 'from the writings by those who have practised this art, he could have learnt, that the daily rotation visible in the heavens, is in reality proper to the earth.' He therefore refers to the medieval philosopher Roger Bacon, as well as to his compatriot, the '*summus philosophus*' Cornelis Drebbel, who should have demonstrated this with his own hand. [266]

The reference to Bacon was second-hand; Lansbergen took it, as he explains, from the German chemist Libavius. [267] The reference to Drebbel is more interesting. Drebbel was born in or around 1572 in Alkmaar and

---

[261] Lansbergen (1629) 59.

[262] Lansbergen's Hermeticism is discussed by Donahue (1981) 128, 155, 243.

[263] Lansbergen (1619) 110.

[264] This explication is given by Polacco (1644) 110 (assertio 180). Note the resemblance to Beeckman's later speculation.

[265] Telle (1992) 239. Crosland (1962) 6.

[266] Lansbergen (1628) 105-106. The reference to Tycho concerns *Epist. Astr.* 1, [1596] 117: his letter to Rothmann, 17 August 1588. See Brahe (1919/1972) VI, 146.

[267] According to Hooykaas, the reference is to Andreas Libavius, *Examen philosophiae novae, quae veteri abrogandae opponitur* (Frankfurt 1615) 58 footnote 16: *Coelestem familiam transferre in globum terrenum, et totam oeconomicus exacte repraesentare, quale quid fecisse dicitur Archimedes, Drebelius, et alii. (...) Arbitratur Rogerus materiam posse invenire, quid quaedam apud nos sic moveatur ut coelum, veluti Cometae, maris aestus esc. Verum haec opinio est. Drebelius aliud videtur excogitam, ad exemplum motus, qui sit in arcanis lapidis vitro inclusi, etc...* (Collection Hooykaas).

developed into a kind of technological wizard. [268] Initially, he tried to make a living as a technician in the Republic. He obtained letters patent for several inventions and was engaged, in 1600 or 1601, in the construction of water pumps for the town of Middelburg. [269] Still, he was no ordinary engineer. Drebbel posed as a kind of miracle-maker who, by his knowledge of the hidden mysteries of nature, could work great things. Thorndike described him as 'probably the most pretentious, secretive and magical figure of the early seventeenth century.' [270] Drebbel's claims can only be understood in the framework of a magical and neo-Platonic worldview. He soon left the Republic for the Renaissance courts of London and Prague, where, probably, he found more scope for such an attitude than in the rather down-to-earth atmosphere of the Dutch Republic. [271]

Still, his compatriots did not forget him and probably felt some pride in his being appreciated by royalty. Some of them appear even to have sympathised with his esoteric ideas. Drebbel's most ardent supporter in the Dutch Republic was his fellow citizen from Alkmaar, Gerrit Schagen, a self-taught man celebrated for his learning. Regrettably, hardly anything is known about Schagen. He published several books to promote Drebbel's fame. The first of these was published in 1607. Its contents are summarised in the rather longish title: 'Miraculous discovery of the perpetual motion, which Cornelis Drebbel, philosopher from Alkmaar, has occasioned by a perpetual moving spirit contained in a sphere. Whose dedication (on the occasion of his offering it to the powerful King James of Great Britain) is rendered here verbatim [*naecktelijck*]. Equally the testimonies offered by Cicero, Claudianus and Lactantius of the perpetual motion allegedly found by Archimedes. Equally from Bartas on Ferdinand, who has sent a perpetual motion to the Turkish emperor at Byzantium. Equally is added a book Pymander, written by Mercurius Trismegistos, who allegedly has been a philosopher, priest and King in Egypt in Moses' time'. [272]

The references to earlier examples of perpetual motion clearly serve a purpose. Schagen's aim with the latter part, which represents the first full publication of the *Corpus Hermeticum* in Dutch, [273] is more difficult to assess. Still, it is by far the most comprehensive: 60 pages as opposed to 10 for the other

---

[268] The fundamental study is Jaeger (1922). For a recent overview, see Snelders (1980) 110-119. English works are Tierie (1932) and Harris (1961) 119-223.

[269] Jaeger (1922) 14-15.

[270] Thorndike, VII, 492.

[271] Cf. Evans (1973) 81, 189.

[272] Schagen (1607).

[273] Cf. Janssen (1989) 233-235. The translation appears to have been made after the 1548 Italian edition by Benci.

parts taken together. Apparently, Schagen thought there was some connection between the wisdom of Hermes and the inventions of Drebbel. The key part of the edition, however, was clearly the dedication by Drebbel to King James. This text is only preserved in the version offered by Schagen, which, however, turns out to be the original. Drebbel, unable to write in either English or Latin, had it translated before offering it to King James. [274]

Central to the publication is the announcement of Drebbel's discovery of perpetual motion. In the literature, one generally denotes as such Drebbel's instrument which could imitate the tides of the sea (modern authors consider it a kind of thermoscope or baroscope). [275] This, however, seems to be based on a misunderstanding, which may have been partly intended by Drebbel. The said instrument was only one of the applications by which Drebbel wanted to demonstrate that he had penetrated to the cause of the *primum mobile*, and thus of all motion and order in the universe. Dedicating the instrument to King James of England, he explains how he has found this cause. But, 'if, My King, I could not prove this with actual [*levendige*, 'living'] instruments as well as with natural reason, I would not have dared to write as much... So, as a proof that I have found the cause of the *Primum mobile*: I make a globe which can be moved eternally after the course of the heavens, all round once in every 24 hours or so much more often as needed, so that it will not fail in a thousand years. Denoting us years, months, days and hours, as well as the course of the sun, the moon, and all planets and stars known to man...' There follows a long list of other instruments Drebbel claims he can produce, with in the end the instrument reproducing the tides, which is offered on the occasion. [276]

In fact, all these instruments should be seen as expressions of perpetual motion. More in particular, however, this name applies to the first-mentioned instrument, the globe continuously turning all round. So it was generally interpreted by Drebbel's contemporaries. Although the dedication does not state that Drebbel actually built this instrument, there exists a later description by William Boswell (who exposes it as a fraud), which presupposes a working

---

[274] The letter was also printed (at least in later editions) in another book edited by Schagen, which contained a tract by Drebbel on the nature of the elements. The first edition of this book seems to have been lost, but there is a German translation from 1608 and a second Dutch edition from 1621, along with a further tract by Drebbel on the fifth essence. Most probably, the first edition was from 1607 as well. (I take as spurious an edition of 1604, which is sometimes mentioned in the literature, but which nobody has ever seen.) For a full list of all editions of Drebbel's works, see Jaeger (1922) 5-6.

[275] Drebbel's perpetual movement is discussed by Jaeger (1922) 63-69; Harris (1961) 152-159; Michel (1971); Drake-Brockman (1994) *passim*.

[276] Drebbel (1607) (n.p.) and (1621) 56.

prototype.[277] In some cases, it was combined with the instrument which reproduced the tides, which at least partly accounts for the confusion. Probably, Drebbel conceived of his *perpetuum mobile* in Holland. It seems to be identical to an invention described in letters patent granted him in 1598 as 'a watch or timepiece which can be used continuously during a time of 50, 60 or 100 years or more, without being winded up or treated in any way, as long as the wheels and the other clockwork are not worn out.'[278] The instrument, or claim, must have been well known in the Dutch Republic. That talk of such a *perpetuum mobile* was common is attested by Mulerius: 'Several artisans are trying with great diligence to invent an instrument [*automatum*], that moves with the heavens both eternally and equally, that is, without any intensifying or slackening of its motion. Given this, one could find geographical longitudes in the following way. (...)'[279] (That is, by using it as a timekeeper.) Apparently, an esoteric world-view could well go with an eye for practical applications.

However, the instrument could also serve cosmological speculation. When Gerrit Schagen published Drebbel's letter of dedication to King James, he dedicated it in his turn to the engineer Adriaen Anthonisz (note that all three – Drebbel, Schagen and Adriaen Anthonisz – came from Alkmaar). Schagen's dedication is dated December 1607. He explains that astronomy cannot be perfectly known without Drebbel's invention. 'Were this science common among astronomers, one would not need so many hypotheses, and calculations of the planets and the other stars. Astronomy would be easy, and Copernicus would flourish: for he proves (by reason) that the earth moves all round every 24 hours. But this philosopher from Alkmaar is able to prove the same not just by reason, but also with actual [*levendighe*] instruments.'[280] The allusion is rather cryptic, but as it seems, the turning globe should be seen as an image of the earth itself. The motion of the globe would therefore prove the motion of the earth.

This is indeed how Lansbergen took it. Lansbergen's reference to Drebbel as substantiating the rotation of the earth clearly concerns the latter's *perpetuum mobile*, the globe turning all round. Lansbergen appears not to doubt that Drebbel's perpetual motion was a reality. As authorities to substantiate this he refers to Libavius, Fanianus and John Dee. Libavius has been mentioned already. Johannes Chrysippus Fanianus was an alchemist from Basel. His work had been included in a collection of alchemical tracts originally pub-

---

277 Jaeger (1922) 66-67.
278 Full text of the letters patent: Jaeger (1922) 119-120.
279 Mulerius (1616) 113.
280 Schagen (1607).

lished in 1602 and re-edited at Strasbourg in 1613 by Lazarus Zetzner.[281] The English mathematician John Dee was deeply involved in the study of Renaissance magic. The reference is to the dedication to the emperor Maximilian of his *Monas hieroglyphica*, originally published at Antwerp in 1564, but also included in Zetzner's collection.[282]

In fact, although these authors do mention Drebbel's claims, it is hard to take them as saying that Drebbel actually carried out the experiment. Moreover, none of these authors was a Copernican and none of them interpreted the experiment as supporting the theory of a moving earth. Drebbel spoke of it as representing the motion of the *heavens* and that is how most authors appear to have looked at it. It seems probable that Lansbergen knew of Drebbel's machine from other sources, and simply looked up some learned references to substantiate his claim. Considering Drebbel spent some time in Zealand in 1600/1601, it is even possible that the two met. It is striking that Gerrit Schagen, who had publicised Drebbel's text (equally in a Hermetic context) and indeed had represented the *perpetuum mobile* as proof that the earth is turning on its axis, is not mentioned by Lansbergen. Yet, Lansbergen holds the same view: 'Those who nowadays exercise this art [alchemy], know that the *terra physica* not only moves all around in a day, but, what is in particular remarkable, that it is moved continuously from west to east. I do not have any doubt that the great earth (*Tellus*) too is moved in a day in the same way, according to the saying by Hermes Trismegistus, '*Sic mundus creatus est*', which I earnestly approve.'[283]

The expression *terra physica* is not quite clear, but seems to denote the element of earth as it allegedly can be procured by alchemical operations. Drebbel's experiment reminds one of the *terrella* of William Gilbert. Froidmont already identified Lansbergen's argument with the more familiar (though hardly less esoteric) argument from magnetism and rejected it as such. However, Lansbergen's son Jacob, in his reply to Froidmont, denied such identification and pointed out that the question was about the 'physical earth' of the chemists, which appeared in chemical processes. After this explanation, Froidmont not unsurprisingly despised the argument still more and joined

[281] *Theatrum chemicum* (1613), I, 25-62. (I found this reference in the Hooykaas collection.) Cf. Ferguson (1954) II, 439. See on Zetzner: Pagnoni-Sturlese (1995), for the *Theatrum* in particular 363-366.

[282] *Theatrum chemicum* (1613), II, 191-230 (*Monas*); 191-204 (dedication). An English translation of the full work is offered by Josten (Dee 1964). The best study on Dee is Clulee (1988), see pp. 77-124 for the *Monas*.

[283] Lansbergen (1619) 106.

Mersenne, who had written earlier that he would leave the argument of the 'physical earth' to melancholics, Paracelsists and Fluddists. [284]

It may well be that Lansbergen himself in some way conflated magnetism and the occult qualities of the alchemists. At another place in the same work, he attributes a magnetic nature to the earth in order to account for the constant position of its axis. [285] However, Lansbergen's *terra physica* takes on quite a different character than Gilbert's *terella* or Stevin's ideas on magnetic force. With Lansbergen it becomes a mystical alchemical concept. The phrase *Sic mundus creatus est* (Such the world has been created) is the tenth (or, according to another division, the twelfth) section of the *Tabula smaragdina* (Emerald Table), an obscure text of rather unclear provenance, which was generally ascribed to Hermes and had a lot of prestige among alchemists. The authoritative commentary on the *Tabula* by Hortulanus explained *Sic mundus creatus est* as: just as the philosopher's stone has been formed, so the world has been created. [286] Drebbel's apparatus thus is thought to mirror the universe as such. The conclusive point seems to be that the *terra physica* turns from west to east, like the earth in the system of Copernicus. The heavens, according to Ptolemy, move from east to west. Taken as an analogy between the philosopher's stone or *terra physica* and the outer world, one should conclude to a motion from west to east in the world, which could be in the earth only.

None of this seems really incompatible with humanist scholarship. Yet, there is a difference of emphasis. To Lansbergen, the opinions of the ancients are still the yardstick for measuring truth and reality, but he feels less bound to established scholarly tradition as to what these opinions really were. In the end, his world is not shaped by the study of ancient authors but by religious notions. Hence his readiness to engage in controversial ideas – Paracelsian medicine, Copernican astronomy, and so on. Lansbergen is certainly aware of the ideas of the humanist scholars, but he uses them much more freely. The results of humanist scholarship are used only in so far as they fit in. It is exactly in this way that the humanist notions could be given a more radical use and turned into arguments for Copernicanism.

---

[284] Jacob Lansbergen (1633) 15; cf. Monchamp (1892), 102-104. Mersenne (1623) 915, cf. 891.

[285] Lansbergen (1619) 114.

[286] Ruska (1926), offers on p. 2 the text of the table, on p. 185 the relevant commentary by Hortulanus, and on pp. 206-224 an overview of the work's authority in the sixteenth and seventeenth centuries.

Conclusion: the Leiden interpretation of Copernicus' theories

It appears that the way people picked up Copernicus' theories in the Dutch Republic was quite different from the 'Wittenberg interpretation' as it has been defined by Westman. People at Wittenberg regarded Copernicus' theories as foremost a mathematical tool for calculating stellar positions. They ignored his arguments that the heliocentric system better maintained the mathematical order of the universe. At Leiden, people appear not very interested in 'saving the phenomena'. The argument introduced by Osiander, that one should use Copernicus' hypothesis for astronomical purposes without bothering about their realness, is hardly met in the Dutch Republic at this time. It appears that people were not very impressed with Copernicus' mathematical-astronomical theories. The Leiden humanists were, on the contrary, interested in the constitution of the universe, and quite impressed by Copernicus' arguments regarding cosmic harmony and '*symmetria*'. I propose to call this approach the 'Leiden interpretation', as opposed to Westman's 'Wittenberg interpretation'. This interpretation thus represents a common way of reading Copernicus' work and building upon, or ignoring, parts of his theories, rather than a well-defined cosmological theory. It covers both heliocentrists like Lansbergen and geocentrists like Mulerius.

This is not to say that people at Leiden interpreted Copernicanism as a physical rather than a mathematical theory. Things were more complicated than that. The 'astronomical axiom' that all celestial motions happen in regular circles appears to have been widely agreed upon. Now, this was largely a physical principle. Mulerius defended it referring to the *cause* of heavenly motions. Therefore, Copernicus' astronomical (mathematical) hypotheses were attractive from a *physical* point of view. On the other hand, the heliocentric system appears to have been attractive mainly because it better maintained the harmonious ordering of the universe; that is, the succession of the heavenly spheres according to their periods of revolution. This, indeed, appears to have been the major reason why Stevin and Lansbergen accepted the system of Copernicus. And whereas most astronomers appear to have rejected heliocentrism in the end, Copernicus' arguments carried enough force to make them modify their world-view. Mulerius was ready to accept the daily rotation of the earth for this very reason, and the heliocentric orbit for Mercury and Venus appears to have been adopted unanimously. Saving the harmonious order of the cosmos appears at first sight to have been a mathematical argument, and in a sense it was. So, here emerges the paradoxical situation that whereas the hypotheses Copernicus used in his mathematical astronomy were found attractive from a *physical* point of view (regular circular motion), his model of the universe was deemed attractive for *mathematical* reasons. The simple distinction between a physical and a mathematical part of astronomy clearly does not hold. It is the belief in a harmonious order of the cosmos which makes such mathematical arguments relevant. People could be convinced by them because reality itself was deemed *essentially* mathematical in character.

While people at Leiden interpreted Copernicus' theories as a view of reality, they had grave difficulties in endorsing it fully. The problem appears to have been how to bring the new view of the universe into line with traditional notions of the world's order – an order which was deemed mathematical, but also more than that: moral and religious. As a result of their humanist education, Leiden scholars had appropriated such traditional notions to a considerable degree. Small wonder that they shrank back from the radical consequences of Copernicus' view of the universe. It is probably no accident that heliocentricity was openly endorsed at this time only by relative outsiders, viz. people who had come into the orbit of Leiden University and had come to appreciate the arguments of the humanists, but who had earlier been formed in a different way of life. They integrated elements of the 'Leiden interpretation' into a different world-view, which in some cases proved more malleable or more compatible with heliocentrism. Lansbergen's biblicism, with Hermetic and occult elements, is an outstanding example of integrating cosmographical and Christian notions. In Stevin's life and works, on the other hand, religion appears to play no part. His is rather a mechanic's view of the universe; but as such, it does offer an integral view. He may well have thought about its moral dimensions, but in that case, he wisely did not put this down on paper.

Part II. The challenge to philosophy

6. Conceptions of the cosmos after 1610

Dissemination of the debate

Until the first two or three decades of the seventeenth century, the system of the world does not appear to have been a pressing problem. Outside scholarly circles probably not many people were aware of the debate. Stevin's work, which sought to bring the issue to a wider audience, drew little attention. Lansbergen's *De motu solis* from 1619 did not sell very well either, as we have seen. But when in 1629 he published his *Bedenckingen*, this work drew considerable attention. As noted, the book was the first popular account of the Copernican system published in Europe and as such undoubtedly played its part in the further propagation of the Copernican system. The publication of his tables and other more scholarly works in the following years certainly enhanced his credibility.

Still, one may surmise that Lansbergen's work was noticed primarily because of a growing interest in cosmography. In the same year (1629), the French philosopher Pierre Gassendi made a tour through the Dutch Republic. He sent an account of his experiences to his patron Fabri de Peiresc. Gassendi visited scholars in The Hague, Amsterdam, Utrecht and Dordrecht. He passed through Middelburg (Zealand) but: 'I failed to remember that this was the home town of Mr. Lansbergen. So, much to my regret, I did not meet him.' Gassendi's regret will have increased as in the following years Lansbergen's main works came off the press. At the States' army, engaged in the famous siege of 's Hertogenbosch (the United Provinces were still defending their independence), he was introduced to a number of the States' engineers, among them Albert Girard. Gassendi concluded this part of his overview with the words: 'For the rest, all those people are in favour of the motion of the earth.' [1]

Probably the most important source of the rising interest in cosmology

[1] *Au reste tous ces gens là sont pour le mouvement de la Terre*. Gassendi to Peiresc, 21 July 1629. Quoted by de Waerdt in IBJ, IV, 153.

were the discoveries Galileo and others made concerning the heavens from 1610 onwards by means of the telescope. These turned the discourse on the heavens from an abstruse mathematical or scholastic debate into a display of undeniable and fascinating wonders. Although none of these discoveries was made in the Dutch Republic, the Dutch could still pride themselves that their country was the cradle of the instrument. They soon learned to use it for astronomical observations themselves. As we have seen, Mulerius and Metius had watched sunspots through a telescope. The telescopes they used were most probably those of Metius' brother, Jacob Metius, one of the claimants of the invention of the instrument and by any account a very able technician. Astronomical observations by means of telescopes were also practised in adjacent territories, as demonstrated by Thomas Harriot in England and David Fabricius in the principality of Eastern Frisia.

However, before 1630, the telescope was not a widely used instrument. It was rare even among the learned. Jacob Metius was a rather unsociable man who did not produce for the market and who took no trouble to communicate his art to others. His telescopes were not widely available and when he died, in June 1628, the secret of his craft was lost. Contemporary instrument makers did not seem able to achieve his high standards. By 1622, Lansbergen appears to have become enthusiastic about telescopes and he urged Isaac Beeckman to build one. [2] Beeckman took the subject to heart, but it was far from easy. As a matter of fact, one could buy telescopes in the Dutch Republic. Beeckman himself bought one in Delft; he used it in 1631, but apparently it did not satisfy him. Lens grinders in the Dutch Republic proved unable to procure the lenses he needed. Beeckman therefore decided to learn the craft himself. For several years, he served apprenticeships with lens grinders in Middelburg and Amsterdam. [3] The poor quality of Dutch telescopes during this period is confirmed by Hortensius. When discussing with Galileo his method of determining longitude – which was based on observations of the moons of Jupiter – Hortensius complained that no instrument maker in the Dutch Republic was able to provide the lenses necessary for this work. They were generally ignorant men, with no knowledge of dioptrics. [4] Hortensius himself used a telescope for several observations from 1625 onward. [5]

Thus it was some time before the Dutch public was informed about, and

---

[2] *IBJ*, II, 294-295.

[3] Van Berkel (1983) 140-142.

[4] Vanpaemel (1989) 122.

[5] Lansbergen (1632), *Thesaurus*, 177, 171, 144 (observations from 1625, 1632, 1627). These observations were made at Leiden and the instrument need not have been Hortensius' own. An observation at Dordrecht in 1630 (*ibidem*, 124) may have been made with Beeckman's telescope.

believed in, the new and undreamed-of face of heaven. But by 1630, telescopes gradually became more common. The regent and poet P.C. Hooft, who was not known for his interest in natural science, was by 1641 familiar enough with the instrument to refer to Galileo and his telescope in a private letter – and in a completely non-astronomical context, one should add.[6] Dirk Rembrandtsz, a shoemaker turned mathematical practitioner, observed the phases of Venus in 1646. He also made observations of Mars, probably at about the same time. He used a telescope four feet long, which he may have obtained (but this is just a guess) thanks to his acquaintance with Descartes.[7]

A quite remarkable booklet, largely inspired by the new telescopic observations, appeared in 1638: 'Oration on the new wonders of the world, and on the use and dignity of mathematics and geometry'.[8] Its author was Jacob Spoors, who announced himself as public notary in the town of Delft, solicitor at the municipal court of justice and sworn-in surveyor. The last-mentioned qualification presupposed some mathematical knowledge. He explains elsewhere that he had passed his exam under Willebrord Snellius. Mathematics was Spoors' real passion. The 'new wonders of the world' of which his title speaks, are mainly the new cosmological discoveries of the time. Spoors appears to be familiar with the works of Stevin and Galileo. He claims, however, to have made his own astronomical observations, with the telescope (in fact, he writes: 'by means of a certain optical instrument'[9]), of sunspots, the phases of Venus, the moons of Jupiter (he indeed calls them 'moons', and believes each of them has its own sphere), and the surface of the moon.

These unheard-of marvels, which had come to light in his days, moved him to a further consideration of cosmological questions. As he argues, all these strange new things confirm the theory of Copernicus concerning the daily and annual motion of the earth. This leads him to a long digression on the history of the ideas on the system of the world. Heliocentrism was known in antiquity, in particular by the Egyptians. However, this science was suppressed by the Greeks, who thought that knowledge of the heavens was divine and unattainable for humans. Hence they refused to investigate it by means of natural reason. Philosophers who knew better and did not hide their views, such as the Pythagoreans, were persecuted. Copernicus at least brought the old theories to light again, not because he chanced upon some

---

[6] Hooft to Barlaeus, in Hooft (1979), II nr 1078: '*Wat dunkt U.E. van zoo een starrekijker, die dat afzien kan? Zoud'hij't Galileo Galileï niet wel te raeden geeven?*' On Hooft see below, p. 218–219.

[7] Van Nierop (1661) 78. On Van Nierop see below, p. 193–198.

[8] Spoors (1638). This book was recently discovered by Eric Jorink (Groningen), who kindly permitted me to include it in this overview.

[9] '*door sekere ghesicht-ghereetschap.*' Spoors (1638) 6.

ancient works preserving the older traditions (an idea Spoors ascribes to Stevin) but because – having found of his own accord that Ptolemaic theory was utterly wanting and that heliocentrism would be more rational – he finally found in old Pythagoraic writings the data that enabled him to bring the theory into agreement with observations.

As a whole, Spoors' book is rather programmatic in character. Spoors claims he is able to give mathematical proofs for his ideas, but the proofs themselves are lacking. One suspects that with this book, which he dedicated to the Delft municipal government, Spoors tried to get support for publishing a full account of his ideas. He indicates that he had written several other works, on astronomy and other branches of mathematics; however, as far as is known, he never published anything apart from his 'Oration'.

How far the Copernican system gained credibility after 1610 is difficult to tell. There is an interesting instance, however. In 1623, the Admiralty Board of Amsterdam decided on a request by a certain Claes Jacobsz Broeck, from Edam. 'Having found, as he says, that not the heavens but the earth is moving and turning', he asked to be put on board a man of war as a pilot, 'so that he could on that occasion test his invention.' The Board did not think this idea at all unbelievable and agreed to the request. A month later, they even awarded a remuneration of forty-two guilders to this same Claes Broeck, for having offered the admiralty 'an instrument of his own invention about the course of the sun and the attraction of the [magnetic] needle, for the use of navigation.' [10] Probably, the two instances refer to the same invention. Nothing more is heard of it and details are lacking. Its alleged use for navigation might allow for the willingness of the admiralty lords to test it, but it is still curious that this aspect is not mentioned in their first considerations.

A second factor which aroused interest in the Copernican question was the trial of the famous astronomer Galileo and the open condemnation of heliocentrism by the Roman church in 1633. This trial drew attention all over Europe and put the question at stake at the centre of public interest. As earlier surmised by Hooykaas, it aroused interest in the Copernican system in the Dutch Republic, too. Galileo's condemnation had no direct consequences for the United Provinces. The Pope did not have jurisdiction there and, besides, most people regarded his decisions with suspicion – though not many went so far as the German theologian J.J. Hainlinus, from Tübingen. According to the Swiss professor Petrus Megerlin in a book published in Amsterdam in 1682, Hainlinus had told him in 1644 that the question of the compatibility

---

[10] ARA, 1.01.46 nr. 1369; I thank Bert Westera (Amsterdam University) for drawing my attention to this source.

of the Bible with the Copernican system had vexed him for a long time. However, when he heard of the Roman Church's condemnation of Galileo, he was in doubt no longer: 'for the spirit of deceit has occupied that see of the great antichrist so firmly, that no truth can ever come forth from it.' [11]

Still, the trial certainly drew attention. From Paris, Mersenne explained the issue to André Rivet, court preacher to the stadholder at The Hague, apparently in answer to a question posed by Rivet. [12] There were also some reactions in the academic sphere. In 1636, Caspar van Baerle (Barlaeus), professor at the Amsterdam Athenaeum, as a kind of introduction to the course of physics he was to teach that year, held a public oration *De coeli admirandis* (on the marvels of the heaven). In it, he presented the heavens as a demonstration of God's greatness. Although the question of the world system was not really his concern, he touched on it when he discussed the velocity at which the heavens move. After all, this makes a great difference in the Copernican and the Ptolemaic system. 'I shall not investigate the arguments within this controversy. Nor shall I say my opinion, although I have no fear of the Pope.' In the remainder of his argument, he presupposed the Ptolemaic system. [13] At the opening of Utrecht University in 1636, the rector Voetius held an oration on the sciences, during which he briefly touched on the Copernican system. As Voetius' role will be more fully discussed in a later chapter, I will not deal with it here.

After Galileo's condemnation, there was also an attempt by a group of Dutch scholars to offer him asylum. Plans were made to invite Galileo to come to the Dutch Republic to continue his work unhampered by the Inquisition. The initiative seems to have come from Hugo Grotius, who had been convicted after the Arminian troubles and was living as an exile in Paris. He suggested the idea to his friend Gerard Vossius in Amsterdam, who in turn discussed it with some other friends. One of these was Martinus Hortensius, whom we met earlier as Lansbergen's pupil and who by then was a professor at the Amsterdam Athenaeum. The others were the cartographer Willem Jansz Blaeu and Laurens Reael, who was an important regent and an amateur scientist. All supported the proposal enthusiastically. That it came to naught was due to Galileo himself. When he heard of the project, he replied that his advanced age and state of health did not permit him to undertake the voyage. [14] Had it not been for this, the project might well have succeeded.

---

[11] Megerlin (1682) 72-73.

[12] Mersenne to Rivet, 8 Feb. 1654. Mersenne, (1932-1986) IV, 37-38. Rivet's letter has not been preserved.

[13] Barlaeus (1636).

[14] Van Berkel (1983) 146-147.

Of course, it would have needed the support not only of some Dutch scholars, but also of the Dutch authorities. However, Grotius and his friends had found an effective means of arousing the interest of the regents: Galileo was to bring his method for finding longitude with him. Indeed, even after the original plan had been dropped, the possibility that Galileo might offer his method to the States-General gave rise to discussions between Galileo and the Dutch. For several reasons, the discussions were protracted and in the end nothing came of them.[15] Yet, these very discussions on a subject the Dutch took to heart may well have contributed to drawing further attention to Galileo's theories generally.

We meet very nearly the same people in another pro-Copernican action in the wake of Galileo's trial. This concerned the publication of *Philolaus*, a Copernican treatise by the French astronomer Ismaël Boulliau (Bullialdus). Boulliau had written it before the trial, but hearing of Galileo's fate, he considered it too dangerous to have it published in France. He therefore sent it to his friend Vossius in Amsterdam in order that it might be printed, anonymously, by Blaeu. Hortensius was asked to recommend the work, which he did enthusiastically, and Blaeu readily agreed to publish it. Blaeu, however, was rather slack in the printing of scientific works in this period and did not live up to his promise. This, however, was nothing new: other scholars had suffered from Blaeu's slackness. The publication of Boulliau's work was delayed for several years: it came out only in 1639, after Willem Blaeu had been succeeded by his son Joan.[16]

The debate reached some of the lower strata of society as well. In the years following the condemnation of Galileo, two full-length pamphlets were published on the issue of the world system. Both rejected Copernicus and upheld the Ptolemaic view. One was by the well-known navigation instructor Sybrand Hansz Cardinael from Amsterdam.[17] It appeared in 1635 in the form of a 'mathematical argument'. The other pamphlet appeared in 1636 and was written by the otherwise unknown Jan Jansz de Lange, 'amateur of the mathematical arts, living in The Hague'.[18] This pamphlet is in the form of a dialogue. A skipper and his first mate discuss the merits of the two systems, viz. that of Ptolemy and that of Copernicus. This setting suggests a navigational background, but The Hague is not really the right place to be if one wishes to keep in touch with navigation. It might rather suggest the type of audience

---

[15] Vanpaemel (1989) 118-124. van Berkel (1983) 147-151.

[16] Blok (1996); Nellen (1994) 64-69; Bots (1974) 30-31. On Blaeu's slackness in general: de la Fontaine Verwey (1979) 29-30.

[17] See on him Wijnman (1933/34).

[18] The preface is dated Haarlem, 1 December 1636.

thought to be interested in these questions. The pamphlets were written in Dutch, and were not aimed at a learned audience but, as De Lange writes, at 'amateurs and lovers of astronomy and navigation'.

The very start of De Lange's dialogue is worthy of note:

Skipper: 'How great a dispute there is among the learned astronomers. Some are saying that the whole of the firmament is turning all round, others that only this earth is moving, turning all round in 24 hours.'
First mate: 'Not only among the learned astronomers, but also among the common people there is every day much dispute about what is turning.'

In his introduction, De Lange says the same. He asks his readers to study the work attentively, 'as these days, much quarrel and discord is rising more and more' on the subject. How far one can take this at face value is, of course, open to question. There is undoubtedly an element of rhetoric in stressing the topicality of the subject. However, the very fact that two pamphlets – both in Dutch, but one in a rather clumsy style – on the issue of the world system appeared in quick succession, suggests that the subject by that time had reached a wider audience. At least in navigation schools, the topic seems to have been a matter of dispute. Both authors appear mainly at issue with the arguments from Lansbergen's *Bedenckingen*. That both feel compelled to defend the traditional view in print suggests the latter work had aroused widespread interest.

The two pamphlets are interesting enough to merit a brief discussion. Cardinael's pamphlet is an extensive, well-organised and nicely illustrated argument running to some 30 pages. He purposely does not treat every aspect of the subject. In particular, he omits arguments from Scripture and from '*hemelmetinghe*', that is, the determination of cosmic distances. The latter he rejects as indecisive – as it begs the question – and the former he leaves to others. Instead, he mainly limits himself to a discussion of falling objects on earth. The argument was a familiar one, of course. According to Aristotelian physics, a body could not move without a moving agent. An object in the air would necessarily remain at rest – rest and motion being interpreted in absolute terms. So, if the earth were moving, objects surrounding it would be left behind. Within the context of Aristotelian physics, this objection was a valid one. The Copernicans were forced to find a way by which the motion of the earth was shared by all objects surrounding it. Generally, they explained that the motion of the earth was natural not only to the earth, but also to the surrounding air. The air communicated the motion to bodies not directly in touch with the earth itself.

The issue of motion was something of a stock argument to repudiate the Copernican system. Few people, however, made as much of it as Cardinael.

He tried to refute the Copernican counter-argument as it concerns falling bodies. To this purpose, he made the Aristotelian distinction between forced and natural motion. He granted that the Copernican counter-argument may be right in the case of bodies moving naturally, but a forced movement can for a time undo the force which makes a body move naturally. A projectile shot from a cannon supersedes gravity: the forced upwards motion undoes the natural downwards motion. In the same way, it would supersede the 'natural' rotation of the earth, provided there is one. Because of its violent motion, a projectile would not turn with the earth. When shot upwards, this should result in a marked aberration. Cardinael concludes that he has proved that the earth stands still, and that it is up to the astronomers now to solve, by means of their circles and epicycles, the question of the heavenly motions. Mathematical theories, thus, have to answer to physical demands. This was a remarkable stance for a man who was foremost a practical mathematician himself.

Cardinael's main motive, however, appears to have been a religious one. Although his pamphlet is certainly not a theological work, scriptural arguments were important. He prefers to take the Bible literally and he rejects what he regards as the forced exegesis of the Copernicans. Actually, this strict biblicism is somewhat surprising. Cardinael belonged to the Mennonite confession, which as a rule was rather lenient in its Bible interpretation. One surmises that his stance derives from a more general conservative attitude – as he explains, if Copernicus were right, nearly the whole of humanity would be wrong.

De Lange's pamphlet is less well organised. The dialogue form allows the author to pass in review a rather diverse collection of arguments. The skipper – who is defending the Copernican system – mainly repeats the arguments offered by Lansbergen, although the argument based upon the tides, which is also discussed, must have been drawn from Galileo's work. The arguments of the first mate – who is refuting the skipper's arguments and clearly is the author's mouthpiece – concern in general simple observations and reasoning as well as elementary spherical astronomy. For instance, if the earth were not at the centre of the universe, at some places one would see a larger part of the heavens than at others. If the earth were moving around the sun, it would not be possible for the magnetic needle to be exactly pointed at the pole of the firmament (magnetic deviation does not refute this argument as, according to the author, such deviation is due to magnets in the sea etc.). The tides are not caused by the motion of the earth, as this fails to explain spring tides; they are clearly caused by the moon. There are also arguments about falling bodies, rising smoke, floating clouds, etc. The arguments are not all nonsense, but on the whole, the scientific level is not high. The author uses his common sense and ignores contemporary physical insights. One should add that arguments from Scripture fail completely, although the work has a certain religious tenor. Lansbergen's third heaven is mentioned approvingly as the goal of human striving.

Attempts to arrive at a new cosmology

The telescope not only drew attention to existing cosmographical debates, but also made a difference to the questions raised. The telescope made people look at the heavens with a different eye. Until then, the heavens had been observed by means of the cross staff or the quadrant. Such observations were confined to measuring stellar positions. They served to study the motions of the heavens, 'to search for the numerous turning spirals of the stars' as Ptolemy said in a famous epigram. [19] In this kind of study, the telescope was of little help. But the new instrument raised questions of quite a different kind, in making people wonder what things up there looked like. Far from introducing people to a heavenly realm of mathematics, eternal motion and harmony, it made them wonder about the physical appearance of the things in heaven, and quite naturally to compare them with the things down here.

An early illustration of the impact of telescopic discoveries is offered by Ernst Brinck, a minor Dutch regent. He started his career as secretary of an embassy to the Ottoman empire and later became burgomaster of the small town of Harderwijk (province of Gelderland). He was a well-educated man with an interest in learning, and translated Blaeu's *Atlas minor* into Dutch. On his travels, he visited many prominent men as well as scholars and had them sign his *Album amicorum*. Passing through Florence in 1614 he visited Galileo, who contributed a little sketch of the moons of Jupiter for the album. Brinck wrote about this meeting: 'At Florence in Italy, I have anno 1614 spoken to the famous mathematician Galilaeo Galileo, who by means of his *tubus opticus* (...) could observe the interior of the moon. He considered it to be full of mountains, hills and valleys, also woodlands and rivers, men and animals, and scattered with towns and villages. He thought that the sun, too, was inhabited. He has also found new stars, which he called Medicean.' [20]

This, then, is what a reasonably cultured Dutchman of the time understood, and found worth recording, about the new cosmology. I would not claim that Brinck's diary is a reliable source on Galileo as a philosopher, although it may give an impression of Galileo as a brilliant and witty conver-

[19] Quoted after Taub (1993) vii.
[20] Van Rappard (1868) 56. Thomassen (1990) 72, where also Galileo's drawing in Brinck's album is reproduced.

sationalist, entertaining his curious and somewhat naive guest with the marvels he wanted to hear. But the point that supralunar and sublunar things are similar certainly came through. As it seems, Brinck found it all very new and thrilling, and had no philosophical or religious objections.

Life on the celestial bodies was not a subject of scholarly study, but the idea certainly fascinated the public. Francis Godwin's well-known fantasy *The man in the moon*, published in English in 1638, was received favourably in Holland. It aroused the curiosity of Constantijn Huygens. His friend, the scholar Johannes Brosterhuyzen, made a Dutch translation as early as 1639. The translation was wanting, however, and it was never printed. Then, in or shortly before 1650, another Dutch translation appeared. Several editions were to follow, and its success was great enough to inspire two anonymous Dutch authors to publish a sequel; both appeared in 1651. One of these is of no interest here, as, although quite fantastic, it did not leave the earthly realm. The second, in its second part, deals with a voyage to the planet Mercury. As science fiction, the book is not very inspired, but the preface clearly plays on the scientific interest of potential readers. The author promised a third part, on the other planets, but this probably never appeared.[21]

The impact of the telescope was experienced differently by different people and different groups. People raised in the tradition of scholarly, mathematical astronomy would not change their world-view overnight. They were undoubtedly interested in the new discoveries, but remained committed to traditional aims and methods. This applies even to the Copernicans among them. It is striking that Lansbergen simply omitted the newly discovered moons of Jupiter from his plan of the Solar system, which was published with his 'Considerations' of 1629; they were included only in the second impression.[22] Still, Gerard Vossius appears to have changed his views on the nature of the heavens, regarding them as fluid rather than as consisting of solid spheres, largely under the influence of Galileo's discoveries.[23]

However, other people went further and abandoned the old Aristotelian theory of the cosmos, as well as the view of a harmoniously ordered universe which should be understood in mathematical terms. The anti-Aristotelianism before 1610 drew mainly on Paracelsist and neo-Platonist ideas. After 1610, there was a marked interest in philosophers who had alternative views on the substance of the heavens, e.g. Giordano Bruno and Johannes Kepler. As the universities offered little scope for radically breaking with the tradition,

---

[21] 'Domingo Gonzales' (1651). Janssen (1985) 30-36.

[22] They appear for the first time in Hortensius' Latin translation from 1630.

[23] Vossius (1641) 1, 520-521.

the most daring attempts were made by independent thinkers and by people outside the scholarly community.

One of the more prominent Dutch scholars of the first half of the seventeenth century is Isaac Beeckman. He was a person of international stature, who corresponded with Mersenne and had a deep and lasting influence on the ideas of René Descartes.[24] For our purposes, he is particularly interesting because of his *Journael*, a kind of diary in which he recorded his private considerations, particularly those on scientific subjects. His ideas are accessible to us not in their finished, polished version (in fact Beeckman published hardly anything), but in their various stages of accomplishment, as he turned them over in his mind. In this way, we can follow the development of his ideas in a much more complete way than would have been possible had we only printed publications.

Beeckman was born in Middelburg in 1588. His father, a close friend of Philips Lansbergen, earned his living as a candle maker and by constructing and maintaining waterworks. Isaac Beeckman studied theology at Leiden, during which time he also took courses in mathematics under Rudolf Snellius. He never entered the ministry. Instead, he followed in his father's footsteps as a candle maker and builder of waterworks. The latter came down to the kind of civil engineering we met earlier in the case of Stevin and Drebbel. Many of the annotations in his *Journael* in this period concern technical problems. In 1616, Beeckman quite suddenly decided to follow a more scholarly career. Two years later, he took his doctorate in medicine at Caen University (France) and went on to make a living as a Latin schoolmaster. He died in 1637 as rector of the Latin school of Dordrecht.

Beeckman started his *Journael* during his student years at Leiden, i.e. from 1607 to 1610. The question of the system of the world is among the first subjects discussed in it. He pondered various arguments for and against the motion of the earth, taken from Stevin, Tycho and Cardano.[25] Apparently, he had not taken a stance of his own yet. When in 1613-1614 he first formulated his principle of inertia, he applied it to, among other things, the motion of the sun, not the earth.[26]

By 1616, however, his thoughts took a new turn when he pondered the influence of the stars on the sublunary world. At first, he seems to have regarded this as a conclusive argument for the centrality, and hence immobility, of the earth. Only at the centre of the universe was the stellar influence from all sides equal. Looking closer into the matter, however, he decided that this

---

[24] See on him, van Berkel (1983).

[25] *IBJ*, I, 1, 2, 3.

[26] *IBJ*, I, 24.

need not be the case. The earth was not necessarily at the centre of the universe, but at the point where all virtues moving the earth had equal force. [27] At this point, Beeckman seems to have been struck by the idea that his argument that stellar influence should be equal in the centre, could also be used as an argument in favour of the centrality of the sun, instead of the earth. The earth and the moon simply need not be affected by the stellar radiation. The sun, however, is. As it is pressed on from all sides, it is pushed towards the place where the various forces counterbalance each other, that is, at the centre of the universe. The sun in its turn emits light which pushes the planets away a certain distance. 'This speculation seems such, that accomplished astronomers can save all the appearances with it.' [28]

This idea gave a physical explanation of the fabric of the heliocentric universe. Apparently, it held some fascination for Beeckman and he elaborated on it in various further notes. The sun obtains its power from the starry radiation it receives. In its turn, it radiates the light back to the eighth sphere. In this way, the sun and the stars entertain each other. It might even be that the sun is not a specific body, but simply the effect of the various light rays from the stars, running together in the midst of the universe. [29]

Apparently, Beeckman had explained the system of the world to his own satisfaction. For a long time, he did not return to the matter. He only referred to it when it might seem to contradict his other theories. So when by 1620 he speculated that the fiery particles emitted by the eighth sphere might be the cause of gravity, he added that the eccentric place of the earth did not refute this hypothesis. [30] It seems clear that by 1616, Beeckman had accepted the Copernican system, even if he did not state so explicitly – after all, he did not need to convince anyone in his *Journael*. [31] Conclusive was the agreement of the heliocentric system with his physical principles, which came down to a mechanisation of traditional ideas concerning stellar influence. Actually, Beeckman has been heralded as one of the first mechanical philosophers of the seventeenth century. [32] His celestial physics was foremost a kind of cosmic economy of power. Beeckman does not seem to have occupied himself very much with mathematical astronomy.

An interesting question is the influence of Lansbergen in all this. By 1616, of course, he had not yet published any astronomical work but, as noted before,

[27] *IBJ*, I, 26, 100.

[28] *IBJ*, I, 101.

[29] *IBJ*, I, 103-104; see also 101, 194-195.

[30] *IBJ*, II, 107, 119-120. See also 138-139.

[31] In my view, van Berkel (1983) 163, is too cautious in this respect.

[32] Van Berkel (1983) 5.

Beeckman's father happened to be a close friend of Lansbergen. In 1616, Beeckman himself moved from Zierikzee to Middelburg, where Lansbergen had been living since 1613, which would have given the two ample opportunity to discuss astronomy. Of course, their respective approaches were quite different. Lansbergen was primarily a mathematical astronomer, whereas Beeckman was mainly interested in the underlying mechanisms of the world's constitution. Still, it is quite possible that Lansbergen raised Beeckman's awareness of the problem of Copernicanism, and may even have hinted at some possible directions. Beeckman's speculations on the sun as the centre of the starry radiation bear a striking resemblance to Lansbergen's 'Hermetic' axiom of 1619, viz. that everything that is actually dispersed in the circumference of a circle should potentially be at its centre as well. As stated, Beeckman's ideas are not Hermetic at all, but this does not preclude him from having developed his mechanistic world-view by using elements from existing theories.

As stated, Beeckman let the matter rest for several years. He only took the matter up again in 1628, when he came across the works of Kepler. He was fascinated by them and studied them avidly. It was not so much Kepler's mathematical-astronomical work which drew his attention, but his attempts at a physical explanation of the universe. In his *Astronomia nova*, Kepler had tried purposely to give a 'physics of the heavens', a project more or less in line with Beeckman's earlier speculations. [33] On the whole, Beeckman could agree with Kepler's ideas. 'These things by Kepler, what he writes about the physical motion of Mars, please me very much, perhaps because long before I saw his work, something similar occurred to me as prerequisite for the restoration of astronomy. This can be seen at various places in this book [the *Journael*], especially where I spoke physically on the motions of the earth. However, now Kepler has forestalled me, I hope to accomplish at some time a work on this matter, with recourse to my meditations which he shall not see.' [34]

Perhaps because he was writing a book of his own (which, by the way, never materialised), Beeckman quite extensively noted his reflections on three of Kepler's works – *Astronomia nova*, the *Mysterium cosmographicum* and the *Epitome*. [35] Beeckman's main aim was to incorporate Kepler's speculations into his own mechanical principles. He wanted to show 'that all three motions of the earth can be accomplished without any fictitious internal force, and that they follow mathematically from the movement of the particles emitted by the sun.' [36]

---

[33] Beeckman's cosmological speculations are discussed by van Berkel (1983) 183-186 and Schuster (1977) 567-578. On Kepler's celestial physics, see Stephenson (1994).

[34] *IBJ*, III, 103.

[35] *IBJ*, III, 68-69, 73-76, 99-109, 115-121, 158, 160, 165-166.

[36] *IBJ*, III, 108.

Kepler had made much use of magnetic forces and virtues. According to Beeckman, he need not have done so, had he realised two basic truths which he, Beeckman], had discovered long before. One was the principle of inertia: everything, once moved, will keep moving. Once the earth is turning round the sun, no further force is needed to keep it moving that way, whereas Kepler had formulated an elaborate theory of how the earth was pushed forward by the rays of the sun. Second was the idea that light was a corporeal substance. Once this was agreed on, it no longer appeared strange that the radiation of the sun and the stars had a physical effect on planets and the earth. Beeckman's basic cosmology was the same as ten years earlier. The planets ('among which I count the earth') are repelled from the sun by the corporeal light particles it ejects (in fact, the sun reflects the light emitted by the sphere of the fixed stars).

In the remainder of the *Journael*, Beeckman kept returning to the system of the world. In 1631, reading Lansbergen's *Uranometria* made him wonder why the greater planets were generally placed at greater distances from the sun. He surmised that the planets had originated from 'exhalations' from the sun.[37] Another work which elicited a lot of reflections was Galileo's (yet to be published) manuscript on the tides, *De fluxu et refluxu maris*. Beeckman had pondered over the tides before. His own ideas were a kind of mechanisation of older ideas on celestial influence. He surmised that moon rays consisted of particles which added humidity. In this respect, he was in line with traditional astrological lore, which held moistness to be a quality of the moon. Initially, Beeckman thought that the pressure of the air decreased after it had been moistened by the moon. The tides were the effect of the variability of the pressure on the waters. Later, he thought that moon rays directly affect the waters by making them swell.[38] Galileo explained the tides from the combined daily and annual motion of the earth. Beeckman approved of the idea and even elaborated on it. He thought that the movement of the winds could be explained in a similar way. Beeckman tried to account for the fact that the tidal movements were synchronised with that of the moon, a fact Galileo simply denied. Beeckman no longer regarded the moon as a cause of the tides, but vice versa: he considered the moon's motion to be a tidal effect. The moon drifted on a large sea of air, which underwent tidal effects just like the ocean. These tidal waves moved the moon with it. Despite some difficulties he found later with Galileo's theory, Beeckman stuck to these ideas.[39] Other reflections on the fabric of the world were elicited by other works, or

[37] *IBJ*, III, 206-207.

[38] *IBJ*, I, 151, II, 167-168, 317-318, 386-388, III, 11. See also van Berkel (1983) 175-176.

[39] Van Berkel (1983) 176. *IBJ*, III, 171, 205-206, 281; see also 253, 345-346.

had no apparent cause at all.[40] All in all, the economy of the world's system and the physical explanation of the universe were among the subjects most intensively studied by Beeckman in the later years of his life.

There were also some people of lesser status who indulged in cosmological speculation. Albert Girard was born in France and in 1626 became an engineer in the States' army. He edited the mathematical works of Samuel Marolois in 1628 and published a work on algebra in 1629. He died in 1633. A year later, his French translation of Simon Stevin's 'Mathematical memoirs' was published. In the part on the Copernican system, Girard had added an extensive note with regard to Stevin's cosmology, a subject in which apparently he took great interest.[41] He appears to agree completely with Stevin's Copernicanism and just adds some further considerations. For one thing, he refers to the recent telescopic discoveries (which of course had been unknown to Stevin), concerning both the face of the moon and the new celestial bodies. Not only had many 'new' planets been discovered, but also it was possible that there were many more yet to be discovered. Girard also maintains that the stars are at variable distances, thus implicitly denying the existence of the sphere of fixed stars.

Moreover, Girard discusses and refutes the arguments meanwhile advanced against Copernicus. He denies that Copernicus' theory contradicts the Bible (he could feel justified in bringing up this topic in his annotations, as Stevin had not touched on it). In his defence of Copernicanism, Girard relies on a rather idiosyncratic cosmology. He considers as ridiculous the argument that the earth is fixed because it is heavy, and therefore tends towards the centre. In his view, different 'primitive (or original) bodies' (such as stars) have different places they tend towards. Moreover, they are moved by sympathy for or antipathy to each other. 'As they are more in need of each other, they will be closer, and vice versa.' A body which is derived from a primitive body, and therefore less noble, must turn around its major, from which is has its origin: 'As continuously looking for everything which is necessary to it by means of appearances, aspects, nearness and withdrawal, figure and harmony...'

This harmoniously ordered universe is governed by magnetic forces. These played already an important part in Stevin's work, but Girard turns them into the cosmic agents which preserve the order in the universe. They are of two different kinds. Some serve to make the bodies cohere, thus causing their natural heaviness, while others are 'for making the bodies turn round, and making them approach and withdraw according to their mutual need.' The

---

[40] *IBJ*, III, 207-208, 216-218, 225, 229, 272-273, 276-278, 279-280, 325-326, 344-345, 350-351, 353, 363.
[41] Stevin (1634) II, 184-185.

perfect balance between the celestial bodies is illustrated with the example of the distance between the sun and the earth: were it greater, all water on earth would freeze; were it smaller, all water would evaporate. The distance between the sun and the earth is maintained such that water can occur in a fluid state. When aged, bodies can change their quality. As this affects their sympathetic relations with other bodies, this will change their natural place. Such a body, falling head over heels into its new place, is seen as a comet. As the earth is getting older, too, Girard speculates 'that one day, the earth will be a comet, that is to say (I do not dare to be specific, as this belongs only to God), will fall and change its place, when our Lord will come to judge it, and bring it into another place. But this I leave for another time.'

Girard puts forward his opinions in a rather self-conscious way. He sniffs at the bookish learning of natural philosophers. In his view, the fall of man has undone his pristine state of natural understanding. Since then, he has had to learn by experience.

An even more curious figure is the miller, Balthasar van der Veen. Van der Veen cannot be regarded in any sense a learned man, but he did have a curious and inquisitive mind. He picked up knowledge wherever he could find it, only to brew his own cocktail from the various elements. One is reminded of his sixteenth-century Italian colleague Menocchio, but with the difference that Menoccio's speculations mainly concerned religious issues, whereas Van der Veen was preoccupied with physics and cosmology. As such, he was one of the first of an impressive array of lay physicists who would come to the fore in the second half of the seventeenth century.

Van der Veen left no writings of his own, but some of his ideas were recorded by Isaac Beeckman in his diary.[42] Copernicanism was one of the elements on which Van der Veen built his cosmology. His interpretation was somewhat peculiar. From the theory that it is the air which keeps things in place on the turning globe, he derived a theory that air becomes denser the further it is from the earth's surface and eventually forms a solid shell, like glass or ice.[43] He also knew that, in principle, the annual motion of the earth could be proved by annual parallax, but was rather idiosyncratic in stating that this could be done in practice, too – and in fact already had been. As he argued, various authors gave different values, which differed by two or three degrees, for the same distances between stars. According to him, this could only have occurred because they observed at different times of the year, so that the outcomes were influenced by parallax. 'For it cannot happen that

---

[42] Van Berkel (1983) 111-112, 248-251.

[43] *IBJ*, III, 140.

of such people the instruments would have been made so badly, or the observation been done so badly.'[44] He also had a theory to explain the various distances of the planets with respect to the sun by means of the varying quality of the air in the universe.[45] Among his more daring speculations is the idea that the earth is hollow and that there are people living on the concave surface inside it.[46]

Other ideas are related to contemporaneous scholarly speculations. Thus, Van der Veen held that there is no sphere of the fixed stars. The sun, the earth and the planets are essentially similar. The sun is not the centre of the universe, but is moving in its turn around some other centre. The stars rotate on their axis, just like the earth does. He may well have borrowed these ideas from Bruno, as Beeckman noted.[47] Van der Veen's ideas on the nature of stars and the planets are akin to Girard's. Van der Veen assumed that formerly the stars had been worlds like the earth, 'which have been clarified by now and therefore are shining.' The earth, too, will one day become a shining star. God is creating new worlds every day.[48]

Beeckman still strove to fit the new discoveries into an old framework, with heavenly spheres, celestial influence and such like. It was less educated people like Girard or Van der Veen whose cosmological radicalism was most outspoken. With hindsight, their theories may seem somewhat fantastic. Still, one might suppose that in some way, the later theories by Descartes on stellar evolution sprang from such a tradition of cosmological speculation.

---

[44] *IBJ*, III, 140-141 (21 Nov. 1629).
[45] *IBJ*, III, 323 (Oct.-Nov. 1633).
[46] *IBJ*, II, 388-389 (4 march 1627).
[47] *IBJ*, III, 208-209 (April-June 1631), 323.
[48] *IBJ*, II, 389 (4 March 1627).

7. Astronomy at the universities

New cosmological ideas kept creeping into the universities, too. Ptolemy's system was increasingly dismissed in favour of that of Tycho Brahe. The elliptical orbits were upheld in the work of Ravensberg and Holwarda, as well as by the philosopher Renerius. Two convinced Copernicans – Hortensius and Holwarda – occupied academic chairs. In this section, we will deal with the universities in the pre-Cartesian epoch. The changes brought about by Cartesianism will be dealt with in the following section. Of course, there is no clear chronological boundary between the pre-Cartesian and the Cartesian era. Rather than establishing a fixed time limit, our division will be determined by whether the various authors show any awareness of the new situation as created by the advent of Cartesian philosophy. This will include quite a number of contemporaries of Descartes among the pre-Cartesian philosophers. By 1650, however, nearly everybody had lost his or her innocence. The debates on Cartesianism of the 1640s made it virtually impossible not to take sides on the issue.

As we have seen, outside the universities the philosophical implications of the new discoveries were quickly recognised. The universities, however, were much more traditional and did not allow for daring speculation. Academic discussion on the system of the world still remained very much the affair of astronomers and mathematicians. Even so, physical issues do turn up in their works. The Aristotelian world-view was no longer a self-evident background. Moreover, by 1640 the system of Ptolemy had run completely out of favour. It simply did not match the astronomical phenomena. On the other hand, the system of Tycho Brahe became a serious alternative. So, gradually, a situation arose whereby a choice between Tycho and Copernicus had to be made.

Tychonians

As stated, the first adherent of the system of Tycho Brahe in the Dutch Republic was Johannes Isacius Pontanus. His was rather an exceptional case. His personal acquaintance with Tycho will surely have been of influence. On the whole, it was only in the 1640s that Tycho's system was seriously considered by Dutch astronomers.

One rather curious figure is Antonius Deusing. He is mainly remembered as a physician, but he started his career as a professor of mathematics and physics at Harderwijk University. He later claimed to have been a student of Golius.[49] Deusing was a rather ambitious scholar, and he published a large number of works. In 1640-1641, he presided over a series of nine physical disputations, which provide some information on the background to his ideas. He found the principles of natural philosophy in the history of Creation as told in the Bible: matter, spirit and light. Having discussed these extensively, he argued that the principles of Aristotle agreed marvellously with those of Scripture.[50] The remainder of the disputations consisted largely of an exposition of the works of Creation, in the classical form of a hexahemeron. (It appears that he encountered some problems because he, a philosopher, had discussed the Creation.[51]) In the corollaries to the fifth disputation, Deusing upheld the Tychonian system and rejected the celestial spheres.[52] Meanwhile, in an elementary work on cosmography and astronomy he published the following year, he simply stuck to the Ptolemaic order.[53]

However, Deusing's ambitions went further. Another corollary of the fifth physical disputation was: 'Having rejected the Tychonian motion, we may demonstrate, saving the phenomena, the simple planetary motion around the sun by epicycles and epicyclepicycles of the orbs.'[54] A few years later, in 1643, he devoted a full-length treatise to the planetary orbits. It had the promising title *De vero systemate mundi* ("The true system of the world'). As it appears, Deusing felt much admiration for Copernicus' simplification of astronomical theory by having the planets revolve round the sun, but he had big problems with the motion of the earth. In particular, he considered the immense cosmic distances this would require as absurd. Consequently, he aimed at transposing Copernicus' theories to an immobile earth.[55] This results in, of course – as Deusing himself acknowledges – a Tychonian system. One should remember that Tycho himself had given only an outline of the geoheliocentric system, and had not formulated the exact theories. This is exactly what Deusing now undertook to do. The whole thing is rather tough reading, as nearly the whole

---

[49] Deusing (1642) 199. On Deusing as a physician, see Ebels-Hoving (1997).
[50] Deusing, disp. Harderwijk, June 1640, quaestio 1 & 2. Cf. Verbeek (1992) 8.
[51] Deusing (1650) preface.
[52] Deusing, disp. Harderwijk, 25 Feb. 1641, coroll. 1, 2.
[53] Deusing (1642).
[54] Deusing, disp. 25 Febr. 1641, coroll. 3.
[55] Deusing (1643) 6-7.

book consists of mathematical calculations. Delambre, who in his days was famous as a diligent astronomical arithmetician, commented on the book: '...this work is tiring [*pénible*] to read. One finds a horrible multitude of lemmata, theorems and corollaries, the quiries of which are long and obscure, and the demonstrations difficult to follow.' [56] Moreover, Deusing nowhere uses observational data or numerical values. His only concern is the construction of geocentric theories which are equivalent to Copernicus'.

It is not quite clear how far Deusing took these theories, which involved complicated models for the various motions, as mere mathematical hypotheses, or as physically true descriptions. In a work on the Creation two years later, he still appears to vacillate between a Tychonian and a Capellan world system. Speaking on the meaning of the word *mundus* (world), he explained that it could refer not just to the whole universe, but also to separate parts: 'So, there is the sublunary world, wherein the moon is rotating; the Solar world, wherein at least Venus and Mercury are turning round; the Jovial world, wherein the stars which are turning round Jupiter...' [57]

Anyhow, all theories should be in accordance with the immobility of the earth. However, having dismissed the annual motion of the earth, Deusing is still prepared to accept its daily rotation. The decisive argument is taken from the immense velocities the revolving heavens should have. [58] The assumption of the earth's daily rotation leads him to an elaborate discussion of the phenomenon of free fall, largely in opposition to Ismaël Boulliau. He even arrives at some quantitative relations, but he keeps to Aristotelian theory and appears unaware of the work of Galileo. None of this is of great scientific importance. Deusing does not make a single effort to match his theories with observations and, as Delambre dryly remarks, 'nobody, I think, will take the trouble'. The astronomers seem to have completely ignored the book. It is preserved in the important libraries, but one is hard pressed to find a single reference to it in contemporary literature. [59]

So, the various discoveries of recent times moved Deusing to enrich his world-view with new elements, but he continued to understand the world in biblical terms. To him, mathematics was a gateway to truth. Astronomy, the contemplation of the heavens, in particular led man to the knowledge of the divine. [60] Like Lansbergen and Mulerius before him, he aimed to reduce

---

[56] Delambre (1821) II 146. Deusing is dealt with on pp. 144-146. Deusing is also succinctly dealt with by Dunin Borkowski, IV, 500-501.

[57] Deusing (1645).

[58] Deusing (1643) 125.

[59] Lipstorp (1653) is one of the few authors to refer to Deusing.

[60] Cf. his orations on these disciplines in Deusing (1642) 181, 219.

all heavenly motions to uniform circular movement. He knew of Kepler's innovations, but deliberately opted for the simplicity of Copernicus' circles.[61] We may conclude that Deusing was a scholar in the Leiden tradition who did his best to grasp in traditional notions a world ever more complex and bewildering.

At Utrecht University (founded 1636), physics and mathematics were taught by Jacob Ravensberg.[62] He had been a student of Mulerius at Groningen, where he obtained his doctorate in philosophy in 1639 on a collection of miscellaneous theses. One thesis was devoted to the question whether the earth was moved or not. He gave a long list of both the ancient and the recent proponents of the earth's motion, and summarised their opponents in a single line (peripatetics, Ptolemaeans, Tychonians). Finally, he simply stated: 'It is difficult to refute the motion of the earth by natural reasons; still, one cannot deny absolutely the motion of the sun.'[63]

After his graduation, Ravensberg moved to Utrecht, where in 1640 he defended a disputation on the system of the world, wherein the Copernican question was tackled much more seriously. Disputations 'on the system of the world' would be rather popular in the second half of the century, when they discussed the question in a set way. Ravensberg's disputation, however, was the first of its kind, and not at all a standard one. In fact, the work is rather puzzling. At the outset, he states that there are only two serious claimants for the true system of the world, viz. the Copernican and the Tychonic. One of them has to be true. He will lay out the system of the world according to the Copernican theory. The theory may seem rather absurd, but it makes astronomy easier, even if it may be wrong in other respects. So he will accommodate his exposition to it. 'So far as it disagrees with the Tychonic system, it will suffice to have it qualified here once.'[64] A similar qualification is also to be found in the dedication of the disputation to the Utrecht rector Schotanus, professor of law. Ravensberg explains that he may seem to be rather too fond of Copernicus; but then he does not strive after strict accuracy, but proposes these things for disputation's sake. He is only concerned with matters fit for a disputation, not with exact astronomical descriptions of motions and magnitudes.

The disputation can be divided into several parts. After a few theses about the world in general, Ravensberg first explains the various motions of the earth according to the Copernican theory (th. 12-19). Still, whereas Coperni-

---

[61] Deusing (1643) 6.

[62] Or Ravensperger. See on him Dibon (1954) 211-214.

[63] Ravensberg, disp. Groningen Feb. 1639, th. 48.

[64] Ravensberg, disp. Utrecht 25 Nov. 1640, th. 2.

cus himself had acknowledged only three motions, Ravensberg came up with nine. He also counted such phenomena as the anomaly of the obliquity of the Zodiac and the variation of the eccentricity as motions. In the end, however, he reduces the whole lot to three again. Next, he devotes a number of theses (19-24) to the magnetic nature and other properties of the earth. The final part (th. 25-36) is devoted to special phenomena which could be derived from the theory of Copernicus, such as stellar parallax, the phases of the planets, their variety in magnitude in different places in their orbits, retrograde motion and other properties of the planets' courses, Galileo's theory of the tides, etc. In some cases, such as stellar parallax, Ravensberg acknowledges that the phenomenon cannot be observed because of the dimensions of the universe, in others he confirms their existence; some phenomena, such as the influence of the earth's motion on falling bodies, he regards as probable, but unproven as yet. Without Ravensberg's warning at the start, the whole could very well be read as an extensive argument in favour of Copernicanism.

Ravensberg was well acquainted with the astronomical literature. Among others, he refers to Kepler, Galileo, Gilbert, Boulliau, Blaeu, Lansbergen and Mulerius. His arguments are indeed largely mathematical. Most of them can be accommodated to the theory of an unmoving earth, as happens in the Tychonic system. On the other hand, Ravensberg does not have a physical view of the universe which would support heliocentrism. His cosmology is a mixture of old and more recent elements. His view that the world (in the sense of the whole cosmos) is spherical is traditional. As such, it must have a centre, where either the sun or the earth is located (th. 1, 3). Still, he argues that the fixed stars may be like suns, each of them with its own planets. The stars were not created primarily for the benefit of man (th. 7). He adheres to Gilbert's theory that the daily rotation of the earth is caused by its magnetic virtue (thus implicitly acknowledging the daily rotation). He supplements this theory, however, with Galileo's idea of inertia (th. 21-22). Thus, he was clearly aware of the physical difficulties of Copernicanism and these may well account for his doubts. Ravensberg certainly was not an adamant adversary of Copernicanism, but in the end, he refused to advocate it openly.

As to the further developments of his ideas, in 1642 Ravensberg published, in the form of disputations, his *Encyclopedia mathematica*, which dealt with the whole of mathematics. The work was a collection of unconnected, for the most part rather short theses, grouped under such headings as spherics, astronomy, 'planetaria', astrology and statics, as well as 'medica' and 'physica'. In some respects, Ravensberg appears more advanced by now: he not only rejects the existence of celestial orbs, but accepts that the planets do not move in circles but in ellipses. He still regards the theory of epicycles and excentres as useful, but he no longer holds, as he had done in 1640, that they really exist.[65] Among Dutch astronomers, he seems to have been the first to defend elliptical orbits in print. However, he had been forestalled by a philosopher, Henricus Renerius, who had recommended Kepler's ellipses in a disputation in 1635.[66] As Renerius was closely connected with Descartes, he will be more fully discussed in a later chapter. He taught at Utrecht University, where Ravensberg defended his theses, but it does not seem very probable that he influenced the latter directly, since he died at about the time Ravensberg arrived there.

Other elements in Ravensberg's 1642 work probably were influenced by his new academic environment. He devoted a rather long section to a discussion of the vortex theory of planetary motions – remarkably, as this theory would not be published until two years later, in Descartes' *Principia philosophiae* of 1644[67] (the *Discours de la méthode* does not discuss it). On the other hand, Ravensberg had become more cautious with regard to cosmographic speculation. On the whole, he stressed the insufficiency of the arguments in favour of Copernicus, while he appears to regard Tycho's work rather favourably. But by now, Ravensberg also was aware of theological restrictions: 'I leave the Copernican, that is, daily and annual motions of the earth, to the theologians in order to investigate them.'[68] Theological pressure was rather heavy at Utrecht, as we will see in a later chapter.

In Ravensberg, Utrecht had a competent mathematics professor with a special interest in astronomy, who seems to have transmitted both his astronomical learning and his predilection for Tycho to a number of students. His influence can be discerned in several doctoral disputations. The later Amsterdam professor of mathematics Alexander de Bie obtained his degree in philosophy at Utrecht on 25 August 1642 on a thesis on spontaneous generation, the sun's motion, and usury. (Utrecht doctoral disputations were triple at this time, each part dealing with a separate part of philosophy – normally physics/metaphysics, mathematics and ethics.) In the second part, De Bie agreed that Kepler's ellipses saved the phenomena, but like Ravensberg, he maintained that the theory of excentres and epicycles remained of practical value.[69] As

[65] Ravensberg, disp. Utrecht 1642, planetaria, th. 32. Ravensberg, disp. Utrecht 25 Nov. 1640, coroll. 18.

[66] Renerius, disp. Utrecht 10 June 1635, th. 36.

[67] Ravensberg, disp. Utrecht 1642, statica th. 4.

[68] Ravensberg, disp. Utrecht 1642, Geographia th. 8: '*Motus vero Terrae Copernicanos, seu diurnum & annuum, Theologis hac vice examinandos relinquo.*'

[69] De Bie, disp. Utrecht 24 Aug. 1642, *De motu Solis*, th. 6.

for Copernicus' theory, 'To investigate it would require an opinion like the one Copernicus had. But although we presuppose from Holy Writ that his hypotheses and his system recede from truth, it will be allowed for us to hear and consider the arguments of the philosophers against Copernicus.'[70] De Bie, too, expressed his preference for Tycho's system.

Also in 1642, Barthold van Wesel graduated at Utrecht on a thesis on prime matter, the celestial orbs, and the affects. In the second part, after providing a historical overview of the several theories, he concluded that it is highly doubtful whether such orbs really exist. The sphere of the fixed stars is known to exist for biblical, not philosophical reasons. Planetary spheres are rather superfluous; according to Kepler, one can use ellipses instead of excentres with epicycles. Meanwhile, Van Wesel does not seem to have been very fond of the more mathematical parts of astronomy. This is not the case with Bernard de Moor's doctoral disputation, which he defended at Utrecht in 1643. It deals with the continuum, the new planets and the Stoic errors concerning the affects. De Moor was well acquainted with the astronomical literature of the day, including Galileo's work and Kepler's *Epitome*. In the corollaries, he expressed his preference for the Tychonic system, as best according with the heavens, and his dislike of the celestial spheres.

Copernicans

Probably the first full-fledged Copernican to hold a Dutch chair was Martinus Hortensius, whom we met before as the champion of Lansbergen.[71] He was one of the most outspoken Copernicans in the Dutch Republic. He was clearly interested in astronomy at an early date; born in 1605, he regularly made astronomical observations from 1625 onwards.[72] In 1634, he was appointed professor of mathematics at the Amsterdam Athenaeum. In 1634 he taught astronomy, in 1635 optics and in 1636 navigation. According to Varenius, who later set his sights on this chair, he had hardly any auditors.[73] It need not have been Hortensius' fault. 'This city has many followers and ob-

---

[70] *Ibidem*, th. 5: '*Haec inquirere ingenium requirebant, quale fuit illud Copernici, Quamvis vero ex sacris praesupponamus ejus hypothesis & ejus systema à veritate aberrare, licebit nobis philosophorum argumenta contra Copernicum audire & expendere.*'

[71] On Hortensius, van Berkel (1997). See also de Waard in NNBW I, 1160-1164; van Berkel (1983) 143-145, 150; Rademaker (1981) 247-250.

[72] Lansbergen (1632), *Thesaurus*, 124, 144, 162, 164, 171, 175, 177, 183. Pingré (1901) 57, 62, 64-66, 68, 72-75, 77-78, 83, 86-90, 103, 105, 114, 123. Pingré (p. 55) also mentions an observation at Dordrecht on 20 May 1621, but this appears doubtful.

[73] Varenius to Jungius, 24 Dec. 1647 and 12 April 1648, Jungius (1850) 379, 381; German translation in Jungius (1863) 328, 331.

servers of Mercury, but none of the other stars,' as Varenius remarked caustically on another occasion.[74] In 1639, Hortensius was appointed professor of mathematics at Leiden, but he died on 17 August, before he could start his courses.

Hortensius was not really an original thinker, although he did contribute to the propagation of Copernicanism in several respects. He made Latin translations of Lansbergen's *Bedenckingen* and of Blaeu's *Twee-voudigh onderwijs*, both of which had a clear Copernican tenor. Hortensius seems also to have advocated Copernicanism in a more private way. In 1635, in the aftermath of the Galileo affair, Gerard Vossius, by that time Hortensius' colleague at the Athenaeum, was asked for his opinion on the Copernican issue by his friend Abraham van der Myle. Vossius did not want to commit himself on the subject and he passed the request on to Hortensius, who apparently was only too pleased to state his views. There is a manuscript known wherein he discusses and refutes some objections Van der Myle had brought forward against the Copernican system. Among other things, Van der Myle thought it improbable that the radius of the earth's orbit would be 'quite nothing' compared to the distance to the fixed stars.[75]

Moreover, he published an extensive introduction with his translation of Lansbergen's book. This introduction was principally an attack on the astronomy of Tycho Brahe. It aimed not so much at Tycho's world system, however, as at his observations. Hortensius made it quite clear that only Lansbergen could claim to have restored true astronomy. As it seems, he regarded Tycho as a rival claimant for this honour. Hortensius did his utmost to show that Tycho's astronomy was built on rather shaky foundations. The points dealt with seem rather trivial to the modern reader – in fact, not just to the modern one. Kepler spoke of 'minute things which are controverted among specialists'. Finally, Hortensius asserted that Lansbergen had discovered the two pillars upon which he was to restore astronomy: the motion of the earth, and the true measure of the celestial spheres.

This introduction provoked a vehement reaction from the Danish astronomer Peder Bartholin Kierul, who came out in defence of Tycho Brahe's honour.[76] Kepler, too, made some rather disdainful remarks about Hortensius' preface, in an appendix to his ephemerides for 1624.[77] In the latter case, Hortensius answered with an entire booklet, in which he replied to Kepler's ob-

[74] Varenius to Jungius, 16 June 1647. Jungius (1850) 378, cf. Jungius (1863) 326.

[75] Rademaker (1981) 249. Leiden University Library, Pap. 2. The attribution of this piece to Hortensius is based on a marginal annotation; van der Myle appears as 'Milius'.

[76] Bartholin (1632). Cf. Moesgaard (1972) 133.

[77] Kepler, *Werke*, IX-1, 204-205.

jections point by point. Behind the discussion of astronomical details stands the question: who is the true restorer of astronomy? Especially the fact that only Lansbergen holds all ancient observations in esteem, whereas Tycho, Longomantanus and Kepler tend to neglect them, is regarded by Hortensius as a clear sign of superiority.[78] In fact, Hortensius points out that he decided to answer Kepler's objections because of Lansbergen's exhortations. 'For he has no other wish than to see, while he is still alive, the security of his hypotheses, which I put forward in my preface, publicly asserted, and confirmed by certain and convincing arguments.' Although Kepler had died in the meantime, the book was still published.[79]

His defence of Lansbergen led Hortensius to attack the work of the greatest astronomers of his age, but he was not in all cases so offensive. As indicated above, he supported the publication of Boulliau's *Philolaus*, although this book contained a rather critical evaluation of Lansbergen's argument from the moon's motion. He also appears to have genuinely admired Galileo. Hortensius played a prominent role in the plan to get Galileo to the Dutch Republic after his condemnation, and in the discussions on the project of determining longitude thereafter. When he died, he was about to go to Italy on a special embassy to Galileo to settle this affair. He was also instrumental in getting two copies of Galileo's *Dialogo* to the Republic, in 1634.[80] As the book had been banned in Italy, it was rather difficult to obtain. Hortensius seems to have welcomed the Latin translation that Matthias Bernegger was about to make at the behest of Galileo and his friends. Bernegger, a German from Strasbourg, was urged by the Leiden professor Boxhorn, on behalf of Hortensius as well, to produce this translation.[81] It appeared in 1635 under the title *Systema cosmicum*.

In a dissertation, dated 1636, explaining the project of a full course in mathematics, Hortensius announced his intention to publish shortly a book on astronomy, *Controversiae astronomicae*.[82] His death, however, left the project unfinished. The fundamentals of his astronomy are however pretty clear. To the end, he remained a firm defender of Lansbergen's calculations. When he published a reaction to the book the French astronomer and philosopher Pierre Gassendi wrote on his observation of Mercury's conjunction with the

---

[78] Hortensius (1631) 9-10, 12.

[79] Hortensius (1631) 4; see also preface. Extracts from the debate were published by Frisch in: Kepler (1858-1871) VIII, 543-547.

[80] Van Berkel (1989) 104-105.

[81] Bernegger to Diodati, 14 April 1635: '*Verum enim est... Leydensem illum Boxhornium suo et Hortensii nomine ad versionem Systematis me adhortatum esse.*' *Briefe* (1889) 936.

[82] Hortensius (1645) 591.

sun, he used Gassendi's observations as confirmation of the accuracy of Lansbergen's tables.[83] In his determination of the diameter of the planets, he based himself on the values for cosmic distances in Lansbergen's *Uranometria*. In his cosmological ideas, he probably also remained strongly influenced by Lansbergen. He dismissed Kepler's speculations regarding cosmic harmony. Astronomy, he stated, should be based on observations and mathematical demonstration, not on speculation.[84] Although he rejected the vain prognostications of the astrologers, he firmly believed in the influence of the celestial bodies on the earth.[85]

The second Copernican to occupy a Dutch chair was a more important astronomer than Hortensius. Jan Fokkes ('Fokkes' is not a family name; it simply means 'son of Fokke') was born in the village of Holwerd (province of Friesland) in 1618. As he entered an academic career, he Latinised – or rather Graecised – his name as Johannes Phocylides Holwarda. He studied at Franeker, took his doctorate in medicine there in 1640, and then taught there as a professor. In 1639, a year before he was awarded his medical doctorate, he had already been appointed professor *extraordinarius* of logic. He became an ordinary professor in 1647, and died just four years later.

Although neither mathematics nor astronomy was part of Holwarda's charge, astronomy was certainly the field he took most interest in. He observed various eclipses and remains known for his discovery of *Mira Ceti*, a variable star in the constellation of the Whale.[86] Moreover, he published two astronomical books. A third one, as well as a book on physics with a large astronomical section, was published posthumously.[87] The *Friesche sterre-konst* ('Friesian astronomy') is a handbook of mathematical astronomy. Its core is a series of astronomical tables with which one can calculate stellar positions. These are accompanied by extensive theoretical explanations of the motions of the various planets. Although by this time many similar books had been published, Holwarda felt justified in writing it for two reasons: he writes in the vernacular, which nobody had undertaken before in the Dutch Republic, and he had new astronomical data. In particular, Holwarda accepts Kepler's innovations of putting the sun as the centre of motion and making the orbs of the planets elliptical rather than circular.[88] Holwarda was one of the first ad-

---

[83] Hortensius (1633) 16-23, 79-81.
[84] Hortensius (1633) 68; see also 65.
[85] Hortensius (1633) 67-68.
[86] Pingré (1901) 122, 128, 142, 148-149.
[87] On Holwarda, see Galama (1954) 91-100; a bibliography of his works on 291-293. Also Terpstra (1981) 65-74.
[88] Holwarda (1652), preface; cf. 237-240.

herents in the Dutch Republic of Kepler's theories. On the other hand, he was highly critical of Lansbergen's work.

He took from Kepler not only the theories we now regard as 'modern'. 'The sun, being the centre of the whole world, causes, by its rotation on its axis and the emanation of its strong rays (as if it were a big magnet), the motion of all further celestial bodies, in proportion to their distances and their particular properties.' The idea of the sun as moving force of the heavens is clearly Keplerian. Holwarda, after some hesitation, declares that the fixed stars, too, are moved by this central force. Their immobility would contradict nature, which is never idle. Precession, then, is a real motion of the fixed stars, not just of the earth's axis. Holwarda's ideas here are akin to Lansbergen's. He, too, thinks that the velocity of the heavenly bodies decreases with their distance from the centre, that is, the sun. [89] On the other hand, Holwarda ridicules the idea of solid celestial spheres. [90]

Holwarda represents the transition to a new era. By the time he occupied his chair, not only had the old Aristotelian system been discredited, but the outlines of a new world-view had become visible. Descartes had published his *Discours de la méthode* in 1637. But if Holwarda was acquainted with Descartes' ideas, there are few traces of them in his work. Although his view of nature appears to have been influenced by the new mechanical philosophy, another mechanical philosopher seems to have been more important to him: Pierre Gassendi, whose work Holwarda praised highly. Like Gassendi, he defended the view that nature consists of atoms. On the other hand, there is also the influence of Renaissance philosophers, such as Julius Caesar Scaliger. Holwarda's atoms are endowed with sympathy and antipathy. He remains a transitional figure and certainly cannot be called a mechanical philosopher.

We might conclude this section on academic Copernicans with a mathematician who remained outside the universities, Bernard Varenius. He was, however, a scientist of note who worked in the academic sphere. Varenius came from Germany. He had studied at Hamburg with the famous Joachim Jungius before starting an academic pilgrimage which brought him first to Koenigsberg and finally to Leiden. Apparently, the scientific climate in the Dutch Republic was much to his liking. He settled in Amsterdam and set his hopes on obtaining the chair of mathematics at the Athenaeum, which had been vacant since the departure of Hortensius and his successor, John Pell, some time before. Despite his poor living conditions, he turned down an offer

---

[89] Holwarda (1652) 230-231; cf. (1640) 218.
[90] Holwarda (1640) 232-233.

from his teacher Jungius to return to Hamburg. His hopes, however, came to nothing. Soon afterwards he died, in 1650.[91]

In order to qualify himself for the chair of mathematics, Varenius undertook the composition of a book on geography, then regarded as a practical application of mathematics. Shortly before his death, it appeared as *Geographia generalis*. The book is considered a milestone in the development of the geographical sciences. Here, however, we are concerned only with Varenius' contribution to the discussion on Copernicanism. The book opened with a general description of the earth, wherein he discussed not only its shape and size, but also its motion. He rather elaborated on this subject, as 'there is no affection of the earth on which there is a greater or harsher dispute, as not long ago it suffered from the censorship of the Roman Church.'[92]

The fifth chapter is devoted to the annual rotation of the earth, and the sixth to its course around the sun. Varenius argues elaborately in favour of the Copernican system, rehearsing the well-known arguments for and answering the common objections against it.[93] He has clearly been impressed by the new cosmological discoveries. Aristotelianism is dismissed and he rejects the traditional distinction between heaven and earth, although he feels that one might still maintain it for practical reasons.[94] However, he makes no use of the newly developed Cartesian physics. He is familiar with Descartes' vortex theory, to which he refers in a section on ocean currents and the tides, but appears rather critical. He promises a further consideration on Cartesian physics, but this never appeared.[95]

Varenius' book immediately became popular and was reprinted many times in the seventeenth and eighteenth centuries. It was exactly what the educated public needed at the time: a thorough but readable overview of an extensive subject, presented in a new scientific spirit. One should not call it a mere book of popularisation, although it certainly helped to spread scientific insights. Its discussion of Copernican theory must have reached a far wider audience than most other books on the subject, in the Dutch Republic as well as abroad. In the eighteenth century, it was even translated into Turkish and so became one of the sources from which knowledge on Copernican theory reached the Ottoman empire.[96]

---

91 His biography in Günther (1905).
92 Varenius (1650) 48.
93 Varenius (1650) 48-62.
94 Varenius (1650), dedication.
95 Varenius (1650) 180-190.
96 Ihsanoglu (1992) 86-87.

Philosophers

The most interesting development in the academic sphere, however, is not so much the shift from Ptolemy to Copernicus and Tycho Brahe: more far-reaching was the fact that cosmological questions gradually became a matter of concern not just to mathematicians and astronomers, but also to physicists and philosophers. In a sense, this was a result of the shift which had been started by Galileo's discoveries: the main question by now was the constitution of the heavens, not the mathematics of celestial motions. It seems quite natural that philosophers increasingly discovered the relevance of the topic. Still, the new context did not leave the discussions unaffected. By being discussed in philosophy, questions other than those hitherto common came to the fore. Moreover, the old questions were put in a new framework. The order and constitution of the heavens became entwined with the doctrines on nature as a whole.

That the new cosmology was gradually coming to the attention of academic philosophers is illustrated by one of the outstanding academic philosophers of the first half of the seventeenth century, Franco Burgersdijk. In his case, we see the gradual shift from complete disregard to cautious acceptance of the new cosmology. Burgersdijk was the dominating philosopher during the first part of the seventeenth century. He held the Leiden chair of logic and also taught ethics and physics. [97] He was not an original philosopher, but was renowned as a pedagogue. His textbooks were often reprinted and widely read, not just in the Dutch Republic, but also in England. He wrote two textbooks on physics, which appear to be collections of previously held disputations. *Idea philosophiae naturalis* was originally printed in 1622 (the dedication is from December), and *Collegium physicum* in 1632.

Idea philosophiae naturalis is very short, composed of very short theses with little explanation. The text is traditional, commenting on the works by Aristotle, not on current cosmological questions. Speaking on the heavens, Burgersdijk mainly discusses their influence on the sublunary world. Orbs and heavenly motions are simply taken for granted; the references are mainly to the Conimbricenses. [98]

In 1627, however, Burgersdijk presided over a disputation, *De coelo* ('On the heaven'), which did display some awareness of new developments in cosmology. It does not just pose the time-honoured scholastic questions about the goal and form of the heavens, but is largely devoted to more down-to-earth questions regarding its shape, division and substance. His astronomical

---

[97] On him: Ruestow (1973) 28-32; Bos and Krop (1992); van Bunge (2001) 27-28, 30-32.
[98] Burgersdijk (1635) 31-35.

knowledge is not always up to contemporary standards (he argues that all heavenly bodies, including the moon, are luminescent, although he admits that it is believable that the sun contributes as well), but he does try to account for new insights. Starting from the assumption that stars are denser parts of their orbs (a point he had made in the *Idea*), Burgersdijk explains the occurrence of novae, like the one of 1572, as the result of a temporary thickening of the heavenly substance. He also discusses the motion of the earth, which he rejects with the familiar argument about the behaviour of falling bodies. He still maintains that the heavens consist of solid orbs and, as a simple body can have but one motion at a time, there are as many orbs as there are different motions. In an earlier chapter, we noted his division of the heavens, on scriptural grounds, into three parts.[99]

Burgersdijk appears to have changed his views even more in *Collegium physicum*.[100] In the eleventh disputation ('On ordinary and extraordinary stars'), he still upholds the view that stars are dense spots of heavenly substance. But he has become much more critical about celestial orbs: these are mere figments, invented to describe the motions of the stars. They suffice as hypotheses to save the phenomena, but it should be regarded an error to think that they exist in reality. Several arguments force us to reject them. It seems much more probable that the planets move of their own in a fluid medium.[101]

As to the system of the world, Burgersdijk has become hesitant. As he says, the cases of both Ptolemy and Copernicus are argued with strong reasons. (Tycho Brahe is not mentioned.) As it seems, Burgersdijk's change in attitude has come about mainly under the influence of the works of Lansbergen, to whom he explicitly refers. He is most impressed by Lansbergen's calculation of the velocity which the fixed stars should have if they, rather than the earth, were turning. 'And if the earth moves with a daily motion, it is easier to believe that it moves with an annual motion as well. And if Lansbergen has not hallucinated in his *Uranometria*, we shall have to think about a solution for the arguments, which argue in favour of Ptolemy.'[102] Had death not prevented him, Burgersdijk might well have turned into a full-fledged Copernican. The real innovation in his work, however, is that here is a philosopher, not a mathematician, discussing the system of the world. In his own time he was an exception, but in the second half of the century, the tables would be turned: discussions on the world system would become the near exclusive domain of the philosophers.

---

[99] Burgersdijk, disp. Leiden 16 June 1627.

[100] Burgersdijk (1637). I use the updated second edition, as I have not seen the first.

[101] Burgersdijk (1637) 106-113.

[102] Burgersdijk (1637) 113; see also 112.

Another philosopher of a somewhat later date is Albert Kyper, who was born in Germany. After his studies at Leiden, he was allowed to teach physics privately and even to preside over disputations. In 1646, he became professor of philosophy at the illustrious school at Breda and physician to the stadholder. In 1650 he returned to Leiden, now as a professor of medicine.[103] In 1645-1646, he published in two parts an introduction to physics. Although this work was published after the important work of Descartes, which transformed the face of philosophy, it hardly mentions him.

In the introduction, Kyper announced his programme. His enemies had insinuated that he had introduced new and dangerous opinions into philosophy, which were apt to undermine the foundations of theology and disturb the academic peace. With this publication, he wanted to defend his honour. He admitted that in many respects he was rather critical of Aristotelian philosophy. But that was not to say that he wanted to banish it from the university, or to frame a new system.[104] The hostility Kyper had met may partly have derived, as he supposed himself, from the fact that he had only recently become a member of the Dutch Reformed Church (originally, he had been a Lutheran), and was suspected of having done so just to further his career.[105]

From the text of his book, there appears no reason to regard Kyper as a dangerous modernist. He appears as a man deeply committed to religion, who refers to the Bible much more often than is usual in philosophical textbooks. His deviations from Aristotelian philosophy appear to derive mainly from this biblicism. Biblical arguments support his view that darkness is not a mere privation, as the Aristotelians have it, but something positive. He also ponders the question whether Christians can use in good faith pagan names for the stellar constellations, but admits that it would be very unpractical to have them changed.[106] Kyper rejected Burgersdijk's view that the stars were dense spots of heavenly substance, on the grounds that Genesis teaches us 'that the stars are placed into the heavens, not made from the heavens, by God.'[107] In astronomy, Kyper appears to have been an adherent of Tycho's world system. Perhaps that is why he preferred the view that the heavens are a homogenous, fluid body, to the view that they are made of solid orbs: in this way, they were best suited for the motion of the stars, as well as for the propagation of the stars' influence. As to the fixed stars, because of the uniformity of their motion, he thought it probable that all of them are at the same dis-

---

[103] On him: Sassen (1962) 323-68; Ruestow (1973) 39-43.
[104] Kyper (1645-1646), preface.
[105] Ibid., dedication.
[106] Ibid., I, 463-464; II, 42. Ruestow (1973) 40.
[107] Kyper (1645-1646) II, 19.

tance from the earth.[108] This argument, of course, presupposes that the diurnal rotation is in the heavens themselves.

Indeed, Kyper rejected the Copernican view that the earth was moving. In accordance with his general stance, his main arguments appear to be biblical. Kyper argues that the Bible expressly teaches that the stars move. Moreover, from the history of Creation in Genesis, it appears that the sun is not the centre of the universe. Physical arguments appear only in the second place; most of them are fairly commonplace. He dismisses the Copernicans' claim that their system is the more 'economic'. The system of the world should be valued according not only to its economy, but also to its necessity and greatness. As long as we do not understand these, we should not make a judgement contrary to Scripture. Interestingly, he also dismisses the argument of 'cosmic harmony', i.e. that the velocity of the spheres should diminish according to their distance from the centre. Kyper argues that the argument is based on a false reading of the original source, viz. Aristotle in his book on the Heavens (*De caelo*, Book 2, Chapter 3). There, Aristotle spoke only about precession. Wondering how the stars are able to make all these complicated motions (earlier, he had rejected not only orbs but also their being moved by intelligences), Kyper looked for a solution in the sympathy of the heavenly bodies both to each other and to the sublunar things.[109]

On the one hand, Kyper's world-view appears rather traditional; on the other hand, however, he is clearly not satisfied with traditional scholastic philosophy. Notably, he feels the need to gain an insight, as a philosopher, into the pressing cosmological questions of the day. Therefore, he starts on new topics and new explanations, although in a much more cautious way than some thinkers outside university. Kyper appears as a kind of belated representative of 'Mosaic philosophy'. The point is, however, that with the shift from a mathematical to a physical consideration of the heavens, the status of the biblical sentences became problematic. No mathematician would turn to the Bible for the solution to a mathematical problem, and no theologian would expect him to. Physics and philosophy, however, traditionally did interfere with theology. It was hard to imagine that the Bible would have no relevance to the explanation of the world. In this sense, Kyper's stance was not just an echo of the sixteenth century, but also a foreshadow of things to come.

---

108 Ibid., II, 56-57, 9-10.

109 Ibid., II, 104-110; see also 57.

Part III. The universe of law

8. Cartesian cosmology

A physics of the universe

Although Aristotelian philosophy had been the subject of attack for nearly two centuries, in 1635 it still stood firm. The strongest argument in its favour, however, was simply the lack of an alternative. Especially in teaching, it was felt that a clear systematisation of knowledge was needed to introduce students to the world of learning. Aristotelianism was the only such system available.

Aristotelianism collapsed in the end, not by a slow crumbling under the accumulation of its problems, nor by gradual adaptations until it had simply grown unrecognisable. Aristotelianism was finally brought down by a deliberate attack. Descartes designed an all-embracing philosophical system which established itself as a viable alternative to Aristotelianism. In most of the literature on Descartes, emphasis is laid on his work in metaphysics and methodology. However, it can be defended that these philosophical ideas served mainly as a legitimation of his new physical world-view. [1]

In the end, Descartes views were compelling because they offered the new intellectual framework into which the many discoveries and insights of the previous century could be fitted. It can be argued that, for instance, Harvey's theory of the circulation of the blood or Sanctorius' static medicine took on their modern 'scientific' shape only under the influence of Cartesian philosophy. Galileo's discoveries had been anomalies in the Aristotelian cosmos. In Descartes' world, they appeared as the logical consequences of his general view of nature. Likewise, Copernicanism was transformed from a mathematical theory of the heavens into the application of general physical principles to the phenomena of the solar system. Copernican cosmology had always been handicapped because it ran counter to accepted ideas on the universe.

---

[1] Gaukroger (1995); van Ruler (1995). The literature on Descartes is of course immense. I restrict myself to the most relevant sources.

Descartes finally changed this. Thus, it was Cartesianism which turned the heliocentric theory into an acceptable and indeed dominant idea. [2]

This is not the place to discuss Descartes' ideas in general, but a brief overview of his ideas on the system of the world cannot be dispensed with. These ideas were put forward mainly in his *Principia philosophiae* (1644). An earlier work on the same subject, *Le monde*, remained in manuscript. Anyhow, it should be emphasised that these ideas were compelling only because they were part of a larger whole. Descartes claimed that his physics had a mathematical character. Mathematical reasoning, not the unreliable testimony of the senses should be our guide to truth – an argument which could well have been framed with the Copernican debate in mind. Descartes maintained that he could prove the basic constitution of the world in a rigorous way from some basic 'axioms' or principles. Descartes himself did not capitalise on the analogy between the physical world and the world of mathematics, but some of his followers did, as we are to see shortly.

Descartes regarded nature as uniform. This entailed a universe which was in principle without limits and without a centre. Neither the sun nor the earth could be allotted a special place in it. The Aristotelian world with its many different spheres and regions was simply incompatible with the world-view propagated by Descartes. On the other hand, Copernicus' world, with the sun in the centre as on a royal throne, would also not do. In Descartes' universe, all the stars were suns. Each star was the centre of a vortex. In the case of the sun, this vortex carried the planets around the sun. The planets in their turn were the centre of other smaller vortices, carried along with them. These secondary vortices caused the planets to rotate on their axis. They also carried with them the planets' moon(s), if they happened to have such. [3]

The various celestial bodies were in a sense equivalent. Like Girard and Van der Veen before him, Descartes felt that over the course of time they could be transformed into each other. The vortices were in a state of equilibrium with their surroundings, but if circumstances changed, they could collapse. Stars constantly became obscured by matter as it coagulated, as evidenced by sunspots. Normally, grosser matter was ejected, but when there was too much of it, the star's surface would be obscured and the star would not be able to keep its vortex turning. In this case, the vortex would collapse

[2] The pivotal role Cartesianism had in many cases in the acceptance of heliocentrism has been recognised before. Brockliss (1990) makes a rather general case for France. In Sweden, according to Sandblad, geocentrism lost favour in the course of the second half of the seventeenth century because of 'the general Cartesian tide of the time': Sandblad (1972) 265 and *passim*. See also Vanpaemel (1995) 116-118, on Louvain, and Moesgaard (1972) 134-140, on Denmark.

[3] For a discussion of Descartes' theories, see Aiton (1972) 34-58.

and the star would be absorbed by a neighbouring vortex. Depending on the circumstances, it would move either to the near centre of its new vortex and become a planet, or to its periphery and become a comet, traversing space at the edge of the various vortices. According to the Cartesian hypothesis, the earth itself had come into being in this way, as a star which had been obscured and absorbed by the vortex of the sun. Descartes himself stated that this theory was a mere working hypothesis which could explain the phenomena, but that it could not be true as it was not in accord with the story of the Creation.[4] One may reasonably doubt whether this was his real opinion, or a matter of caution with the fate of Galileo in mind. His followers generally preferred to forget about this qualification.

A convincing argument in favour of Descartes' vortex theory was that it could also give a plausible explanation of the tides. It had long been remarked that the movement of the sea was synchronous with the course of the moon (although Galileo rejected this, apparently because he found it smacked of occult forces), but the exact mechanism was unclear. According to Descartes, the tides were caused by the pressure of the vortex around the earth. As the vortex had to pass through the relatively narrow space between earth and moon, its sideways pressure increased, pressing down the sea level. At the same time, it slightly pushed the earth aside, increasing the pressure of the vortex, and hence lowering the sea level, on the other side as well. The waters naturally flowed towards the other sides of the earth. As the moon's orbit was not circular but oval, the distance between moon and earth – and thus the space through which the vortex had to pass – varied over time. The narrower the opening, the higher the pressure of the vortex and the lower the sea level would fall. Thus even the alternation of spring tide and slack water could be explained by Descartes' theory.[5]

There is one final element which should be explained in some detail, as it directly affects Descartes' interpretation of Copernicanism. This has to do with his theory of motion. Descartes rejected the Aristotelian absolute distinction between motion and rest and upheld a relativistic definition: motion can only be defined as such with respect to something else. For a philosophical definition, however, he felt such a random choice would not do. The motion of an object, then, should be defined with respect to the object's immediate surroundings. A ship drifting on a swiftly flowing river should be said to be at rest, as it is not moving with respect to the water. Applied to the case of the earth, it is moving with respect to the sun. But if we are to answer the

[4] *AT*, VIII-1, 99-100, 203.
[5] *AT*, VIII-1, 232-238.

question of the earth's motion or rest in a more philosophical way, we must look at its immediate surroundings. Now, the earth is drifting in a large vortex of celestial matter, just as the ship is drifting in water. As the earth is simply drifting, it will not be moving with respect to the vortex. Properly speaking, therefore, the earth is at rest. The argument seems deliberately framed to attenuate religious misgivings about the Copernican system. It is hard to see why otherwise Descartes would have stated this so explicitly. Indeed, it is introduced for the first time in *Principia philosophiae*.[6] When writing *Le monde*, Descartes had not yet contemplated the problem. Still, the argument, without being flawless, follows quite logically from Descartes' general ideas on motion.

Under the influence of Descartes' ideas, many people came to regard the theory of the annual and daily motions of the earth as an established fact, rather than as an explanation which was plausible at best and anyhow open to discussion. Heliocentricity seemed to follow inevitably from his view of nature. How this came about is analysed in the following sections. Three figures, all advocates of Copernicanism, will be discussed in some detail. All three formulated their ideas in a clear and personal way. Two of them were minor figures who at an early date became convinced Cartesians. The third was one of the greatest scientists of the age, and who showed the stamp of the new physics throughout his career. In all cases, their Copernicanism appears to follow from the general ideas which had been forced upon the age by Descartes.

Daniel Lipstorp: mathematics as philosophy

Daniel Lipstorp made a great contribution to the identification of Cartesianism and Copernicanism. Lipstorp, a German, was born in Lübeck in 1631. He studied for some time at Rostock University, where he obtained the title of master in 1651, and at Leiden, where he matriculated on 4 July 1652. In 1653 he went to Weimar as a court mathematician. In 1656 he returned briefly to Leiden to matriculate (26 September) and to be awarded his doctorate in law (2 October). His further career carried him to Uppsala and The Hague, and then back to Lübeck, where he died in 1684.[7]

During his first stay in Leiden, Lipstorp published a *Specimina philosophiae Cartesiana* (1953) with a sequel called *Copernicus redivivus*: 'Copernicus revivified, or On the true system of the world'. The *Specimina* consist mainly of a discus-

---

[6] AT, VIII-1, 89-91.

[7] Günther in: *Allgemeine deutsche Biographie*, XVIII (Leipzig 1883) 746.

sion of air, describing among other things various pneumatic instruments. Pertinent to our subject are the two introductory chapters, which discuss generalities. In the first – 'On the certitude of Cartesian philosophy' – Lipstorp argues that one should demand mathematical proof not just in mathematics, but also in physics. Of course, it is only Cartesian philosophy which can give this kind of proof.[8] In the second chapter, he shows how one should take this. Here, he gives Descartes' rules of motion, demonstrating them in – what he claims to be – a mathematical way. Lipstorp starts with a number of definitions and postulates. These are followed by twelve 'axioms or general rules', which comprehend Descartes' three laws of nature as well as some other generalities Descartes had presupposed rather than formulated; for instance, some rules concerning the transmission of a quantity of movement from one body to another. These axioms are followed by nine 'theorems', which correspond to Descartes' nine rules of percussion.[9]

Copernicus redivivus, the second part of the book, is a vigorous defence of the Copernican world system. It was largely based on a series of six disputations Lipstorp had held the previous year (1652) at Rostock University (Germany): 'Physico-mathematical discourse on the Copernican system of the world'. In these disputations, he had shown 'that that Pythagorean opinion is not so improbable and paradoxical as is commonly reputed, but that, as far as it is defended by mathematical experiences taken from heavenly appearances and by physical arguments, it is in its way preferable to, and more probable than, the Ptolemaic system.'[10] Still, even when defending that Copernicanism was not contrary to natural reason, in 1652 he had rejected it in the end on biblical grounds. The second half of the sixth disputation was largely devoted to a discussion of the arguments from Scripture. Scripture, he stated, clearly stated that the earth stood still, and this argument should be decisive. Although discussing the Copernican counter-arguments in some details, they failed to convince him.

His position appears to have changed by the time he wrote *Copernicus redivivus*. Although largely based on the earlier work, it has been reworked up to the point of being unrecognisable in many places. The whole work is divided into two books. In the first, Lipstorp, after some general remarks on scientific progress, starts with a rather traditional account of the theories of Copernicus and their history. In the fourth chapter, he explains Descartes' idea of the 'true motion' of the earth, that is to say, that it rests in the celestial matter. The other chapters are also rather traditional. Lipstorp discusses the various systems

---

[8] Lipstorp (1653) I, 1-27.

[9] Lipstorp (1653) I, 28-58.

[10] Lipstorp, disp. Rostock 1652, title of the separate disputations.

which try to compromise between Ptolemy and Copernicus, in particular the one by Tycho Brahe, and points to the various absurdities of the Ptolemaic system. The second book is mainly a rehearsal of all the arguments put forward against the Copernican system, along with their refutation. The eighth chapter, the final one, deals with the arguments from Scripture. Although these are not our main subject in this chapter, it is important to note that Lipstorp had completely revised his earlier opinion. Whereas one year earlier he had deemed the scientific arguments insufficient to decide that the biblical passages should not be taken in a literal way, in *Copernicus redivivus* he quite emphatically argued that Scripture cannot decide questions of mathematics or knowledge of nature.

As said, Lipstorp presented his defence of Copernicanism together with a work on Cartesian philosophy. So, how far was his Copernican conviction in 1653 based on Cartesian philosophy? On the whole, the Cartesian element in *Copernicus redivivus* is not very marked, apart from the fourth chapter of book one. Partly, this is undoubtedly because it concerns a reworking of an earlier, non-Cartesian work. Besides, Lipstorp of course had had a largely non-Cartesian education, and it seems clear that originally he had entered the discussion on the system of the world because of his interest in mathematics, not because of Cartesian physics. His work is rather an exercise in learning. Instead of developing an argument of his own, he mainly discusses those of others. He clearly likes to show off his erudition and refers to as many famous and obscure authors as he possibly can. Characteristically, he closes the book with a phrase in Arabic, 'It has been finished with the help of God, all glory be to Him!', a common conclusion of Arabic works. [11]

Still, his contemporaries had little doubt. To the Leiden minister Du Bois – one of the most prominent clerical opponents of Copernicanism in the Dutch Republic – Lipstorp served as a living example of the pernicious effects of Cartesian philosophy. He had seen the six earlier disputations, and knew that in the first five and the beginning of the sixth, Lipstorp, using natural reasons, argued in favour of Copernicus; but that, when he finally arrived at the arguments from Scripture, he abandoned his pro-Copernican stance. That, Du Bois thought, is how it should be: natural reason yielding to revealed truth. But then he found that as a Cartesian, Lipstorp thought he could dismiss the arguments from Scripture. In Du Bois' view, this served to prove that Cartesianism was undermining religion. [12]

In the preface to the *Specimina*, Lipstorp himself is quite explicit. In 1652, he

[11] With thanks to Dr J. Hogendijk, Utrecht, for the translation and explanation.

[12] Du Bois (1655) b, 25-27.

had not been convinced of the truth of the Copernican system; since then, however, he had dedicated himself to mathematics, optics and Cartesian philosophy at Hamburg under the guidance of Johann Adolph Tasse (Tassius) and at Leiden under Frans van Schooten. While studying optics, he came to admire Descartes. He then found that in Descartes' system, all objections to Copernicanism disappeared. [13] Of course, we have to regard such a 'spiritual autobiography' with some caution. It is not inconceivable that Lipstorp remodelled his life after a convenient model. One might wonder whether Lipstorp was not already a convinced Copernican at Rostock, and that it was outward pressure rather than inner conviction which led him to the cautious conclusion of his disputations. In that case, his change of mind would have been only apparent and not caused by philosophical considerations.

This matter is hard to decide. Lipstorp was not a die-hard opponent of Copernicanism in 1652, and if he was not already convinced of its truth he must at least have been pleased to find a reason to become so. However, if Lipstorp thought it expedient to present his Copernicanism as an outcome of a conversion to Cartesianism, that means that such a presentation made some sense. Cartesianism could be seen as the foundation of Copernicanism as a real, physical theory, instead of a mere mathematical hypothesis. How did Cartesianism, in his view, strengthen the Copernican hypothesis? One might suppose that Descartes' theory of the 'real' motion of the earth, which after all was posited prominently in his book, played some part. However, it is striking that this theory is not used in the final chapter to reconcile the difference between the physical model and the biblical text. Apparently, Lipstorp had other, better reasons not to doubt the Copernican system.

Most probably, this was the mathematical structure of Cartesian philosophy itself. His Cartesian conviction is summarised by the sentence: 'Mathematics itself is the true and best philosophy'. [14] What he meant by this is shown by his discussion of the elliptical shape of the planetary orbits. Credit for demonstrating this elliptical shape he allotted to Boulliau, not to Kepler, who had shown it (in the case of Mars) only by calculation. To Lipstorp, such a proof clearly was not enough. Boulliau had demonstrated it 'from general and known dispositions of motion.' [15] The theory of elliptical orbits is not in Descartes, nor in most later Cartesian authors. Lipstorp himself speaks of it only in his preface. Still, the passage shows what he considered vital for a scientific theory. In this respect, we should not regard *Copernicus redivivus* in isolation from the preceding *Specimina*, with its mathematical presentation of

[13] Lipstorp (1653)a, preface.
[14] Lipstorp (1653)a, preface.
[15] Lipstorp (1653)b, preface.

Cartesian physics. Cartesianism set new, mathematical standards for theories on the world. Many older theories were dismissed in this way, but for theories which answered these, Cartesianism claimed mathematical – that is, absolute – truth.

Christophorus Wittichius' decisive argument

Another German who studied in the Dutch Republic and was won over to Cartesian philosophy was Christoph Wittich (Christophorus Wittichius). Wittichius was a theologian who, after professorships at Herborn, Duisburg and Nijmegen, ended his career as professor of theology at Leiden. In 1653, the same year Lipstorp published *Copernicus redivivus*, Wittichius published in Amsterdam two dissertations (*Dissertationes duae*), which had earlier been defended at Duisburg University under his presidency. These disputations are primarily a theological work, concerned with biblical exegesis. Wittichius' main aim is to demonstrate that the theory of Copernicus is compatible with biblical revelation. As such, the work will be more fully discussed in a later chapter, on Copernicanism and biblical exegesis. At this point, it is important to note that the confidence with which Wittichius tackled the issue was founded in Cartesian physics.

That Cartesianism was really the foundation of Copernican cosmology was made clear already in the introduction: 'About the most important of the physical propositions of Descartes, still fouled by fools' mud, concerns the annual motion of the earth around the sun and the daily motion on its axis. Descartes, having expounded this hypothesis according to Copernicus, who had resuscitated it after it had long remained buried, has defended it by means of a very certain and evident mathematical proof which no one (except those who do not understand it) has so far been able to refute; nor will anybody ever be.' It was this new Cartesian insight which moved Wittichius to discuss the question of Copernicanism and the Bible afresh: 'Until now, Copernicus' defendants were not able to demonstrate the theory of the double motion of the earth in such a clear way, as Descartes has done after having laid surer foundations of physics.' [16]

The first of the two dissertations is completely devoted to biblical exegesis, which does not interest us here. It is complemented, however, by the second dissertation, which is primarily a physical work. As the title states, it 'deals with the disposition and order of the universe as a whole and of its main bodies, and defends the opinion of Descartes on the true rest and true motion

---

[16] Wittichius (1653) praefatio.

of the earth'. First of all, Wittichius discusses the size and form of the universe. As humans do not know the outer limits of the universe, we have to take it as indefinite. As a consequence, we cannot speak of a centre of the universe (a centre can only be defined in relation to a circumference). Of course, we can speak of a centre of the movement of the planets, 'centre' to be taken here in a physical, not a geometrical sense. This centre is the sun: both Copernicus and Tycho agree on this point (Ptolemy is no longer a viable alternative). Only the earth has so far remained a matter of dispute. [17]

Wittichius next turns to an elaborate discussion of the world's system. This part is little touched by Cartesianism. Despite his declared intention to leave most arguments untouched, Wittichius' argument is mainly a rehearsal of a large number of ancient and modern authorities supporting heliocentrism. Most attention is paid to respectable, but in modern eyes highly spurious, ancient authorities such as Pythagoras and Numa Pompilius. Cartesian physics turns up only in the fourth chapter, where Wittichius defends Descartes' definition of motion against his opponents.

In the fifth chapter Wittichius finally presents his 'mathematical' proof of the earth's motion. As he explains, he will not refer to simplicity, harmony or the velocity of the heavens, but will use a 'demonstrative and most evident argument taken from our philosopher.' [18] This argument appears to come down to just a summary of the relevant passages in Descartes' *Principia philosophiae*. As Lipstorp did at about the same time, Wittichius put them in a synthetic-mathematical form. As he had explained in the preface, Descartes' physics is a coherent system, in which one cannot understand the later propositions without having understood the earlier ones – he explicitly makes a comparison with the work of Euclid. So, he is compelled to start from first principles. He cannot discuss all of Descartes' physics, of course, but a clear idea of Descartes' idea of motion is essential. [19]

Wittichius gives Descartes' theory in the form of 28 propositions. The first is that all corpuscles in the world are made of one and the same kind of matter. The second is the impossibility of a vacuum. This entails that any displacement of a corpuscle affects the corpuscles in the environment as well. From this, the theory of vortices is deduced and hence follows the whole of Descartes' cosmogony (which Descartes himself had presented as a mere hypothesis). Finally, he arrives at the motion of the earth in the solar vortex, wherewith the truth of the Copernican system has been proved. [20] Wittichius, it

[17] Wittichius (1653) 171-172.
[18] Wittichius (1653) 225.
[19] Wittichius (1653) praefatio.
[20] Wittichius (1653) 226-244.

seems, regarded this as a definitive argument – as mathematical arguments should be, of course. Some years later, he returned to the subject after his two dissertations had been the target of vehement criticism. Although much more elaborate, this new work is really just a repetition of the former; so, we re-encounter exactly the same 28 propositions. [21]

Wittichius, then, at the same time and in roughly the same way as Lipstorp, found in Cartesian philosophy a 'mathematical proof' of the motion of the earth, convincing up to the point that it should decide a theological debate. Both, be it in different ways, reformulated Descartes' principles as a mathematical argument. How this coincidence comes about is difficult to say. One suspects the influence of academic teaching at Leiden, but clear indications are lacking. Nor would their cases remain unique. A work which would undertake the same kind of reformulation in a more systematic and comprehensive way is Spinoza's *Renati des Cartes principiorum philosophiae pars I & II, more geometrico demonstrata* from 1663. Later in the century, of course, the *'mos geometricus'* became a rather popular argument for non-physical discourse as well, Spinoza's *Ethica* taking pride of place among the literature of this kind. This geometric way of reasoning seems to have its origin in the mathematical reformulations of Descartes' physics. [22]

Out of Descartes: Christiaan Huygens

The significance of Descartes' ideas for the reception of Copernicanism was not limited to minor philosophers and theologians, nor to Cartesians in the strict sense. Descartes' fundamental ideas determined the scientific discussion of the second half of the seventeenth century. They also dominated the thought of the truly great scientists of the age. In order to demonstrate this, we shall finally have a look at Christiaan Huygens.

Huygens grew up at a time when Cartesianism was much in vogue. His father was the diplomat and virtuoso Constantijn Huygens, one of Descartes' main protectors in the United Provinces. He obtained his main scientific education from the mathematician Frans van Schooten, a close friend and admirer of Descartes. Still, his relation to Cartesianism remained ambiguous. In his youth, Huygens was much impressed by Descartes' system, but over the course of time, he came to acknowledge various weaknesses in it, particularly in its physics. Huygens was too much of a mathematician and too little

---

[21] Wittichius (1659) 80-339; for the 28 propositions, see 241-339.

[22] As is well known, Descartes himself had already put his proof of the existence of God and the distinction between soul and body in the form of a mathematical argument, at the suggestion of Mersenne in the latter's objections to Descartes' *Meditationes de prima philosophia*. Cf. Dear (1995).

of a philosopher to feel truly at home in the Cartesian world. Rather, he continued the tradition of men like Kepler and Galileo in their mathematical description of nature. If Descartes aimed at supplanting Aristotle, Huygens figured as a new Archimedes.

Still, Huygens owed a lot to the new physics as put forward by Descartes. His critical stance can be largely explained by the fact that Cartesianism was to him a point of departure, not a new theory which solved old questions. It offered him a frame of reference for his scientific investigations. To Huygens, the system of the world was no longer a matter of debate. The truth of the Copernican system was simply an established fact. Huygens never bothered to discuss the relative merits of the Copernican, Ptolemaic and Tychonian systems. Instead, he pondered the various alternatives within Copernicanism. In a letter written in 1656 to the Polish astronomer Hevelius, he explains his preference for Kepler's elliptical orbits, above Copernicus' construction by means of epicycles. [23] Only when directly attacked on his Copernicanism did he comment on the relative merits of Tycho and Copernicus: 'Which of these two I apply hardly matters as far as phenomena are concerned. But the truth of the matter is explained only by following Copernicus.' [24]

Copernicanism, then, was an integral and important part of Huygens' world-view, self-evident up to the point that it needed no special defence. Still, there are also some instances wherein he clearly gave his stance on the subject. Huygens was aware that Copernicanism had not been accepted generally and he thought the matter important enough to propagate it when the occasion presented itself. The points where Huygens' work is part of the history of Copernicanism and cosmological theory are his *Systema Saturnium* and his *Kosmotheoros*. As a third point, one might mention the Copernican planetary he designed and built in 1680-1682. In this case, his primary aim seems to have been to construct an accurate planetary, which could be used for predicting planetary conjunctions. It must have seemed only natural to him to use a Copernican model. But at the same time it also had a didactic purpose, as Huygens himself acknowledged. [25]

In the case of the *Systema Saturnium*, the situation is somewhat different. Huygens earned his first public success with his telescopic observations. In 1655 he discovered a moon of Saturn, which later was named Titan. It was the first really new discovery in the solar system since the time of Galileo, and as such was a scientific sensation of the first order. At the same time, Huygens also gave the first correct interpretation of the curiously varying shape of

---

[23] Huygens to Hevelius, 25 July 1656. *OC*, I, 463-464 (no. 318).
[24] *OC*, XV, 459 (*Assertio systematis Saturni*, 1660).
[25] Seidengart (1982) 210.

Saturn as observed by a telescope: the planet was surrounded by a ring, slightly inclined with respect to the plane of the ecliptic. These discoveries were momentous in their own right. But Huygens also perceived that they offered additional evidence for the truth of the Copernican system. Already in 1656, in a letter to van Schooten, he called them a 'confirmation of the Copernican system: another earth, another moon'.[26] In the official publication, *Systema Saturnium*, which appeared a few years later (initially, he had only announced his discoveries in a four-page pamphlet), the point was more explicitly made. In the dedication to Leopold de Medici, prince of Tuscany, Huygens emphasised that his system of Saturn confirmed most strongly 'that beautiful general order of the world, which has its name from Copernicus'. The fact that Saturn proved to be so akin to our own earth, having, like the earth, one single moon (instead of four, like Jupiter, or none at all, like the other planets), and being, as apparent from the position of its ring, slightly inclined with respect to the ecliptic (again, just like the earth), made it most probable that the earth was just a planet.[27]

So, Huygens used his *Systema Saturnium* to take sides in the Copernican debate, which was still going on in Europe and had just reached its peak in the Dutch Republic, as we will see in Part IV. In fact, Huygens was ready to go out of his way to strengthen the case of Copernicanism. One might ask how he could have been so sure that Saturn had only one moon. In fact, before the seventeenth century was over, Cassini would discover another four. Huygens had apparently considered the rejoinder and countered it in advance by means of an argument that has rather bewildered most historians who have dealt with it.[28] As he explained in his dedication, it seemed improbable that more planets (moons included) would be discovered, as his latest discovery had brought the number of secondary planets to six (one moon of the earth, one of Saturn, and four of Jupiter), which was equal to the number of primary planets. That is, both kinds of planets equalled the perfect number of six (in mathematics, a 'perfect' number is a number of which the addition of the factors gives the number itself: $1 + 2 + 3 = 1 \times 2 \times 3 = 6$.) This could be no coincidence, but apparently had been ordained on purpose by the all-wise Architect. The use of this kind of number mysticism to explain the frame of the universe had of course been quite common in the first half of the seventeenth century, but it is astonishing to encounter it with Huygens, who had been brought up with the new mechanical philosophy. Indeed, it has no parallel anywhere else in his work, nor does he repeat it to a more scholarly corre-

[26] Huygens to Van Schooten, 10 March 1656. *OC*, I, 389 (no. 269).

[27] *OC*, XV, 215.

[28] I.B. Cohen (1978).

spondent. His prime reason for introducing it here was probably rather ad hoc, viz. to strengthen his argument of the analogy between Saturn and the earth.

The second argument, too, is rather far-fetched. Huygens found that Saturn's ring was inclined to the ecliptic at a constant angle. As this was more or less similar to the constant inclination of the earth's axis, he *assumed* that the inclination of Saturn should be the same as the inclination of the earth, i.e. 23.5 degrees. [29] Proving from this assumption the similarity between Saturn and the earth is something of a circular argument, of course. Still, this was not just an argument for the good of the cause. Even after Huygens realised that his theoretical values did not match observations, he sought the fault with other magnitudes (such as the width of the ring) rather than with the angle of inclination itself. It was many years before Huygens was ready to abandon his 23.5-degree angle for Saturn's inclination in favour of a more realistic value of 31 degrees. [30]

Thus, it appears that Huygens' interpretation of his discoveries was guided by a strong intuition of the similarity between the earth and Saturn. From this analogy, he argued the truth of the Copernican system. The analogy, of course, was based in a belief in the basic similarity of the celestial and terrestrial worlds. This was a deep-rooted conviction Huygens entertained not only in his youth, when he wrote *Systema Saturnium*, but also later in life. The last work he accomplished was his *Kosmotheoros*, which was published posthumously in 1698. Huygens' subject here is the probability of life on the celestial bodies. Most modern historians have regarded it as scientifically uninteresting, rather as some kind of playful entertainment, a cosmological fantasy. [31] But Huygens obviously did not think of his book as a work of the imagination. He referred to it shortly before his death: 'At present I am working on some philosophical subject.' [32]

Huygens indeed pretended to give a logical argument. Our observations teach us that the planets are basically the same as our earth, and the stars to our sun. We may surmise that they will be similar, too, in those respects which defy observation up to now. Consequently, we are allowed to draw conclusions about the different planets on the analogy of our own earth. Hence the

---

[29] OC, XV, 309 (French translation 308). Cf. van Helden (1980) 153.

[30] OC XV, 476-477, 483-484. See for the whole episode (and a sequel): D'Elia (1985) 99-122, who also points out the significance of Huygens' discoveries in corroborating the truth of the Copernican system.

[31] OC, XXI. The work is discussed by Seidengart (1982). See also Snelders' postface in the reprint of the Dutch translation (Huygens 1754).

[32] Huygens to Gregory, 19 Jan. 1694, in: Vermij and van Maanen (1992) 512, 523.

seemingly paradoxical but in fact very apt title: 'Conjectures on the heavenly earths'. The argument from analogy led Huygens to suppose that life on the planets is similar to that on earth. After all, there is no reason why there should be a greater difference between various planets than between various continents. Consequently, the animals on other planets are built like those on earth, procreate like those on earth, and some of them will be intelligent, just as on earth. The latter's intellectual and spiritual life will be akin to that of man, and they will live in a similar society.

It is easy to see that the whole book is built upon one basic idea: the universality of nature. In the second part of the book, Huygens deals with the stars and the various planets in a more rigorous astronomical manner and gives information about their size, distances and appearances, and what it would be like to live on them. Nature is the same, always and everywhere. The differences between the sun and the other stars are differences of size and measure, not of principle. The same holds for the differences between planets. Life itself, even intelligent life, has no specialised position in nature. Huygens just drew a leading idea of the scientific revolution to its (seemingly) logical conclusion.

The argument appears to have appealed to contemporary readers. Within a few years, the book had been translated into English, French, German and even Russian. As for the Netherlands, it was enthusiastically reviewed by the journalist Petrus Rabus, who shortly afterwards, in 1699, published a Dutch translation. In part, the book will have pleased because of its second part, an elementary introduction to the universe. But its first part was approved, too. In a poem appended in front of his translation, Rabus praises Copernicus for establishing the fact that every planet, 'in vastness and size, equals our earthly living place'. He endorses Huygens' view that on the planets would live 'animals, no lesser in soul and reason than you. Or would the intellect, which teaches us to ascend gradually to the sciences, have been given only to us, and to no others? Such has never been written. Reason itself denies it.' [33]

It is this basic belief of Huygens in the universality of nature which forms the foundation of his unqualified support of Copernicanism. If there is no basic distinction between heaven and earth, between sublunary and superlunary nature; if the universe is indeterminate and has no centre, then no body occupies a special place in the world. The sun is but a star, and the earth but a planet. This belief Huygens clearly owed to Descartes. Copernicanism received its ultimate support from Descartes' view of a geometricised nature.

---

[33] Huygens (1754).

It is this background to Cartesian philosophy which makes Huygens' work different from that of Kepler or Galileo.

New theories of motion

It seems that the principles of Cartesian physics indeed were a turning point in the reception of the Copernican system. Copernicanism not only fitted in with the view of the universe put forward by Descartes, but was even felt to be a necessary consequence of the constitution of nature. First there was the new view of the universality of nature, which had been in the air since Galileo's discoveries but was now used as a starting point for natural philosophy. And second, this new natural philosophy claimed a mathematical character, and hence mathematical exactness.

It was such general changes in world-view rather than more specific arguments which won most people over. After all, the old arguments against the motion of the earth were as valid as ever. One might presume that the acceptance of the motion of the earth was facilitated by the waning of the Aristotelian concept of motion, and its replacement by the modern, mathematical view of motion in the work of Galileo and Huygens. The concept of inertia offered a solution for the common objections against the motion of the earth, wherein a cannonball, a bird or anything which was above rather than on the earth was supposed to remain behind if the earth moved. Theoretically, this solution may have been available. In practice, however, the role of these new insights appears to have been very limited. It is possible to find the concept of inertia fully developed in Descartes' writings;[34] however, he appears unaware of its full consequences and his ideas on motion are actually rather inconsistent. His followers interpreted his work in a variety of ways and their ideas of motion (or rest) as a rule place these concepts in the realm of metaphysics rather than mathematical physics.[35] In understanding the motion of the earth, these ideas were of little help.

The way the theory of inertia was ignored can be illustrated from the theory of ocean currents. An ancient theory attributed the ocean currents to the general motion of the world (that is, the universe) from east to west. Because the heavens turn round in 24 hours, the elements of air and water still had a remnant of this movement left, thus creating the trade winds and ocean currents. Only the element of earth stood motionless in the centre. This theory is propounded (for the case of ocean currents) in an anonymous Latin manu-

---

[34] *AT*, VIII-1, 54-55.

[35] Vermij (1996) 277-281. Ruestow (1973), *passim*.

script on astronomy, which dates from after 1610 and seems to be of Dutch provenance. In order to calculate the dimensions of the earth, the author uses the difference in latitude between Franeker and Deventer. He probably was teaching in one of these towns. So, in this case ocean currents were seen as supporting the movement of the heavens and the rest of the earth. [36]

With the advent of Copernicanism, this theory was simply reversed. According to one Dutch author, the ocean currents (from east to west) in the tropics are caused by the daily rotation of the earth, 'because the fluid water cannot follow the earth apace, but continuously lags behind.' [37] (Still, the wind may also contribute.) This theory is not found in the work of Descartes, whose ideas on motion were probably too sophisticated. It was defended, however, by Galileo (in the case of trade winds) and Kepler (for ocean currents). [38] In the Dutch Republic, it was first upheld by Descartes' follower Regius. [39] Another Cartesian who upheld the theory was the French philosopher Régis; his ideas on the subject appeared in Dutch in 1700. [40] The argument appears to have been quite popular, not only among academic philosophers, but also among mathematical practitioners. [41] Their Copernicanism thus went hand in hand with a traditional idea of motion.

Huygens, however, knew better. But in his case, it is particularly evident that the plausibility of the Copernican system was not a result of the better understanding of the physical processes involved, but vice versa. Taking the Copernican system for granted, Huygens used this presupposition to solve questions in physics. Galileo had argued for the Copernican system on the ground that the motion of objects on earth should be seen relative to the earth; for an observer on earth, they appear the same whether the earth is moving or not. Assuming the basic soundness of this idea, Huygens used it to gain a better insight into the mechanics of motion. By applying it to percussion phenomena, he found the rules by which they were governed and incidentally showed the fallaciousness of Descartes' rules. Thus, it was not by applying the new rules of mechanics that the old arguments from projectiles etc. against the motion of the earth were refuted; by assuming that the

---

[36] Utrecht UL, VI G 14, f 26^v. On f 24, the author refers to a proof which '*in demonstrationibus nostris ad Euclidem tradidimus*'. On the explanation of ocean currents, cf. Burstyn (1966) 170.

[37] Van der Moolen (1702) 23-24.

[38] Burstyn (1966) 170.

[39] Regius (1646) 93.

[40] Hartsoeker (1700) 241-242. (The translator, Ameldonk Blok, expanded the book from other sources.)

[41] Apart from van der Molen (1702), mentioned above, de Graaf (1659) 15 also explains that the daily rotation of the earth results in a continuous motion of the waters from east to west.

earth moved and that thus the ancient objections had to be fallacious, new rules of mechanics were found.[42]

Other Cartesians and Copernicans did not even feel the need for a new theory of motion. Huygens' rules of percussion met with incredulity on the part of his teacher Van Schooten, who could hardly believe that Descartes could have erred in this matter. And most Dutch Cartesian natural philosophers ignored Huygens' inventions completely.

---

[42] Mormino (1993) 10-18.

9. Cartesian cosmology at Dutch universities

A new programme of learning

Descartes lived in the Dutch Republic for 20 years. It was here that he published his major books. The Dutch poet Janus Montanus, in his epic poem 'The war of the philosophers, or the celebrated battle fought between the famous Aristotle and the great Descartes', tells us how he got here. Aristotle, having conquered the whole world, reigned with an iron fist. Descartes, having heard the complaints of the people, took counsel with some friends concerning what was to be done. His friends advised him to start a revolt against the Greek tyranny. First of all, he should try to establish a bridgehead in the United Provinces, as this country offered the best chances of success.

> Descartes agreed: 'You are quite right.
> To Holland I shall turn my stallion
> In order to stir up rebellion
> Against the Greek and all his might.'[43]

Modern historians often regard Descartes' settlement in the United Provinces as a kind of retreat.[44] However, the word 'retreat' presupposes that the Dutch Republic was some isolated Arcadia, where one lived far from the great affairs of the world. Hardly any simile could be less apt. The Dutch Republic took a prominent position at that time, economically, politically and intellectually. French scholars were fascinated by it. The philosopher Gassendi visited the Republic in the summer of 1629, and the following year, Marin Mersenne followed his example. It seems more probable that Descartes went there as he found the place intellectually stimulating, and because he felt that his new ideas would fall upon fertile soil among the Dutch. He did not live as a hermit in the Dutch Republic, but was in contact with a large number

---

[43] Montanus (1701), 57-58: '*Descartes seyd je hebt gelijk./ Kom aan! ik sal naar Holland trekken,/ En eens een rebellie verwekken/ Tegen den Griek en 't Griekse Rijk.*'
[44] Gaukroger (1995) 187-190.

of Dutch scholars and men of society: Constantijn Huygens, secretary to the stadholder; Isaac Beeckman, rector of the Latin school at Dordrecht; the mathematician Jacob van Wassenaer; the physician Cornelis van Hoghelande; professors Jacob Golius and Frans van Schooten at Leiden, Henricus Renerius and Henricus Regius at Utrecht, and Tobias Andreae at Groningen. The list could easily be extended. He seems to have used these contacts as a means to spread his thoughts in the world of learning. According to Van Nierop, who had it from Descartes himself, Descartes purposely went to Cardinael to discuss the latter's refutation of Copernicanism; according to the story, Cardinael proved reluctant to explain his views.[45]

It is significant that many of Descartes' Dutch friends were university professors. Descartes could have found friends and admirers among the educated of any European country. But only in the Dutch Republic could he have hoped to see his ideas established on the university curriculum. As a matter of fact, Cartesianism was presented not just as a set of philosophical ideas, but as a new programme of learning. This programme came to replace not only the old scholasticism and Aristotelianism, but also the humanist-philological approach which up to then had dominated scholarship. Knowledge should no longer come from the study of ancient texts, but from the investigation of nature. Philology did not disappear on the spot, but faded away to become a field for antiquarians rather than a gateway to learning. At the core of the new programme was Descartes' new physics, which strove to explain everything in the universe from the workings of minute material particles, acting mechanically under the influence of some universal laws of nature.

This affected not only the philosophical faculty. Under the influence of the Cartesian programme, Leiden's faculty of medicine became a centre of physiological experiments. This led to Sylvius' and Walaeus' experiments on the circulation of the blood, Swammerdam's experiments on respiration, Reynier de Graaf's discoveries concerning the ovaries, and a lot of other work. Foreign students, such as Nicolaus Steno, then demonstrated these new scientific methods abroad. At the 'Duytsche mathematique', the younger Frans van Schooten, a friend of Descartes, undertook to make Descartes' new mathematical methods accessible to a larger public. Just as Scaliger had formed the cream of Dutch philologists, Van Schooten gave rise to the flowering of Dutch mathematics in the second half of the century, as becomes apparent in the work of his students Johannes Hudde, Johan de Witt, Hendrick van Heuraet, and – most eminent of all – Christiaan Huygens.

---

[45] Van Nierop (1677) 11, also quoted in Thijssen-Schoute (1954) 87. On Descartes' stay in the Republic, see G. Cohen (1920) 357-685; Dibon (1990) 459-470.

What did Cartesianism bring to the teaching of cosmography that was new? For one thing, at the core of the Cartesian programme was its new physics. As for the discussion on the system of the world, this meant that Descartes' work accomplished a development which had been going on for a long time: the replacement of mathematical arguments by physical ones. Until then, the discussion on the system of the world had been largely an affair of mathematicians: Lansbergen, Mulerius, Hortensius, Holwarda, Ravensberg, Deusing and others. Aristotelian philosophy simply cold not cope with this kind of problem. The efforts to explain the constitution of the world by new physical principles had been the work of independent thinkers, like Beeckman or Girard. The discussion on celestial and mundane worlds left some traces in academic Aristotelianism, but only very few academic philosophers, Burgersdijk being the main example, discussed the system of the world.

After Descartes, this changed completely. At the universities, the system of the world became a stock subject in natural philosophy. This transformation of physics appears most clearly from the changing way philosophers were dealing with the 'world', that is, the universe. Traditionally, disputations *De mundo* ('on the world') followed an Aristotelian argument. Typically, they dealt with the definition of the world, its causes, including its final cause, its form and its attributes. They discussed the opinions of various ancient and some modern philosophers; for example, they refuted the Stoic opinion of the existence of a world soul [*anima mundi*] and the idea that there could be a plurality of worlds. Cosmographical questions, regarding the order of the heavenly bodies, were generally not dealt with at all. They were only incidentally touched upon in the corollaries. On 14 May 1653, however, the Cartesian professor of physics Johannes de Bruyn presided at Utrecht over a *Disputatio physica de mundo* ('Physical disputation on the world'), wherein he completely ignored the traditional philosophical questions. The disputation came down to a description of the physical universe. It was followed by many other physical disputations which ignored the old scholastic questions and instead discussed empirical reality.

Philosophy, rather than being an uninteresting propaedeutic subject, became a true science with a considerable status. Descartes certainly contributed to this development, but it would be going too far to attribute it solely to him. Even traditional philosophers by this time had apparently lost their taste for rehearsing the Aristotelian corpus. They looked for other subjects and their choice was sometimes rather surprising. The Groningen professor Martinus Schoock remains particularly known for the wide variety of his topics; he held disputations on, among other things, butter, peat, nothing, aversion to cheese, and the chicken and the egg. The Amsterdam professor Arnoldus Senguerdius held physical exercises on such subjects as 'the antipathy between sheep and wolf' and 'why we move the right foot first, and then the left one'. Despite these, the ancient questions were not abandoned overnight. Scholastic disputations *De mundo*, discussing Aristotelian physics, were still defended. But after 1650, we find a rapid proliferation of new topics. Pertinent to our subject are the disputations *De mundi systemate*, 'on the system of the world'.[46] Soon, even relatively conservative philosophers were discussing cosmographical questions in their textbooks and disputations. Of course, they did not offer the Cartesian world, but a more traditional one.

Academic mathematicians, on the other hand, largely ceased to engage themselves in the subject. The point seems to be that until then mathematics had been closely linked with classical philology and the humanist programme of learning; after 1650, however, this programme quickly lost ground to the new natural philosophy. Philology remained a respected discipline, but it was no longer regarded as the key to a general understanding of the world. Consequently, philologists lost their interest in mathematics and astronomy. Philosophers, on the other hand, had a quite different approach to obtaining natural knowledge and did not feel the need to master what for them was a new and complicated discipline. There remained but a few mathematicians, most of them outside academia, who seriously occupied themselves with the world's system. Of these, Christiaan Huygens and Dirk Rembrandtsz van Nierop are the most important. However, they too were no longer inspired by classical texts or ideas on cosmic harmony, but rather by the new physical ideas.

Philosophers probably felt less incentive to apply themselves to astronomical calculations, as such calculations by that time had become the field of practitioners, many of whom did not even know Latin. Simple descriptive astronomy became vulgarised to such a degree that it was no longer attractive to scholars. So, hand in hand with the rise in status of natural philosophy, the study of mathematics lost prestige.[47] In the second half of the seventeenth century, few university professors continued to bother about the exact mathematical description of the planetary orbits, or tried to put the new theory to use in order to draw up accurate predictions of stellar positions.

Resistance to the Cartesian world-view

However, not everybody welcomed Descartes' ideas. From the very beginning, Cartesianism met with vehement opposition. Montanus' image of a

---

[46] Ravensberg's earlier disputation *De mundi systemate* had been a mathematical disputation.
[47] Cf. Vanpaemel (1991).

war was quite apt. Cartesianism was opposed on the one hand by more conservative philosophers, medical doctors, and so on, who felt committed to more traditional notions, be they Aristotelian or other. On the other hand – and this opposition was by far the most dangerous – it aroused suspicion within the theological faculty. Leading theologians were concerned not only with the effects of Cartesianism on theology itself, but also with its progress in the philosophical faculty. They did not hesitate to call upon the authorities to forbid the public teaching of these unheard-of new ideas. It was their denunciations which made Cartesianism a subject of public debate.

Conflict arose especially during philosophical disputations. Both Cartesians and anti-Cartesians used them to pronounce their respective views. In some ways, disputations served the same function as learned journals do today: they enabled professors to take a public stance on current events or ideas. Disputations were not only disseminated in printed form, but were also public events. They were usually attended by a number of professors from different faculties as well as a large crowd of students, who did not hesitate to vent their approval or disapproval. 'It is usual in this academy that the audience reacts with foot-stamping if either the president or the opponent brings forth something that does not seem to the point', Varenius explained to his teacher Jungius in Germany. [48] On several occasions, the curators forbade these disturbances during disputations, apparently to no avail. [49] In 1661, the Senate of Utrecht proposed to protect defendants by means of a fence or grille from over-enthusiastic participants. [50]

Personal feelings may have had some part in making these debates so acerbic. However, there was more to it than that. The hostility towards Cartesianism was not an individual peculiarity, but an attitude widely found among theologians. Nor were the attacks limited to the Dutch Republic. All over Europe, they dominated the intellectual scene of the second half of the seventeenth century. It should be remembered that Cartesianism was not just a set of ideas, but represented a complete intellectual programme. As such, it ran counter to established programmes. By the middle of the seventeenth century, Protestant theologians had come to value Aristotelian philosophy and scholastic method, both of which Descartes rejected. Originally, Protestants had been quite wary of Aristotle's philosophy, if not of philosophy in general. Aristotelianism was thought to have exercised a nefarious influence over medieval theology. One should follow the precepts of Christ rather than

[48] Varenius to Jungius, 17 May 1647 and 12 April 1648, in: Jungius (1850) 258, 382; German translation in Jungius (1863) 322, 332. At Utrecht it was no different, cf. *Querelle* (1988) 80.
[49] Resolution of curators, 30 Nov. 1655, in: BGLU III, 110-111.
[50] Kernkamp, I, 148.

those of a heathen philosopher. With the division of Protestantism into well-defined creeds and Churches, however, the Church leaders increasingly felt the need for a solid underpinning of theological learning. The many polemical attacks necessitated a thorough acquaintance with the opponents' arsenal. The academic philosophy to which this gave rise is commonly called 'scholastic', or 'neo-scholastic' (to discern it from the scholastic philosophy of the Middle Ages). Reformed theologians at Dutch universities systematised their learning into an Aristotelian and neo-scholastic mould. This Reformed neo-scholasticism was, somewhat ironically, mainly based on the works of the Spanish Jesuit neo-scholasticism of Pereyra, Suárez and the Conimbricenses. [51]

Its main proponent and architect in the Dutch Republic was Gijsbert Voet (Gisbertus Voetius) [52], a professor at Utrecht. Voetius was a very militant man, whose ideas had been formed by the fierce controversies at the beginning of the Reformation. His grandfather had died a martyr to the Reformation, and his father had fallen as a supporter of William the Silent. Voetius' first ministry was in Vlijmen in 1611, where he built up the Reformed community from scratch in a hostile Catholic environment. Later, he acted as an army chaplain, and as such was instrumental in imposing Reformed religion on the newly conquered town of 's Hertogenbosch. Meanwhile, Voetius took an active part in the Arminian troubles and was one of the delegates to the famous synod of Dordrecht. Finally, he became professor of theology at the newly-founded Utrecht University. He not only completely dominated ecclesiastical life in that city, but was also one of the leading theologians of the country in general. [53]

There can be no doubt that Voetius introduced neo-Aristotelianism mainly as a system to protect the purity of Reformed dogma, and as a fence to ward off the attacks of all kinds of heretics and unbelievers. Attacking the philosophical presuppositions of the whole system was therefore tantamount to attacking the purity of faith itself. Descartes denied the value of bookish learning and claimed that in principle any layman could feel justified to frame his own opinion on divine or mundane matters with complete disregard for received views or established authority. This could not be swallowed by a man like Voetius, who saw himself as the defender of an eternal, well-established and vital truth which nobody should question or even doubt. Even if Cartesianism did not affect truth itself, it undermined the very means of defending it. [54]

---

[51] Eschweiler (1928); Petersen (1964); see also Kusukawa (1995).

[52] The spelling Voët (with diaeresis), which is sometimes found in older non-Dutch literature, makes no sense at all.

[53] See his biography by Duker (1897-1915).

[54] On Voetius' scholasticism, see van Asselt and Dekker (1995); van Ruler (1999).

However, resistance to Cartesianism was based not just on didactic grounds. Cartesianism was offensive not only because it ran counter to received philosophy, but also because it appeared problematic from a religious and theological point of view. Many Reformed were wary about a philosophy that proclaimed its own method without explicitly referring to evangelical truth. As they experienced their belief as a continuous struggle against the powers of hell, they had little understanding for a method which wanted to achieve truth by putting all in doubt. Doubt, after all, was a temptation put there by the devil. Only by sticking to revealed truth, against all odds so to speak, was man prevented from going astray.

As demonstrated above, to people like Lipstorp or Wittichius physical reality was explained by Cartesian philosophy in a mathematical way. Hence, the Copernican system should be deemed an absolute certainty. Such claims were naturally dismissed by Aristotelians (and not just by them). To theologians like Voetius, philosophy in the end offered no certainty at all. It was the Bible that was the touchstone of all knowledge. Pertinent to the background to his view is what he stated in an oration on the usefulness of arts and sciences, delivered at the opening of Utrecht University in 1636. On astronomy, he remarks 'that not without reason some true masters in this art have confessed that it is from Scripture that one can and should prove the truth of all astronomical foundations, so that Scripture is the safe harbour from whence one departs on the sea of astronomical speculations.' [55]

So, Cartesian philosophy led people to claim absolute certainty for a doctrine – Copernicanism – which leading theologians regarded as being contrary to the Bible. In a sense, the Cartesians deliberately provoked their opponents, for ever since Galileo, nobody could be unaware that the heliocentric theory was problematic from a theological point of view. Voetius later asserted that he was a convinced anti-Copernican already in his student days. [56] Be that as it may, he displayed such feelings quite early in his career. He was the first professor of theology at the *illustere school* in Utrecht; the school was founded in 1633, and three years later its status was raised to that of a university. In September 1634, he showed his credentials, so to speak, with a large disputation, *Assertationes theologicae de praejudiciis Verae Religionis*. [57] It was a militant, polemical piece against the detractors of the Reformed faith. For the most part, it deals with purely theological issues. However, so shortly after the sensational condemnation of Galileo, Voetius could not resist touch-

[55] Voetius (1636) 29.

[56] Duker II, 45.

[57] The respondent was David van Boxtel. The whole disputation is reprinted in Voetius (1635) [335]-354.

ing upon this issue as well. One of the added philosophical-theological corollaries reads: 'That the heavens stand still and the earth turns all round each day cannot be proven from Scripture; Scripture even denies this in a rather clear way. See Psalm 19: 9, 7, Psalm 104: 5, and Ecclesiastes 1: 4, 5. The objections put forward by some are vain and absurd.'[58]

The thesis might have gone unnoticed were it not for a reaction by the Remonstrant minister Batelier. Batelier felt urged to defend the honour of the Remonstrants against Voetius, and published (anonymously) an extensive reply.[59] His main concern was Voetius' attacks on the Remonstrants, of course, but he did not leave it at that. He systematically reviewed all Voetius' arguments, and so arrived at the thesis on the earth's motion as well. He started with the confession that he was completely ignorant about astronomy – 'perhaps just like you', he added with some sarcasm.[60] Batelier then argued that the Bible does not want to teach the wisdom of the world, but the fear of the Lord. In matters not pertaining to faith, the Bible adapts itself to common ways of speaking. In all this, he closely followed the arguments of Lansbergen. Batelier accused Voetius of playing a nasty trick on the honourable old man.

Voetius did not leave Batelier's attack unanswered: he rejoined with a comprehensive work with a weighty title, *Thersites heautontimorumenos* ('Thersites the self-mutilator'). All points are passed in review again. The issue of the earth's motion, which in his first writing was a simple thesis, this time has nearly 30 pages devoted to it. Voetius apparently felt stung by Batelier's suggestion that he spoke of matters beyond his competence, for he started with a five-page justification of his occupation with astronomy.[61] Astronomy, he asserted, is relevant for theologians, so they should have a judgement on it. He did not deny that he was not competent in technical astronomy. But should that mean that we must accept the authority of one or two self-styled experts, who uphold Copernicanism? Most other astronomers, from antiquity to the present day, do not share Copernicus' opinions, and Batelier himself does not hesitate to reject their assertion. Voetius, in short, does not regard it necessary to judge a discipline on factual grounds. Fully in line with the scholastic method, he feels that the discussion can be reduced to the question what authority we should follow. The ultimate authority, of course, was divine revelation. He could dismiss all non-biblical arguments as irrelevant to the sub-

---

[58] Dukęr II, 31-32; de Vrijer (1917) 206.
[59] Batelier (1634). See also Duker, II, 32.
[60] Batelier (1634) 57.
[61] Voetius (1635) 256-261.

ject. Probably he did not even understand them; but the point is that he did not feel the need to understand.

The first conflicts over Cartesianism at Dutch universities

As for the reception of Cartesian physics, Leiden for once lagged behind. The honour for having first introduced Cartesian physics into its programme goes to Utrecht University. This university had been founded in 1636, a year before Descartes published his *Essais* and the *Discours de la méthode*. Among its first professors was Henricus Renerius (or Reneri). Renerius had made Descartes' acquaintance around 1629. The friendship became closer in the early 1630s, when Renerius was professor at Deventer and Descartes temporarily took up residence in this small town. When in 1634 Renerius moved to Utrecht, Descartes also took up residence there. Renerius was not a Cartesian philosopher in the modern sense of the word. Nevertheless, he was very interested in geometry and the investigation of nature. He occupied himself with microscopical observations and optical experiments and had great expectations for Descartes' new method. It is difficult to say, however, to what degree exactly he propagated Descartes' ideas. In his thoughts, Cartesian and Aristotelian elements appear to be inextricably intermixed.[62]

The problems of interpreting his ideas become clear in his cosmology. In 1635 Renerius presided over a disputation on the world and the heavens, one of a series of six physical disputations. The part on the world (i.e. the universe) follows a traditional Aristotelian pattern. In the part on the heavens, however, he is more innovative and touches upon questions traditionally dealt with by cosmography. He shows himself familiar with the telescopic discoveries by Galileo and also argues in favour of the system of Copernicus. His arguments are not really mathematical, but seem to derive from a certain vision of the cosmos, not unlike the views of Wittichius or Huygens discussed above. Renerius deems it probable that the fixed stars indeed are immobile. If so, the sun should be immobile, too. 'If this is admitted, the motion of the earth follows necessarily, both daily on its axis, and annually around the sun. And as these statements are more in agreement with reason and the celestial phenomena, and are not refuted by Scripture if correctly explained, we undertake to defend them.'[63] We mentioned earlier his defence of Kepler's ellipses, but probably this was not inspired by Descartes.

Renerius' influence was anyhow limited, as he died at an early age, in 1639.

[62] On him: Verbeek (1993)b; Sassen (1941)b; van Bunge (1999) 34–36; de Haan (1993) 45-58; Dibon (1954) 197-203; Verbeek (1992) 96-97.

[63] Renerius, disp. Utrecht 10 June 1635, thesis 33 (see also 27). See also de Haan (1993) 53.

He was, however, instrumental in having Henrick de Roy (Henricus Regius) appointed professor of medicine at Utrecht in 1638. Regius came from a distinguished Utrecht family and had spent part of his student years travelling in France and Italy. He obtained his doctorate in medicine at Padua in 1623. He was very interested in natural philosophy and was rather critical of Aristotelianism. When Renerius introduced him to Descartes, the latter's ideas fell on fertile soil. One should note that Regius never became a slavish follower of Descartes. He fully credited Descartes as the restorer of philosophy, but felt free to diverge from his opinion.[64]

At the beginning of his career at Utrecht, Regius obtained permission to preside over physical as well as medical disputations. He used this opportunity to propagate the theories of Descartes. Among the topics he dealt with were the circulation of the blood and the motion of the earth. Voetius thereupon started a campaign against these new and dangerous ideas. In 1638, he presided over a series of disputations on the Creation. Herein, he clearly rejected the theory of the earth's motion. He also criticised Kepler's idea that the three parts of the universe corresponded to the persons of the Trinity. Voetius later reissued these disputations in a volume of his selected disputations (1648), using the occasion to add some commentary.[65]

A metaphysical question finally led to the conflict coming out into the open. Regius in one of his disputations touched upon the delicate subject of the union of mind and body, and Voetius openly accused him of offending religion. This led to a protracted academic war, which was rooted partly in the incompatibility of ideas, and partly in an incompatibility of characters, as both Regius and Voetius were stubborn, dominant men who were unwilling to give way to external pressure. Descartes was naturally annoyed to see his philosophy compromised by the taint of heterodoxy. In his *Lettre à Dinet*, printed in 1642, he came out openly on the side of Regius and attacked Voetius bitterly. This only made things worse. The affair grew from an academic quibble into a public issue. Voetius and a former pupil of his, the Groningen professor of philosophy Martin Schoock (Schoockius), launched an elaborate attack on the philosophy of Descartes. Eventually, the authorities felt compelled to impose silence on all participants. After some further appeals, they gave their final verdict in 1645. To prevent further disturbances, they prohibited further publications on the matter, be they in favour of or against Descartes. Moreover, they confirmed that all philosophy teaching at Utrecht

---

[64] On him: Verbeek (1994); de Vrijer (1917). One should note that the spelling Le Roy (instead of De Roy), sometimes used in the literature, finds no support in the sources.

[65] Voetius, disp. Utrecht 21 Sept. & 6 Oct. 1638. Voetius, (1648-1695) 1, (1648) 608, 637. Voetius' later elucidations to these theses: ibid. 597-598, 869, 880-881.

should be based on the works of Aristotle. Regius was ordered to stick to his proper subject, medicine.[66]

At Leiden, Cartesianism became an issue only a few years later. It was controverted within the philosophical faculty itself, but, as at Utrecht, the main attacks came from outside, from leading theologians. The reason for their concern is not immediately clear. Descartes had friends and followers at Leiden, most notably the mathematicians Golius and Van Schooten, but nobody taught his philosophy there. The opponents of Cartesianism were probably alerted by the events at Utrecht rather than by events at their own university.

At Leiden, Voetius' part was played by the theologians Jacob Trigland and Jacob Revius. Nowadays better remembered as a poet, Revius was also a militant theologian. As a minister at Deventer (Overijssel), he had been active in combating Arminianism. In 1642, he became regent of the College of Theologians – the States' college – at Leiden University. In this quality, he had disputations defended by his pupils. In 1643, when quarrels at Utrecht were rife, he had a series of twelve disputations 'on the works of the first day' defended. In the second of the series, some theses were directed against the 'Pythagorean' opinion about the motion of the earth. Revius referred to the usual biblical texts. Moreover, he asserted that the Copernicans inverted the article of faith of the ascension of Christ. Christ ascended during the day, with the sun above him. If the sun were the centre of the universe, this should have been called descent.[67]

At the other side, Descartes' defence was taken on by a rather provocative character: Adriaan Heereboord, professor of logic. Heereboord had been appointed extraordinary professor in 1640, at the age of 27, and ordinary professor in 1644. He was quite a controversial figure, both because of the clamorous way in which he vented his anti-scholastic feelings, and because of his private life, which did not remain very private in the end; it was even discussed in printed pamphlets. From his early career onward, he denounced the slavish following of Aristotle, but it seems that he discovered Cartesianism only in 1644. From then on he became an adamant supporter of Descartes, although he does not seem to have grasped the real novelty of the latter's ideas. Descartes was just one of the moderns, alongside with Bacon and Gassendi, who led the way to the modernising of philosophy.[68]

---

[66] The episode has drawn considerable attention. The best account is by Bos in Descartes (1996) 3-35; see further van Bunge (1999) 297-304; Verbeek (1992) 13-33 and (1991); McGahagan (1976) 165-204; van Berkel (1984); Bizer (1958) 308-314; G. Cohen (1920) 535-578, 595-601; de Vrijer (1917) 16-43; Duker, II, 136-199. The documents relating to this episode have been published, in French and with an introduction, by Verbeek in *Querelle* (1988) (but see also Descartes (1996)).

[67] These theses were reprinted in Revius (1650) 318-319.

[68] Sassen (1942-1943); Thijssen-Schoute (1954) 97-101, 114-125; Ruestow (1973) 49-60; Stewart (1994) 37-38; van Bunge (1999) 308-311.

Heereboord wrote a large number of textbooks on the various parts of philosophy. His work on physics was posthumously published in 1663 as *Philosophia naturalis*. In it he gave an rehearsal of Aristotelian physics, accompanied by the views of some modern philosophers – Claude de Bérigard, Henricus Regius and Descartes – on the issues under review. His own opinion is not always clear. On the whole, Heereboord seems to feel quite at home in such a commentary on Aristotle. He seems to have made the step to a truly mechanistic philosophy – be it Cartesian or otherwise – only halfway. His overall world-view is rather a modified Aristotelianism.[69] One should even consider it doubtful whether Heereboord was a Copernican. On the subject of the system of the world, he remained uncommitted. One might surmise that this has more to do with the exigencies of this kind of textbook and that Heerboord might have spoken his mind more openly in the explications in his courses. This, however, is not confirmed by his disputations, which discuss traditional questions in an Aristotelian framework. A thesis like 'The heaven is not moved by intelligences, but if it were, this motion would still be natural', clearly presupposes a geocentric world-view.[70] It can hardly have been Heereboord's 'Cartesianism' which gave offence at Leiden. Still, his avowed dislike for scholastic philosophy in general, and for the Leiden theologians in particular, may have functioned as a trigger.

At Leiden, too, the dispute was waged by means of public disputations. In 1646 the professor of theology, Jacob Trigland, publicly denounced a corollary which recommended the method of doubt, upheld in a disputation by Golius, as religiously dangerous. In January 1647, Heereboord replied with an address wherein he extolled Descartes to the stars. This led to a reaction from Jacob Revius. In five disputations, defended in February and March of 1647, he discussed such items as the method of doubt and Descartes' proof of God's existence.[71] Revius' and Trigland's attacks led Descartes to write a protest letter to the curators of the university. The curators took the same decision that had been taken by their confrators of Utrecht: it was decreed that all professors of theology and philosophy should refrain from referring to Descartes or discussing his ideas. Heereboord had to remain within the limits of Aristotelian philosophy.[72]

Revius was prevented from holding any more disputations on Descartes' ideas, but the very same year he started a series of disputations on nature and

---

[69] Cf. Ruestow (1973) 55-57.

[70] Heereboord, 22 June 1661, th. 43.

[71] Verbeek (1992) 40.

[72] On events at Leiden, see Verbeek (1992) 34-51; Ruestow (1973) 44-45; van Bunge (1999) 304-306; G. Cohen (1920) 654-667; Wiesenfeld (1999) 39-50.

the Creation. They concerned 'some questions on heaven and earth', the waters under and above the firmament, the question whether the stars are animated bodies, judiciary astrology, and so on. Two disputations, defended on 14 and 18 September 1647, are 'on the firmness of the earth'.[73] Herein, he directly opposes the Copernican system. The choice of topics does not seem to be fortuitous. These disputations clearly served an anti-Cartesian purpose as well, even if Revius does not mention Descartes' name. Heereboord understood very well what Revius was up to and accused him of using the Creation as a pretext to discuss matters of physics. Revius replied in his *Statera philosophiae cartesianae* ('Balance of the Cartesian philosophy') in 1650: 'How could I not? What did God create but physical things?' As to the motion of the earth, Revius pointed out that it was the very advocates of Copernicanism (he referred to Lansbergen, Galileo and Regius) who had started to squeeze this theory out of Holy Scripture. No theologian can allow that 'in the wake of an astronomical theory are produced new explanations of Scripture and new doctrines of faith. Especially if thereby some other true article of faith comes in danger.'[74]

Revius did not uphold traditional Aristotelian doctrine in every detail. He considers it unproven that the heavens are solid, rejects the sphere of fire, and believes that the region of the air stretches up to the stars. Still, he is firmly convinced that the earth does not move. He abstains from a full treatment of the question and ignores all mathematical and philosophical arguments. As a theologian, he considers only the arguments taken from the Bible. In the end, however, these biblical arguments are decisive. The principles of biblical exegesis are purely theological. The Bible should be explained by the Bible, and not in the way of a false analogy. From the fact that the Bible does not mention the true ratio of circumference and diameter of a circle (as Lansbergen had pointed out), one cannot conclude that the Bible does not speak about the world system either. This would amount to saying that, since the Bible does not speak of scurvy, it does not mention leprosy either. In nature and the Creation, Revius apparently felt he had a sure field where theology should reign supreme and where he could legitimately resist the Cartesian onslaught. The prominence the Copernican issue soon would gain in the

[73] Revius, disp. Leiden 14 and 18 Sept. 1647.

[74] Revius (1650) 270-274. As an appendix to the book, Revius added the two disputations on the firmness of the earth, the first only in an extract, the second in its entirety (320-326). He also included extracts from the earlier (1643) disputation wherein he defended the stability of the earth, as well as three disputations on the triple soul. On the accusation that he disputed on subjects that were not theological, cf. Verbeek (1992) 117, note 61.

debate on Cartesianism may well have risen, at least in part, because of such considerations.

Now, what is the place of Copernicanism within these debates? Initially, the controversies regarded mainly metaphysical questions. It is generally held that indeed it was Descartes' metaphysics which was deemed dangerous by Voetius, Revius and their allies, and that later the conflict spilled over into other areas. Copernicanism, then, was only a secondary issue, made prominent because of its symbolic value. Whereas it was difficult to reach a wider audience on the nature of the soul or the theory of substantial forms, anybody could understand the question whether or not the earth moved, and whether or not the Bible spoke the truth.

This view cannot be disproved. Still, it is far from obvious. The Copernican issue was there from the beginning. Even at Leiden, where Heereboord did not really push the issue, Revius used it to attack the Cartesian world-view as early as 1643. If initially this debate was only a side show, that may well have been because traditionally this was not regarded as a philosophical question at all, not because it was seen as a minor or secondary point. Descartes' world-view was not attacked in a direct way by investigating its claims on the natural world, but indirectly by aiming at the methodological and metaphysical underpinnings. In this kind of debate, theologians felt at home. The investigation of empirical reality was simply not their domain. However, as their first strategy failed, the Copernican issue was forced upon them. So, it remains quite possible that what annoyed Voetius et al. in the first place, was the new scientific world-view, which was no longer based on the Bible, and that the metaphysical issues were only secondary.

Aftermath of the conflicts

The first attempts to introduce Cartesian ideas at a European university had ended in a resounding defeat, or so it seemed. Descartes, disappointed, left the Dutch Republic for Sweden in 1649, where he died the following year. In fact, however, the further propagation of Cartesian philosophy was only slightly hampered by the events described above. The authorities had acted in the interest of public peace. They were not against Cartesian philosophy as such, but wanted to prevent uproar, whether caused by Cartesians or Aristotelians. As long as Cartesians did not give offence, they were left in peace. Curators saw no harm in appointing to the professorship people of known Cartesian convictions.

At Utrecht in 1652, a few years after the conflict, a new professor of philosophy was appointed: Johannes de Bruyn. He was born in 1620 and had studied at various places in the Dutch Republic, among others with Heereboord at

Leiden. He had obtained his doctorate in philosophy at Utrecht in 1644, defending, among other things, Descartes' theory of the tides.[75] He was a great admirer of the works of Descartes but, as in the case of other early followers of Descartes, his 'Cartesianism' did not entail a slavish following of Descartes' theories. He defended the existence of atoms (not just corpuscles, as Descartes had it, but indivisible entities) which were moving in a vacuum.[76]

On most issues, however, De Bruyn appears to have sided with Descartes, and this became clear in his teaching. I mentioned earlier his 1653 disputation *De mundo*. Here, we will have a closer look at it. De Bruyn explained that the world consists of three classes of things. The first comprised the sun and the fixed stars, which were regarded as of the same nature: a very hot and intense fire. As such, they consisted of very subtle, moveable particles. The second class comprehended dark, opaque things. These were the earth and the planets. De Bruyn advanced several arguments to demonstrate that the earth and the planets were indeed of the same nature. These included the telescopic observations of the moon's surface and of the phases of Venus and Mercury. The third class of things comprised the ether, which filled the space between the stars. This was a fluid and translucent medium. It had to be much thinner than air, as demonstrated by the atmospheric refraction. Discussion of the nature of this ether, he postponed until a later disputation. This division of natural things into three different kinds is clearly drawn from Descartes' threefold division of matter. In fact, the correspondence of the three Cartesian elements to three kinds of bodies – luminescent, translucent and opaque – had been made already in the French translation of Descartes' *Principia*, of 1648.[77]

De Bruyn's admiration for Descartes becomes clearer still in a disputation on human free will he presided over in 1657. The disputation comprised ten theses, and all but one of these ended with references to the works of Descartes. Indeed, at the end De Bruyn explained that the whole disputation had been aimed 'at informing the students entrusted to my care about the opinion of Descartes, and to arm them against the scruples with which they may be injected by its adversaries.'[78] In one respect, however, De Bruyn maintained a neutral stance: he took care right from the start to stipulate that he dealt only with the philosophical aspects of the issue, and left aside those which concerned theology.

---

[75] Verbeek (1992) 88, where one finds also other information on de Bruyn.

[76] De Bruyn, disp. Utrecht 13 Oct. 1655 and 14 Nov. 1655; see also 14 May 1653, coroll. 2.

[77] *AT*, III, 52. A similar division into luminous, opaque and translucent bodies is to be found in De Bruyn's disputation on the visible world, disp. Utrecht 14 April 1660.

[78] De Bruyn, disp. Utrecht 25 April 1657.

This keeping aloof from theology is in line with De Bruyn's defence of Cartesian physics in his other disputations. Most striking is the fact that at no place does he speak out openly in favour of the Copernican system, not even in *De mundo*. Of course, his stressing of the sun's similarity to the stars, and the earth's to the planets, would contradict any other system. But here, he simply ignores the question of the world's system. Nor do the defendant's corollaries touch upon the issue.

The question of the system of the world is discussed in some other disputations, such as one from 1656 on the heavens and one from 1661 (claimed to have been written by the respondent) 'containing a choice of the most important subjects from physics and mathematics'. In 1656, he explained the motion of the planets (not the earth) around the sun, and of the sun around the earth; this would come down to a Tychonian configuration. Finally, however, he explained that the motions as they had been described so far were apparent motions. The question whether such motion took place in the heavens or in the earth, 'is nowadays much disputed among scholars. We permit that the natural reasons which by both parties use to be adduced are examined, as it happens that there are as yet no demonstrative ones. However, we will not touch the argument from Holy Scriptures.'[79] A similar caution is found in the other disputation, five years later. The author explains the three different world systems, viz. of Copernicus, Ptolemy and Tycho. Both Copernicus and Tycho 'save the appearances'. Copernicus' system is the most simple. The arguments advanced against Copernicus taken from natural reason, are not valid. As for the arguments taken from Holy Scripture, 'this is not the place' to discuss them.[80]

It is the same story in the corollaries to several other disputations. Most of them show sympathy for the Copernican theory, but abstain from discussing the theological issue and limit themselves quite emphatically to arguments drawn from nature and reason. One disputant offered in 1662 to defend the Copernican system by natural arguments. (He also stated, by the way, that 'Sane philosophy does not contradict theology').[81] Corollaries from 1655 affirmed that the Copernican hypothesis agreed with natural reason,[82] and that the motion of the earth could be defended by natural reason.[83] In 1656, it was stated 'that it is more in agreement with reason that the earth moves, than that

---

79 De Bruyn, disp. Utrecht 19 April 1656, quotation from th. 10.

80 De Bruyn/van Rhee, disp. Utrecht 11 Dec. 1661, th. 27-35.

81 De Bruyn, disp. Utrecht 7 May 1662, coroll. 20, 1.

82 De Bruyn, disp. Utrecht 20 June 1655, coroll. 9: *An Hypothesis Copernicana cum ratione naturali conveniat? Ass.*

83 De Bruyn, disp. Utrecht 14 Nov. 1655, coroll. 6: *Terrae motus ratione naturali defendi potest.*

it rests',[84] and that 'the natural arguments advanced against the motion of the earth have nothing conclusive.'[85]

Thus, the question of Copernicanism was a hot topic in 1655-1656. Why this was so will become clear in the next part, on Cartesianism and theology. It seems that the students added these corollaries on their own initiative, although De Bruyn will have made sure that their wordings contained nothing offensive to the theologians. Some of them indeed are in a more neutral mood. So, equally in 1655, one of De Bruyn's disputations was defended by a student who apparently had first been educated at the Amsterdam Athenaeum; he dedicated the work to, among others, the Amsterdam philosophy teachers Arnold Senguerd and Alexander de Bie. He added a corollary that 'nothing certain can be stated from natural reason on the motion or rest of the earth.'[86] In 1660, one disputant (who dedicated the work to, among others, three professors at Harderwijk) even added a corollary wherein he defended that 'it is taught in Holy Scripture that the stars are rotating all round, and the earth is resting in the centre'. Interestingly, it is this disputant who gives the clearest testimony on De Bruyn's own position. He added this corollary only 'with the kind permission of the President, who upholds the opposite.'[87] Evidently, De Bruyn's caution with respect to scriptural arguments was not inspired by pangs of conscience, but was due to outward pressures. Voetius' anti-Cartesian campaign simply did not allow him to speak out openly on the Copernican system. He could defend the new ideas in physics to a large extent, but there were some limits set.

As to Regius, after the conflict with Voetius he was confined to teaching medicine, but remained free to publish books on any subject he liked. In 1646, he published his *Fundamenta physices*, wherein he put down his ideas on physics. The book was well received, and a much extended version appeared in 1654 under the title *Philosophia naturalis*; it was reprinted in 1661. The Utrecht publisher Gisbert van Zyll announced a Dutch version as being 'in press' by 1667, but this apparently never materialised.[88] The first edition appeared only

---

[84] De Bruyn, disp. Utrecht 21 May 1656, coroll. 16: *Rationi magis consentaneum est terram moveri quam quiescere.*

[85] De Bruyn, disp. Utrecht 27 Feb. 1656, coroll. 16: *Contra terrae motum argumenta naturalia producta nullam habent efficaciam.*

[86] De Bruyn, disp. Utrecht 13 Oct. 1655, coroll. 6: *De motu vel quiete Terrae ex ratione naturali nihil solide statui potest.* Two years later, the same student responded to a rather outspoken Cartesian disputation by De Bruyn.

[87] De Bruyn, disp. Utrecht 1660 (n.d.) coroll. 4. *Astra circum rotari, & terram in medio eorum quiescere in S. Scripturis doceri, cum bona venia Praesidis contrarium statuentis, defendam; si tamen eam seponere opponenti placuerit quodlibetice hanc questionem tuebor.*

[88] *Catalogus* (1667): *Ejusdem Physica getranslateert in Nederduyts met al sijn Figuren en aenwijsingen 4° sub praelo*; also in the special section of Dutch books: *Physica ofte Natuer-konste door Henricus Regius 4°.*

two years after Descartes' *Principia philosophiae* (1644). True, there are some differences between the two books. Descartes started with a long part on the principles of our knowledge; Regius skipped all this and started with the principles of natural things, explaining his (or rather Descartes') mechanical conception of nature. Moreover, Descartes had dealt only with inanimate nature. Regius, however, continued to treat plants, animals and humans as well. Regius may have added these parts because of his medical background. A year after his publication of *Fundamenta physices* (1647), he published a kind of companion volume, *Fundamenta medica*. Herein, he explicitly aimed at giving a medical theory based upon the new philosophical insights that had recently enlightened the world. Still, a large part of *Fundamenta physices* – in fact anything dealing with the principles of natural things, the visible universe and terrestrial physics – appears simply a rehearsal of Descartes' book. And this up to the point that many of the illustrations in Regius' book came directly from Descartes'. (As the books had the same publisher, they did not even need to be copied.) However, this does not mean that Regius is simply repeating Descartes, as he based his book on his earlier disputations. Descartes tried to prevent the publication, and started a quarrel when Regius did not comply. He alleged that Regius distorted his views, but one wonders whether it was not the similarities rather than the differences between their respective books which provoked his anger. One should add that Regius gave Descartes full credit in his dedication.[89]

Now, what about the issue of Copernicanism, about which De Bruyn, as we have seen, was very cautious? There can be no doubt that Regius was a convinced Copernican. Copernican corollaries were one of the issues over which the conflict with Voetius broke out. In his *Fundamenta physices*, there is a long discussion on the matter. It is quite unoriginal: Regius simply repeats Descartes' point of view. However, in the second edition of this work, *Philosophia naturalis* of 1654, the section on the world's system is much enlarged. The 30-page chapter 'On the fabric of the visible world' has more than doubled in length and is now organised as a separate book. Most of the information added is of a purely astronomical nature: an explanation of the phases of the moon, of the retrograde motions of the planets, and such like. The last chapter is devoted to 'the usefulness of our system of the whole world, as described'. It compares the world systems of Ptolemy, Copernicus, Tycho Brahe and Regius' own, Cartesian scheme. He concludes that his and Copernicus' models save the phenomena best, and that one needs no longer to introduce

[89] The question of the relation between Regius' and Descartes' books is rather vexed. See in particular Verbeek (1994) for a discussion. On the relation of Regius' physics to Descartes, see also Rotschuh (1968).

absurd figments (epicycles) as in the systems of Ptolemy and Tycho.[90] Now, writing that the Copernican system describes or saves the phenomena best is less offensive than writing that it is true. One wonders whether Regius has chosen this formulation deliberately. There is one intriguing clue. In 1661, he published a new edition of this *Philosophia naturalis*. This re-edition is on the whole unchanged compared to the 1654 edition. However, to the chapter 'on the usefulness of our system of the whole world', a final sentence has been added: 'And so it appears, that this our system of the world, as it has been described before, is the true and only one.'[91] The corresponding place in the 1654 edition is left blank and one is led to suspect that here, too, the sentence was about to be printed, but was removed at the last moment from both the text and the index, which lists the paragraphs. Assuming some corruption of the text in this respect, it is also tempting to speculate that the curious chapter title was some kind of an emergency solution, and that it originally read: 'On the truth of our system of the world', which sounds more logical. Unfortunately, there is no way of proving these suspicions or of knowing how exactly the alteration was brought about. We can just be sure that Regius added a sentence in 1661 which he apparently did not dare, or was not allowed, to publish in 1654.

De Bruyn's and Regius' position was particularly difficult, as Voetius' domination at the theological faculty turned Utrecht into the main focus of resistance against Cartesian philosophy. At Leiden, people went much further in ignoring the ban on Cartesian philosophy. The obligation to remain within the limits of Aristotelian philosophy was renewed in 1656 and 1674 (there will be more on this in a later chapter). But this very fact attests that, despite all official prohibition, Cartesianism was very much alive.

Shortly after the ban on Cartesianism had been proclaimed, the chair of philosophy was occupied by a man who was known as a adamant supporter of Descartes, and who would be the main propagator of Cartesian physics at Leiden in the following years. This was Johannes de Raei (or De Raey). De Raei studied philosophy and medicine at Utrecht, where in 1641 and 1643 he defended some medical disputations under Regius.[92] Later he moved to Leiden, where in 1647 he obtained his doctorate, in both philosophy and medicine. De Raei was a close follower of Descartes, who presented his views in a sharp, polemical form. In 1651, Leiden allowed him to read on Aristotle's *Problemata* in public. In 1653 he was appointed extraordinary professor. As the subject of his oration in 1651, he chose 'the degrees and defects of vulgar

[90] Regius (1646) 47-76; Regius (1654) 71-141. Cf. de Vrijer (1917) 206-207.

[91] Regius (1661) 162: '*Ac proinde patet hoc nostrum mundi systema, jam descriptum, esse verum & unicum.*'

[92] Herborn, Evangelisch-Theologisches Seminar.

notions in the contemplation of nature, and the task of the philosopher in this respect.' Herein, he depicted the views opposing his as 'vulgar errors'. In 1654, despite the ban on mentioning Descartes' name, he published a textbook on physics entitled 'Key to natural philosophy, or Aristotelian-Cartesian introduction to the contemplation of nature'. De Raei was impudent enough to dedicate the book to the curators, and to turn the dedication into a eloquent apology of Cartesian philosophy. It started with a short history of philosophy, which was foremost a polemic against Aristotelianism. Nowadays, according to De Raei, Aristotelianism is mainly upheld by Popish superstition. Teachers who 'in the clear light of this century still want to play at blind man's buff and use other's eyes instead of one's own' abuse their profession and harm the common good. Still, according to the order of the curators, he will be using the works of Aristotle, but the contents will be brought in line with the ideas of Descartes, 'the proponent of true freedom and the felicitous instaurator of philosophy, who went ahead to show others the way.' Curators rewarded De Raei for breaking their rules with a gratification of 100 guilders, but then partially withdrew in the face of vehement protests by the anti-Cartesian party. One wonders whether any of them had actually read the dedication. Still, they continued to look on De Raei with favour. His provocative tone might have irritated his colleagues, but it attracted large numbers of students – something which counted for university governors then as now. De Raei became an ordinary professor in 1663.[93]

On the Copernican issue, De Raei continued to defend an uncompromising stance. He dismissed the whole Aristotelian world-view as mere figments and defended a Copernican-Cartesian system, wherein he followed Descartes' arguments closely. The planets, including the earth, are moved around the sun by vortices of celestial matter.[94] High and low tide are caused by the vortex of the moon. De Raei also took from Descartes the theory on the earth's 'true' motion: '... So the earth is said to be at rest, as it obeys the motion of the heaven which envelops it, wherein it floats. However, it is said to be moving with regard to the sun, or other stars whereto it does not keep fixed distance. This is called motion in an improper way, and is more optical than physical.'[95] De Raei started in 1661 a series of disputations on 'the system of the world and the elements'. Only the first of them is known. This one just discusses the three Cartesian elements and gives no information on the system of the world.[96]

---

[93] De Raei (1654), dedication. On De Raei, see Thijssen-Schoutte (1954) 125-139; Verbeek (1992) 71-73.

[94] De Raei (1654) 62-63; de Raei (1651) 21.

[95] De Raei, disp. Leiden 6 March 1655. On De Raei's views, see also Ruestow, 69.

[96] De Raei, disp. Leiden 1661.

As can be seen, the theological opposition did not succeed in banning Cartesian philosophy from the universities. Even at Utrecht, where opposition was strongest, Cartesianism was freely taught, be it with a disclaimer: philosophers expressly limited themselves to arguments from natural reason and left biblical arguments to theologians. Hence, they could not claim absolute truth for their sayings, but in practice, nobody will have been fooled by this concession. But not all philosophy professors were professed Cartesians. The above demonstrates only that professed Cartesians were not hampered. It may still be that the theological attacks were effective in that they created an atmosphere in which it became rather unattractive to pose as a Cartesian or a Copernican. This cannot be decided from a single case. Let us therefore try to get an impression of philosophy teaching in general.

Academic discussion on the system of the world in the second half of the seventeenth century

In the second half of the seventeenth century, the system of the world became a standard topic for academic discourse. It was dealt with in disputations *De mundi systemate*. The title had earlier been used by Ravensberg, but for a mathematical disputation, whereas by now, disputations under this title were physical. As a standard topic, it was dealt with in a standard way. Generally, any argument exhibited the three conflicting world systems (i.e. the Ptolemaic, the Tychonic and the Copernican) and discussed their pros and cons. The Ptolemaic system was in any case quickly dismissed, as it did not save the phenomena. The real choice was between the Copernican and the Tychonic system. The tradition of comparing different world systems may have been inspired by Galileo's *Dialogo*. Other examples may have been offered by Ravensberg and by Gassendi's astronomical textbook of 1647, which had been re-edited in Holland (using the original plates) in 1656. In this book, after the mathematical part wherein Gassendi followed the traditional Ptolemaic system, two non-mathematical sections followed in which Gassendi discussed the Copernican and Tychonic systems, respectively.[97]

Some of the participants in the discussion were professed Cartesians. The first generation of 'Cartesians' had comprised Descartes' own contemporaries. Such people had come across Descartes' philosophy later in life, when their ideas had largely already been formed. Even if they admired them greatly, there always remained some distance. They accepted Cartesian theories only in so far as they fitted in with their own preconceived ideas. This is the generation of Golius, Heereboord, Renerius, Regius, Sylvius and, to some ex-

[97] Gassendi (1647, 1656).

tent, De Bruyn. This was no longer true for the second generation of Cartesians, those who grew up at a time when Cartesian ideas were the talk of the day. Descartes' philosophy imposed itself with such force that it quickly pushed aside any competing anti-Aristotelian philosophy. Younger students became acquainted with Cartesian ideas during their studies. To them, there were only two viable alternatives: the old Aristotelian philosophy and new Cartesian philosophy. In these circumstances, many accepted the new philosophy of Descartes as the one and only true philosophy. This applies to people like De Raei and, probably, Van Schooten. Others include the Leiden physician Theodoor Craanen and the philosophers Tobias Andreae Jr. at Groningen and Johannes Clauberg at the reformed university of Duisburg (Germany).

At Leiden, Cartesianism was propagated by – in addition to De Raei – the professors Arnout (or Arnold) Geulincx and Burchard de Volder. In a sense, they already belong to a 'third generation' of Cartesians, who started to find shortcomings in Descartes' theories without knowing of an alternative to his system. As for the system of the world, however, there is little indication that they diverged from his ideas. Geulincx was mainly interested in metaphysics, but had to teach physics as well. In his *Physica vera* and in his annotations to Descartes' *Principia*, he simply explained Descartes' ideas on the universe. [98] De Volder presided over several disputations on the world's system. In 1682 he started a series on this very topic. The first of these is known, but could not be found. [99] A disputation *De mundi systemate* from 1694 exhibited the three major world systems and argues for the truth of the Copernican system. Tycho is rejected mainly on philosophical grounds: he posits a centre of the universe, but – as the universe is unbounded – there can be no centre. De Volder goes on with a description of the world according to Descartes, who is called 'the prince of philosophers of all times'. [100]

Students' disputations closely follow the views of their professors. Samuel Köleséri, from Hungary, was awarded his degree at Leiden on a dissertation *De systemate mundi* in 1681. He extensively discussed the various arguments and concluded to the truth of the Cartesian world model, the earth resting in its vortex. As De Volder was to do in his 1694 disputation, Köleséri argued that

---

[98] Geulincx (1968) II, 428-439; III, 455-497.

[99] *Disputatio philosophica de systemate mundi*, 1682. It is mentioned in Walther, (1863) 153 (no. 928), as well as in the catalogue of Van der Woude at the university library of Amsterdam. On my request, the library at Saint Petersburg (where Walther found it) replied that they could not find the disputation.

[100] De Volder, disp. Leiden, March 1694. See on De Volder in general: Wiesenfeld (1999) 51-60, 102-124 and passim.

as the universe has no centre, the earth cannot possibly be there.[101] Some other inaugural disputations at Leiden present some loose theses defending the Copernican system, in all cases in a Cartesian framework.[102] In one case, the defence of Copernicus is undertaken in the corollaries.[103] The Cartesian theory of the tides was defended at Leiden in an inaugural disputation in 1662, by Philippus de la Fontaine, from Amsterdam.[104]

On the other side of the spectrum, there are the convinced anti-Cartesians – not just theologians, but also members of the philosophical faculty. Like-minded philosophers include Adam Stuart and Henricus Bornius at Leiden, and Daniel Voet (a son of Voetius) and Gerard de Vries at Utrecht. The last mentioned would in the last quarter of the century turn Utrecht, where Cartesian ideas had first been introduced, into something of an Aristotelian bulwark. These people unhesitatingly clung to the principles of Aristotelian philosophy. As a consequence, they rejected the Copernican world system. As the system of the world is no real issue in Aristotelian philosophy, however, they seldom held elaborate arguments on the subject. In 1671, before he was made professor at Utrecht, Gerard de Vries was awarded his degree in philosophy at Leiden on a disputation *De mundo*. This disputation is still largely Aristotelian. Only in the last two theses (of 30) is the problem of the world's system touched upon. He briefly mentions the systems of Ptolemy, Copernicus and Tycho; Descartes', according to De Vries, may be regarded as a fourth one. De Vries is not really interested in the astronomical merits or demerits of the various systems. He quotes Mulerius (as a man who knew the Copernican arguments) to the effect that they remain mere hypotheses. 'In my view, it is better to remain ignorant about things the Holy Spirit does not speak of, than to impugn what He has said. A pious soul prefers to believe divine authority only, rather than fallacious human judgements.'[105] In his corollaries, too, the biblicist tendency in his philosophy becomes apparent.[106]

De Vries returned to the subject when he edited the second volume of Daniel Voet's *Physiologia*. Voet rejected both the theory of solid orbs and the vortex theory of planetary motions. He briefly explained the world's system. His ideas appear to be Tychonic. That the sun turns around the earth is, according to Voet, taught by Holy Scripture. In the edition he edited, De Vries

---

[101] Koleseri, disp. Leiden 1681. On Köleséri, see Zemplén (1972) 334-335.

[102] Uythage, disp. Leiden 25 April 1662. Coppenol, disp. Leiden 30 June 1664.

[103] Sevenhuysen, disp. Leiden 13 May 1664.

[104] Fontaine, disp. Leiden Leiden 8 Nov. 1662.

[105] De Vries, disp. Leiden 28 May 1671, th. 29.

[106] Annexa ex philosophia in genere, 9: *Licet Philosopho sua Philosophemata etiam ex S. Litteris confirmare*; 11: *'Veritas revelata Rationis naturalis examini non est subjecta'*.

added some notes wherein he supported the author's view. He gave a short review of the three main world systems: the one by Ptolemy (which he deemed to be against nature), the one by Copernicus (which was against Scripture) and the one by Tycho. De Vries rejected the ways the Copernicans tried to accommodate Scripture to their theories; in the end, it all came down to subjugating Scripture to our feeble understanding. 'But who is moved by real piety towards God's word, would prefer to err (if it is permitted to say so) with Moses, rather than to know with Copernicus.' [107]

In 1699, Johannes Leusden became doctor of philosophy at Utrecht on a nearly completely Aristotelian disputation *De mundo*. Only in its very last paragraph, where he listed some subjects he was to skip, did he mention the question of the world's system. 'We dislike the Ptolemaic system, as it does not agree with nature, as well as the Copernican and Cartesian, because they are contradicted by Scripture, and beset by various difficulties from reason. We prefer the Tychonic system over the others, as it saves the phenomena as well as the Copernican (according to our opponents), and agrees with Holy Scripture as well as reason.' [108]

We should mention here the Groningen professor Martinus Schoock, Voetius' ally against Regius. Despite championing Aristotle on this occasion, Schoock later in his career may have been more willing to accommodate new views than the philosophers mentioned so far. In his many disputations, he generally presupposed an Aristotelian universe, but rarely mentioned the question as such. In 1648 he presided over a disputation on physical problems. Among the questions discussed were 'Is the heaven or the earth moved?' and 'Is it possible to take the sun for the centre of the universe?' Both questions were answered '*ut lubet*' ('as it pleases'); that is, the defendant was ready to argue either side of the argument. Maybe this should be taken as an expression of scepticism regarding the possibility of obtaining certainty on the matter by natural means. [109] Schoock was less ambiguous in his *Physica caelestis*, a collection of 15 earlier disputations, published in 1663. In the fifth disputation, on the planets, he discussed the Copernican system, referring mainly to Kepler. He propounded the view that the system might serve as a convenient hypothesis for representing the celestial motions, but that it was inadmissible in physics. In his view, Copernicus and Galileo had thought the same. [110]

---

[107] Daniel Voet (1678) 133, 145-148. On De Vries' anti-Cartesianism see also Dekker (1986) 93-94 (on his oration *De lunicolis*).

[108] Leusden, disp. Utrecht 1699, thes. 17, p. 13.

[109] Schoock, disp. Groningen 1648, th. 28, 51. See on Schoock also below, p. 250–251.

[110] Schoock (1663) 106-108. I owe gratitude to Han van Ruler for sending me his extracts from this rare work.

In the end, however, both professed Cartesians and die-hard Aristotelians were but a small minority among their colleagues. The great majority of philosophy professors in the Dutch Republic were prudent, cautious men, who thought that Aristotle had said much of value, but that Descartes too might be worth reading. They did not particularly like the old scholastic philosophy, but nor did they particularly like revolutionary change. They read what old and new philosophers had written and picked up what they found useful. Therefore, regarding philosophical contents, it is often difficult to draw clear lines. 'Aristotelian' professors, too, felt compelled to modify their views in accordance with recent insights. For instance, few of them continued to uphold the doctrine of solid planetary orbs. In cosmology, they defended the (rather un-Aristotelian) system of Tycho rather than the Ptolemaic. On the other hand, admirers of Descartes might decide to diverge from his ideas on certain points, or simply misinterpret him. After all, clamorous professions of one's adherence to Descartes offer no guarantee that one understands his theories correctly. It is not the correspondence of one's own ideas with those of Descartes that turns one into a Cartesian, but feelings of loyalty and allegiance.

Still, it is important to note that Cartesianism set the tone of academic discussions. One might reject Descartes' views, but one could not afford to ignore the issues raised by him. It would clearly be going too far to discuss all professors at all Dutch universities in this period, but a few examples may be illuminating. At Franeker, Abraham de Grau tried to find some middle way between Cartesianism and the learning of the ancients. De Grau held the chair of mathematics, but his main interest was philosophy and he managed to do some courses in this discipline as well. His mediating did not come down to stuffing an old framework with new discoveries. De Grau's aim was to find the Cartesian propositions in the writings of the ancients.[111] His cosmology is completely Cartesian, although without mentioning Descartes' name. His views become clear from a disputation on the '*delatio*' of the planets – the typical Cartesian term which denotes the drifting of the planets in the celestial matter, as opposed to motion in its 'proper' sense. The disputation gives a general exposition of the vortex theory of planetary motions and also defends the Cartesian theory of the tides.[112] In a disputation on comets, De Grau also defends the Cartesian theory on the subject. Unlike more radical Cartesians, however, he leaves some room for regarding comets as divine signs or warnings.[113]

---

[111] Galama, 109-115.

[112] De Grau, disp. Franeker 1683.

[113] De Grau, disp. Franeker 1665. Th. 4 par. 7: '*Nos consideramus Cometas, ut sidera densis maculis intecta, quorum vortices absorpti sunt ab aliis validioribus, quique ab uno vortice in alios transeunt.*' Th. 14.

The Franeker professor Johannes Schotanus is generally regarded as a convinced Cartesian. He even rendered Descartes' *Méditations* into Latin verse. Still, his posthumously edited *Physica coelestis et terrestris* is not so much a full and extensive exposition of the physical universe according to Descartes, as a collection of rather heterogeneous physical questions, partly drawn from ancient authors. Schotanus, too, compares the three main world systems. He defends the Copernican system and the fluidity of the heavens, but discusses these points largely within a framework of classical learning.[114]

An eminent example of a philosopher hesitating to accept the new physics is Wolferd Senguerd, professor of physics at Leiden.[115] He was an enthusiastic supporter of the experimental physics introduced by his colleague De Volder. He gave experimental demonstrations with the air pump and published several of his experiments. In 1680, he published *Philosophia naturalis*, a textbook on physics. Five years later, a much reworked reprint of it appeared. It starts with a part on the general principles of things. The general argument seems largely inspired by Descartes. Motion ('the translation of a body from one space into another') is the cause of all change in the world. Like Descartes, Senguerd starts with the simplest rules of motion and continues with the rules of percussion, the movement in a plenum, etc. Despite all this, Senguerd does not mention Descartes' name and it is clear that he would not regard himself as a Cartesian. His whole corpuscular philosophy is put in a decidedly non-Cartesian framework, as he starts out by stating that the basic principles of things are the old Aristotelian concepts of matter and form. Only in the course of the argument does it become clear that this 'form' should be reduced to matter in motion.[116]

In the remainder of the work, Senguerd is mainly concerned with a description of the physical world. The second part is devoted to the world's system and the heavenly bodies. Here, too, various elements seem to be taken from Descartes, for instance, his description of the stars.[117] The third part speaks of the non-living sublunary things, the fourth of living beings. The discussion on the system of the world takes up about a quarter of the book. Senguerd discusses extensively the systems of Ptolemy, Copernicus, Tycho Brahe and – but only in the second edition – Descartes.[118] Descartes' ideas

---

[114] Schotanus (1700); the discussion on world systems is on pp. 37-45.

[115] Ruestow (1973) 78-81, 100-104; Wiesenfeld (1999) 124-173.

[116] The relation between the old concept of form and the new mechanistic concepts had been discussed in a similar way by Regius. In fact, Francis Bacon had already written something similar.

[117] Senguerd (1680) 128.

[118] Senguerd (1680) 114-121; (1685) 155-185. See also Ruestow (1973) 79-80.

are dealt with in great detail and with references to Descartes' work. He rejects Ptolemy's system as being contrary to experience. Copernicus does save the appearances, but his system should be rejected as it presupposes a moving earth, which is contrary to the Bible and to reason. Senguerd rejects not only the annual motion of the earth, but its daily rotation as well. He also states that the earth is the centre of the universe. He considers Descartes' idea of an unbounded universe to be absurd.

Senguerd had given vent to his anti-Copernicanism already in his doctoral disputation of 1667.[119] It seems that what prevented him from accepting the Copernican system were religious rather than philosophical arguments. Although he gives some physical arguments as well (in the second edition, he gives a whole list of them against the Cartesian model), the biblical passages appear decisive. In the first edition, Senguerd refers to Herbinius' work on Copernicanism and the Bible (on which more below) as a refutation of the contrary arguments by those who try to accommodate Copernicanism with the Bible.[120] In the second edition, the argument is wholly revisited and the reference to Herbinius is omitted, although Senguerd still refutes the accommodators. The Bible cannot be held to say the very opposite of reality, even if it speaks metaphorically or *ad captum vulgi*.[121]

Religious arguments also play a prominent role in the writings of Alexander de Bie, professor of mathematics at the Amsterdam Athenaeum. Earlier, we discussed his doctoral disputation from 1642, which was probably written under the influence of Ravensberg and wherein De Bie expressed his preference for Tycho's system. As a professor, De Bie seems to have wavered in his views on nature, as can be seen in his various disputations on comets. In 1653 he wrote that 'the rainbow is a sign of God's favour, a comet of His anger,' but in 1662 and 1666 he refuted the view that comets could cause or portend disaster.[122] Still, the biblical miracles should be regarded as supernatural, as he made clear in his discussion of the darkness during the Crucifixion. He refuted Kepler's suggestion that this darkness was caused by an eruption of Mount Vesuvius: 'But Kepler, this is detracting from the miracle!'[123]

As to the system of the world, De Bie was cautious. In his 1653 disputation on the motion of comets, he spoke about the motion and velocity of the ce-

---

[119] Senguerd, disp. Leiden 7 Dec. 1667, *corollaria ex physica speciali*, no. 4: *Terram quiescere, coelum, stellasque moveri statuimus.*

[120] Herbinius will be discussed in Part IV of the present volume, p. 268–269.

[121] Senguerd (1680), 119; (1685) 172-174.

[122] De Bie, disp. Amsterdam 26 Nov. 1653, th. 26: *iris favoris Cometa, summi Numinis irae indicium est.* Idem, disp. Amsterdam 1 April 1662 1 and 11 Sept. 1666.

[123] De Bie, disp. Amsterdam 12 April 1659, th. 15 and *passim*.

lestial bodies in general. He maintained that the earth was but a point with respect to the universe as such, irrespective of whether one followed the Copernican or the Ptolemaic system. Only on one occasion did he speak his mind openly, viz. in a disputation *De mundi systemate*. He argued that both the geocentric and the heliocentric model save the appearances. In itself, the movement of the earth is not an absurd idea. It cannot be disproved by reason. 'Who surely knows by reason, whether the heaven or the earth is turned around? I say, nobody.' Finally, he admits: 'Were I a heathen, I would keep to Copernicus. But as I am a Christian and know the sacred [writings], I say that those things which have been adduced are quibbles, and I say that the earth rests.' [124]

Another example of the same trend is offered by the Amsterdam professor of mathematics, later Utrecht professor of philosophy, Johannes Luyts. Luyts published a textbook on astronomy in 1689. The book was reissued in 1692; apart from a new title page and an added frontispiece, the second edition is identical to the first one. In it, Luyts offers a standard account of the three major world systems. First of all, he lays down the Ptolemaic system and rejects it as not agreeing with nature. Second, he explains the Copernican system, which is neither confirmed nor refuted by nature, but should be rejected as it is contradicted by the Bible. Then, he comes to Tycho's system, which agrees with both Revelation and nature. [125] Consequently, throughout his book Luyts presupposes the Tychonic system as true.

The tendency to reject Ptolemy on natural grounds and Copernicus on scriptural grounds, thus ending up with the Tychonic system, is also found in a large number of doctoral disputations, especially from Utrecht. The system of the world was one of the subjects on which Nicolaus Shepheard, from Rotterdam, graduated as doctor of philosophy in 1661. It is rather a standard account. Shepheard reviews the systems of Ptolemy, Copernicus and Tycho. A decision between Copernicus and Tycho on physical grounds is difficult. But he declares himself Tychonian because of the reverence philosophers should hold for Holy Writ. [126] In 1675, a large inaugural disputation summarised the same view in one corollary: 'We prefer the Tychonic system to the Copernican and Ptolemaic ones, as it is in full accordance with Scripture and not contrary to nature.' [127]

An inaugural disputation of 1698 on the earth devotes a section to the question whether the earth moves or rests. The disputant presents the three

[124] De Bie, disp. Amsterdam 21 Jan. 1659, th. 56, 61.

[125] Luyts (1689) 202-217. See also the preface, page **2 verso.

[126] Shepheard, disp. Utrecht 22 April 1661.

[127] Steenbergh, disp. Utrecht 3 Dec. 1675, *Annexa ex astronomia* no. 1.

world systems, rejects the Ptolemaic out of hand, and explains that Copernicus and Tycho account for the phenomena equally well. Physical objections are of minor importance. One cannot decide between Copernicus and Tycho from natural reasons. However, the biblical passages, which are infallible, settle the matter. The earth is at rest in the centre, or somewhere near the centre, of the universe.[128]

A similar tenor can still be discerned in the disputation *De systemate mundi*, with which Johannes Blomhert took his degree in philosophy at Utrecht in 1716. The biblicism is even more prominent here. Blomhert starts in the usual way by dismissing the system of Ptolemy. He then discusses the system of Copernicus, which in any case is not contradicted by nature. Still, Descartes' vortex theory is refuted at length. He then moves on to the system of Tycho Brahe. This explains all the phenomena just as well as the Copernican system. The objections of the Copernicans against Tycho's system are of no value. It should be preferred because of the infallible authority of Scripture. True, the purpose of the Bible is not to teach physics, but we can trust its passages in this field, too. The source of truth cannot produce anything untrue. If one were to admit any exception, the whole authority of the Scriptures would collapse. The last part of the disputation is entirely devoted to the explanation of Scripture. Blomhert discusses some texts which, as claimed by adherents of Copernicanism, show that the Bible should not be taken literally in mathematical and astronomical matters. He argues that we should take the Bible literally even there.[129]

At Leiden, too, not all students were Cartesians. Such non-Cartesian students were unwilling to defend Copernicanism. But, as the Tychonic system apparently was not taken seriously in this academic environment, they had recourse to a more general scepticism. In 1671, the Englishman Johannes Annesley took his degree at Leiden on a disputation 'containing various considerations on common astronomy'. It came largely down to a discussion of the three world systems. Having dismissed the system of Ptolemy (partly on biblical grounds), he rejected the Copernican system as well, as being contradicted by the Bible. Who denies such, turns God into a liar, he stated. So, the system of Tycho remained. However, Annesley acknowledged that this had some problems, too. His argument tended to show the uncertainty of all systems, or even of all philosophical knowledge. He rejected the temerarious attempts of humans to scrutinise the hidden mysteries of the Creation. A rather traditional inaugural disputation *De mundo*, defended in 1703, at the end

[128] Bert, disp. Utrecht 1698, thes. 13, pp. 19-22.
[129] Blomhert, disp. Utrecht 16 March 1716.

of a very long rehearsal, mentions briefly the discussion on the system of the world and the three main positions. This disputant prefers simply to suspend judgement and states no preference for either the Copernican or the Tychonic system. [130]

Conclusion: character of the debate

This overview clearly reveals some general tendencies. A first conclusion is not very amazing: all professors who declared themselves followers of Descartes, were Copernicans. This holds true also for people who in our eyes are not 'real' Cartesians, as they interpret Descartes' work in a way different from ours and mix their self-professed Cartesianism with more traditional insights (e.g. Schotanus). A second conclusion comes more as a surprise: all professors who did not regard themselves as followers of Descartes, irrespective of their further attitude to the new science, rejected Copernicanism. On closer inspection, some exceptions may turn up, but in our material the correlation is 100%.

Copernicanism as it existed at Dutch universities was founded upon the reading of the works of Descartes, not upon any astronomical considerations. Only one disputation appears to spring from a deeper involvement with the subject. In 1671, Daniel Arens took his degree at Utrecht on a disputation on (among other things) 'the true system of the world'. He seems to have had his main education elsewhere, as he dedicated his thesis to professors Tobias Andreae (Groningen) and Johannes de Raei and Alexander de Bie (Amsterdam) [131] – two professed Cartesian philosophers and one mathematician. Arens emphasises that the Copernican theory is not just a hypothesis, but a mathematically proven truth. As he maintained in a corollary, it did not contradict reason, and still less Scripture. [132] Arens' views appear firmly embedded in Descartes' physics. Descartes is even mentioned as having accomplished the Copernican system as its 'consummator'. Vortices are mentioned, be it in passing, and Arens appears to support Descartes' theory of planetary '*delatio*'. [133] The work itself is a learned piece, and it refers to a lot of authors, astronomers as well as philosophers: Boulliau, Gassendi, Galileo, Kepler and Gilbert. Cartesianism is not prominent here. The precession of the equinoxes is, following Copernicus, explicitly referred to a motion of the earth's axis. Arens discusses astronomical authors rather than laying down Descartes' theory.

---

[130] Annesley, disp. Leiden 8 July 1671; van der Ghiessen, disp. Leiden 20 Sept. 1703.

[131] De Raei had moved from Leiden to Amsterdam in 1668.

[132] Arens, disp. Utrecht 1671. *Mantissa physica*, no. 14: *Motus terrae nec rationi, multo minus scripturae repugnat.*

[133] Th. 9. *Mantissa physica*, no. 15: *Non propriè movetur, sed quietè circa solem defertur* [sc. Terra].

On the whole, however, the professors of the second half of the seventeenth century had only the haziest idea of astronomy. Important astronomical discoveries were completely missed. Huygens' 1655 discovery that Saturn was accompanied by a moon, and that the planet's puzzling shape was caused by a ring around its main body, was completely ignored in the textbooks written by academic philosophers. In their works, Saturn is always spoken of as having two satellites instead of the one Huygens had discovered. This is not an anticipation of new discoveries, but just a word by word repetition of what Descartes had said about Saturn in his *Principia* in 1644.[134] Like most people before Huygens, Descartes explained the curious shape of the planet as being caused by two satellites. De Raei, in commenting Descartes' *Principia*, at one point appears to have got wind of Huygens' work: 'There are people who think that Saturn is circled by only one planet, just like the moon around the earth.'[135] However, in 1673 another, anonymous commentator referred to a theory even older than Descartes': 'Saturn too has its [vortex], if it is true that it has two satellites. But others deny that these are true planets, distinct from Saturn's main body. They call them Saturn's *anses* [ears], which have their origin from the refraction of the solar light, which is reflected from Saturn to us; about the same way that comets (...) are seen with a tail.'[136]

On the other hand, the anti-Copernicans at the universities were not very sophisticated either. They invariably made their choice on religious rather than philosophical grounds: they regarded the system of Copernicus as contrary to the Bible and as a rule did not bother to defend their choice with any other argument. The physical arguments against Copernicus have not been solved, but are simply put out of sight. Even professed Aristotelians agree that his system cannot be disproved from nature. Senguerd is quite exceptional in stating that Copernicanism is against reason as well. The general consensus that the Copernican and Tychonic systems saved the phenomena, and the Ptolemaic system did not, is rather striking. Undoubtedly, most professors would be hard pressed to demonstrate this view other than by repeating others. But as in the end the arguments from natural reason were not decisive, it simply was the easiest way to deal with the issue.

The anti-Cartesian offensive of Voetius and his theological allies appears to have been quite successful in the end. They had not been able to suppress Cartesianism completely, but they had succeeded in imposing their own bib-

---

134 Senguerd (1680) 116 and (1685) 162-163 (the second edition adds '*juxta Nonnullos*'). de Bruyn, disp. Utrecht 11 Dec. 1661, thes. 29. Cf. AT, VIII-1, 195 (part III, par. 146). See also Borgesius, disp. Groningen 25 Sept. 1658.

135 Johannes de Raeij, Leiden, university library, BPL 907, 62v.

136 Leiden, university library, BPL 908, II, 15v.

lical view upon the majority of Dutch academic philosophers. They had turned the debate on the system of the world into a choice between Descartes or the Bible. This was brought about not just by the academic debates mentioned, but also by anti-Cartesian agitation outside the university, which will be dealt with in a later chapter. Still, Cartesianism may have gained some benefits from the polarisation. The theological opposition which had arisen since the 1640s made it very difficult to uphold heliocentrism without being backed by an alternative religio-philosophical stance. This was only to be found in Cartesian philosophy. So, paradoxically, by vigorously propagating their biblical world-view, the theologians rendered Cartesianism – which according to them was the root of all evil – ever more indispensable to their opponents.

10. The impact of Cartesianism outside Dutch universities

Intellectual life in the second half of the seventeenth century

The first means by which Cartesian ideas obtained a foothold in the Dutch Republic had been the presence of Descartes himself. By means of personal interviews, he succeeded in winning many people over to his new ideas. The main channel for the further propagation of Cartesian ideas were the universities, where the works of Descartes were read, studied and commented upon. But Cartesianism did not remain a purely academic affair. Only very few of the students who during their studies became imbibed with Cartesian ideas remained at university; most of them became a physician, lawyer, regent or minister. Descartes' ideas thus became widely known among the educated elite.

Cartesianism spread even further than that. Descartes himself already had started to disseminate his ideas among people without a university schooling (e.g. mathematicians like Jacob van Wassenaer and Dirk Rembrandtsz van Nierop). Also some of his more zealous followers made an effort to spread his ideas among wider circles of the population. The dissemination of Descartes' work was for the most part carried out in small intellectual circles. A prominent one centred on the Amsterdam bookseller and publisher, Jan Rieuwerts. His clientele appears to have consisted of religious dissenters rather than of people interested in science for its own sake. Rieuwerts is particularly known as the publisher of Spinoza's works. But during the years 1656-1661 he published the complete works of Descartes as well, translated into Dutch by Jan Hendriksz Glazemaker. Dutch was the only modern language, apart from French, in which Descartes' full oeuvre appeared in the seventeenth century.

Around 1680-1690, another Amsterdam bookseller, Jan ten Hoorn, was the nucleus of another Cartesian cenacle, dominated by the learned physicians Cornelis Bontekoe, Stephanus Blankaart and Heidenrijk Overcamp.[137] Bon-

[137] See on these people, Thijssen-Schoute (1954) 276-341 and *passim*; Geyer-Kordesch (1982).

tekoe was the author of various learned works, wherein he developed the Cartesian theories, including metaphysics. Blankaart was more interested in practical subjects like medicine or chemistry. He published most of his work in Dutch, popularising Cartesian theories. Overcamp had written on the Cartesian theory of motion. On the whole, Ten Hoorn was more interested in the publication of purely scientific works, both translated and original, than Rieuwerts had been. The Cartesianism of Bontekoe, Blankaart and their friends implied a general interest in science. They translated the work of such scientists as Lémery and Mayow. Ten Hoorn's son, Nicolaas ten Hoorn, published in 1702 a Dutch translation of Fontenelle's *Entretiens sur la pluralité des mondes*. Descartes, however, offered the framework which made such work relevant. In 1690-1692, Ten Hoorn republished the complete works of Descartes in Glazemaker's translation, supplemented with some new translations (by Blankaart) of some formerly unknown works.

Other people worked in more isolation. A Dutch translation of a work by Descartes, the *Traité de l'homme*, which was not included in the Rieuwerts edition, was published in Middelburg in 1682. The translator, the physician Jacob Copper, explains in the preface that he had undertaken the work because some friends in Den Briel (a very small town in the province of Holland) were not able to read the original. [138] On the whole, philosophical and scientific literature in the vernacular proliferated in the second half of the seventeenth century. Many ancient and modern philosophical works were translated, e.g. Lucretius, Hobbes, Machiavelli, Tschirnhaus, Burnet, etc. Publishers partly responded to market demand, but on the other hand some people wanted to propagate their ideas by means of translations or by writing original works. It would be going too far here to give a full overview of these publications. However, something should be said about the character of intellectual life in general at the time.

The second half of the seventeenth century saw the emergence of a new group of intellectuals who took identity not from their status in society, but from what they regarded as their intellectual superiority. To a large degree, such lay philosophers were inspired by the 'democratic' element in Cartesianism. Descartes had shown that true knowledge was not obtained from books, but by applying one's sound reason. Traditional learning and authority were hampering scientific progress rather than promoting it. The new intellectuals felt that they had gained a true insight into the workings of nature. This encouraged them to rebel against the old intellectual authorities, as represented

[138] For a bibliography of Glazemaker's translations of Descartes, see Thijssen-Schoute (1967) 235-246, 257-259; see also Glazemaker (1982) 1-11. On Ten Hoorn's additions: Thijssen-Schoute (1954) 324-325. On Copper's translation: Thijssen-Schoute (1954) 354-355.

by the universities or the Church. They regarded it as their duty to provide their fellow citizens with a real understanding of things. They were active in propagating the new scientific ideas and applying them to all kinds of pressing technical, political and religious problems.

This part could be played by people within the traditional intellectual elites, as shown by the physicians mentioned above. But the new ideas reached a wider audience. Other people had humble occupations, but still felt themselves elevated above their fellow citizens because of their intellectual standing. Such self-proclaimed philosophers often discovered that there was an abyss between their high ideals and their day-to-day life. Seventeenth-century society cared very little for intellectual credits. Intellectual roles were restricted to academic occupations, mathematical practice and the pastime of rich amateurs. It was difficult for the new, self-styled intellectual to find a role which suited him – unless he was rich. The only way to get some legitimation for one's claims was to gain entrance to some official learned society, as they sprang up in this time. However, no such society was founded in the Dutch Republic itself, probably because of the lack of a court tradition. Some people strove after membership of foreign institutions, such as the Royal Society, but this was attainable only for a selected few.

Sometimes, a position as private teacher would offer temporary relief. Moreover, the second half of the seventeenth century is the period in which the first stirrings of learned journalism occur, but the first enduring efforts in the Dutch Republic were the work of the French Huguenot refugees, Pierre Bayle and Henri Basnage de Beauval. Only in 1692 did there appear a learned periodical in Dutch which lasted longer than just a few issues, viz. *Boekzaal van Europa* ('European Library') by the Rotterdam notary, Pieter Rabus.

Intellectual life was therefore very unstable. Apart from the traditional occupations, which were transformed in their own way, there was an army of new men with new ideals, who tried to gain a foothold in society in one way or another. The second half of the seventeenth century saw the emergence of a kind of literary underground. Johannes Duykerius, who had studied theology but remained without a ministry because of a speech defect, tried to make a living as a man of letters. He undertook many ventures and enjoyed some notoriety after the publication of his satirical novel 'Life of Philopater', the sequel of which was banned; however, he died in great distress. Pieter van Gent, an admirer of Spinoza who called himself doctor of medicine, tried to earn a living as a schoolkeeper, besides relying on the support of rich lovers of learning. He died burdened with debts. Daniel Gabriel Fahrenheit eventually settled down as an instrument maker and public lecturer, which enabled him to earn a living more or less in a way he wanted; but his position was a humble one and he never became rich.

Cartesianism was an important background to the emergence of this groups. But as they lacked the formal schooling of university, their ideas could diverge in unforeseen directions. Especially in the last quarter of the century, when people increasingly realised the weaknesses in Descartes' ideas but did not have an alternative, they tried to reject Descartes in the way Descartes had rejected all earlier philosophers. Some of these people, such as Spinoza or Leeuwenhoek, despite their lack of a university training, worked on a level that few, if any, academics could hope to reach. But most lovers of philosophy, despite their high pretensions, accomplished very little in practice. Their pet problem was how to establish longitude at sea, or 'to find east and west', as it was commonly called. (Although the squaring of the circle was never really forgotten.) Its hold on the imagination could be compared only to that of perpetual motion in a later age. The list of solutions proposed for this problem is a long one and the proposals came from persons ranging from renowned scholars such as Christiaan Huygens, navigational practitioners and instrument makers, to the completely obscure. The problem, after all, was of acknowledged relevance. The inventor could hope not only for a material reward, but also for high social prestige.

By this time, however, also the system of the world had become a well-known problem and, as will be seen, several lovers of learning tried to find a definitive solution for it. Antoni van Leeuwenhoek nearly exclusively specialised in microscopical investigations, the stronghold from which he could withstand any challenge from the army of learned men. Still, he also felt committed to the cause of the earth's motion. He expressed sympathy for the Louvain professor Van Velden, who had got into trouble because of his public stance in favour of heliocentrism. Leeuwenhoek himself did not present any theoretical speculations on the issue, but made a mechanical model intended to demonstrate the motion of the earth by analogy. The model consisted of a glass filled with water, a lead ball and pieces of sealing wax. By turning the glass, the model was intended to demonstrate how the rotation of the earth keeps the clouds drifting; if the earth were to come to a standstill, according to Leeuwenhoek, all clouds and other materials drifting in the air would fall to earth.[139]

At the universities, as we have seen, opinions were rather clear-cut. There was a Cartesian programme and an Aristotelian programme, and a lot of people were either vacillating between the two or trying to accommodate them both. Only the Cartesian programme offered scope for defending heliocen-

[139] See Leeuwenhoek's letter to Nicolaas Witsen, 10 July 1696, in: Leeuwenhoek (1983) 292-299 (with English translation). See also Snelders (1982) 73-76.

trism. Outside academia, people were not bound by teaching programmes and were much more free to defend any opinion they liked. Their views do not just move between the two extremes of Cartesianism and Aristotelianism. It will have to be investigated how far here, too, Copernicanism needed the support of Cartesian philosophy.

The mathematical practitioners

The preponderance of physics in the debate on the world system after 1640 becomes especially clear in the work of a group of men who normally had little use for it: the mathematical practitioners. As explained before, mathematics in the Dutch Republic was closely linked with Dutch shipping, in particular the development of intercontinental trade. Increasingly during the seventeenth century, the large trading companies – the United East-Indian Company and the West-Indian Company – as well as the admiralties, required their pilots to undergo a theoretical training in navigation. From 1619 onwards, a system of exams was set up, which had to be passed in order to obtain a qualification for the function. [140]

Consequently, mathematical training was much in demand in the towns and villages of the Dutch coastal areas, especially those in northern Holland, where many ambitious young men wanted a career in the merchant navy. Mathematical education was offered at many levels, from the simple village schoolmaster who supplemented his income by boarding pupils, acting as verger and organist, keeping some cows, and providing some private tuition in mathematics, to prominent mathematicians who ran a school of navigation, held an official admiralty appointment as examiner of pilots, and published a lot of books. [141]

Most of the more prominent mathematicians were to be found in the big towns, and especially in Amsterdam. The first professional instructors there were active around 1600, and after then, there was never a shortage of them. For the majority of the seventeenth century, they were locked in fierce competition. They advertised their abilities on public broadsheets at strategic points, preferably at the *Nieuwe Brug* ('New Bridge') near the IJ, a meeting point for Amsterdam sea captains and other people involved in navigation. At the same places, they challenged their colleagues by posing problems which they should be able to solve. Such problems were a kind of advertising for their respective schools and as such were intended to impress the laity, not

[140] Davids (1985) 294-298.

[141] For an elaborate discussion of the Dutch navigational instructors, see Davids (1985) 312-333.

to achieve mathematical progress. The practitioners stuck to traditional problems and never engaged in the analytical geometry of Descartes. [142]

One of the most prominent Dutch mathematical practitioners of the second half of the seventeenth century was Dirk Rembrandtz (Dirk, son of Rembrandt) from the village of Nieuwe-Niedorp – or Nierop – in northern Holland, where he lived all his life. [143] He mostly signed his name simply as Dirk Rembrandtz; the toponym 'van Nierop' was added to bring it in line with the tripartite form (Christian name + patronym + family name) common among the higher and urban classes. (In Holwarda, we met a similar case.) Dirk lived from 1610 to 1682. He came from a humble Mennonite family and as a mathematician, he was largely autodidact. Originally, he was a shoemaker by trade, but by applying himself to mathematical studies he became so proficient that soon he was able to make a living by teaching navigation and writing almanacs. He was a prolific writer who (apart from almanacs) from the 1630s onwards published a large number of books. They testify to an impressive command of mathematical and, in particular, astronomical learning, even if Dirk never spoke or read any other language than his native Dutch. It should be added that his works are rather a bibliographical mess. He did not hesitate to give corrections to a work in the appendix of the sequel to another work. His first works were purely mathematical, concerning sundials, problems of navigation and astronomical calculation, and such like. However, somewhere in the 1640s he became acquainted with the philosophy of Descartes and this introduced a new element into his work.

In fact, Van Nierop appears to have been introduced to the new philosophy by Descartes personally. The latter lived from 1643 to 1649 in the village of Egmond, some 25 kilometres from Nieuwe Niedorp, and Dirk used this opportunity to make his acquaintance. Baillet, on the authority of Hartsoeker, tells a touching story about how Dirk was sent away twice by the philosopher's servants, who took him for a beggar, but finally gained access to Descartes who immediately recognised his qualities. Subsequently, Dirk Rembrandtsz became a close friend and follower of Descartes. [144] His introduction to Cartesian philosophy opened up new horizons. As he wrote later, Descartes 'had engendered for me a great light, especially in exhibiting the visible world.' [145]

---

[142] Polemics among the practitioners are discussed in Crone (1962) and by Wolthuis in his introduction to Langendijk (s.a.) 11-43. (NB This introduction is a word-to-word repetition of Wolthuis (1936), with in a few places a somewhat extended annotation.)

[143] See on him: Vermij (1996)b; Smit (1992).

[144] Baillet, (1691) II, 553-555.

[145] Van Nierop (1674) 63: '*de welcke voor desen in my een groot licht geteelt heeft| en dat voornamenlijck in 't stellen der* zichtbaere Werelt.'

It may also have opened up new horizons on the social level, offering entrance to the world of learned scholars. Dirk came into in contact with such people as Huygens, Hudde, Van Schooten, Hartsoeker and others.

Dirk Rembrandtsz was an ardent supporter of the Copernican theory. Being a widely renowned mathematician and astronomer on the one hand, and still very much a commoner on the other, he was an ideal intermediary between the learned world and the population at large. Many people – fellow mathematicians, regents and craftsmen – turned to him in order to discuss mathematical problems. On several occasions, Dirk felt the need to take a public stance against the detractors of Copernicanism. Thus, in one of the almanacs calculated by him, there is a comment on Copernicanism: it is asserted that most astronomers of the day were of Copernicus' opinion. [146] In 1661, Dirk published a refutation of the cosmological ideas of Jacobus Coccaeus. Also in 1661 he published on the subject of Copernicanism and the Bible, which had been much debated since about 1655. In 1677, he published a sequel, in which he discussed, among other things, the anti-Copernican ideas of Sybrand Hansz Cardinael and Hendrik Stevin. The original book was reissued in 1683. These episodes will be dealt with later in this book, along with the works of his respective opponents.

His book 'Some exercises in divine, mathematical, and natural things' – a collection of several unconnected essays – is revealing. [147] The first part deals with the Creation. As Dirk explains, he had found that many people found a contradiction between the Copernican theory and the story of Moses. Therefore, he had decided to investigate the matter. He then proceeds to quote all the verses of the Creation story from Genesis, and explain how these should be understood in the light of physical theory. By doing so, he goes decidedly further than Descartes, who had always refused to relate his theories to the Bible. In this respect, Dirk is more in line with the 'biblical physics' of the sixteenth and first half of the seventeenth century. In some other respects, too, he keeps to older notions. He speaks of the four elements of classical natural philosophy, instead of the three proposed by Descartes, although the four elements are reinterpreted in a mechanical way. Rather curious is his remark that the four elements dissolve from chaos to obtain their place in the universe as cheese and whey, or butter and buttermilk, separate from milk.

Here, however, it is important to note that Dirk's Copernicanism was embedded in a Cartesian world-view. His 'Dutch astronomy' (a textbook of as-

---

[146] Salman (1999) 90. Although Dirk calculated the tables, it is not absolutely sure that this commentary is also by him.
[147] Van Nierop (1674).

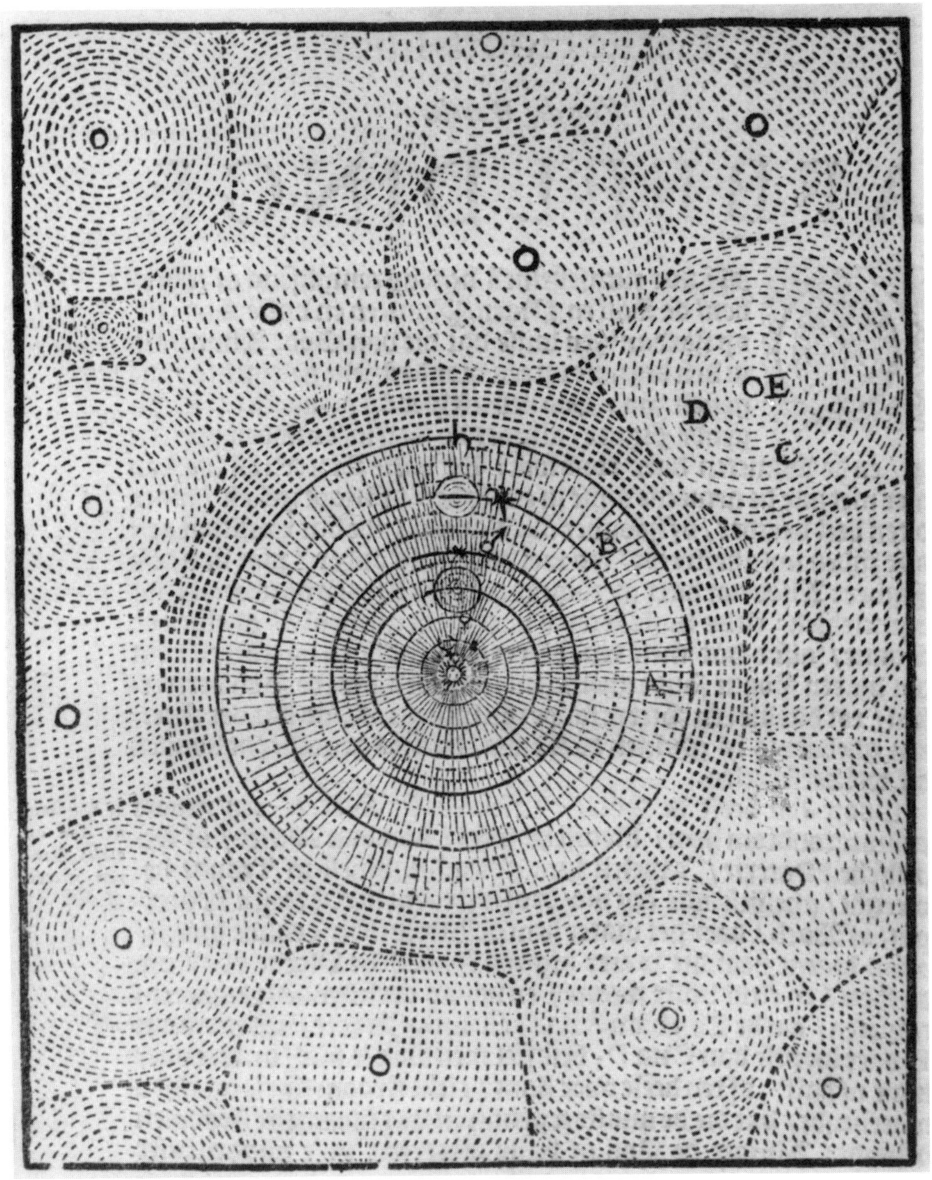

Figure 4: The world system according to Dirk Rembrandtsz van Nierop, *Nederduytsche astronomia*, 1658: the planetary system in the solar vortex. The background of stellar vortices was taken from an illustration in Descartes' *Principia philosophiae*. (Leiden University Library.)

tronomy written in the vernacular, and first published in 1653) started with an elaborate exposition of the Cartesian world system with its vortices.[148] His dependency on Descartes is evident not only from the general world-view, but also from the details. Straight from the *Principia philosophiae* come discussions of the relative turning speed of the moon and the earth, of the question why the moon always turns only one side to the earth, and the explanation of the tides. He also included an illustration of the solar system, adapted from Descartes' *Principia philosophiae.*

This physical introduction was not just an irrelevant extra, unconnected to the mathematical main body of the work. To Dirk, it clarified the way the heavenly bodies moved. The closer to the sun, the greater the velocity of the heavenly matter which made up the vortex. Therefore, the planets were moved faster in their apogee, near the sun, than in their perigee, further away. Before Descartes, this had not been understood correctly. The astronomers had made hypotheses on the assumption that the planets had a uniform circular motion. Dirk, on the contrary, reintroduced the old Ptolemaic device of the *punctum equans*: planets moved in circles, but their velocity was not constant. This mathematical model was legitimised by the physical description.

This idea is not to be found in Descartes' writings, although it remains possible that it was discussed by Descartes and Dirk during their personal discussions. The main asset of the construction was that Dirk found he did not need epicycles, a construction he thought did not fit the Creation – 'Just as though God almighty at the Creation would have worked in a human way, turning different wheels, which should be deemed strange and does not agree at all with the Divine Creation.'[149] All motions were construed from eccentric circles and a *punctum equans.* Finally, by using non-uniform movement, the eccentricity of the planet's orbit was reduced.[150]

The only exception Dirk admitted was the moon. Descartes himself had granted the moon a non-circular orbit, in order to account for the phenomena of spring tide and slack water. In a supplement to the work, published in 1677, Dirk was forced to give an elliptical orbit to Mercury, too. He thought that the reason for the irregular shapes of the moon's and Mercury's orbits lay in their relatively large angle to the ecliptic. These planets periodically dwelled in regions with a different celestial matter, which naturally affected their motion. In no case could one conclude that the other planets, too, moved in

---

[148] Van Nierop (1658). This is the second edition. The preface is dated 1650. The first edition was pubished in Harlingen in 1653, acccording to Dirk's own attestation in van Nierop (1677)a, preface, and (1680), preface, but I have have not been able to find a copy.
[149] Van Nierop (1658), *tot den leser.*
[150] Van Nierop (1661)a 35-37.

elliptical orbits. Elliptical orbits, just like epicycles, were nothing more than mathematical devices, useful as long as one does not understand the real causes of the heavenly motions, but to be rejected thereafter. 'Kepler was not able to work this out but by means of an ellipse. But now this has been obtained by natural reason, as has been proved, I do not understand why one would remain with the unnatural ellipsis (...) Why should one not accept the natural motion according to the aforementioned Descartes? Although he has not done any calculations on this subject himself, this has been done by now in this my *Astronomia*...' [151]

Dirk's views were criticised by a fellow practitioner, Claes Jansz Vooght. In 1662, Vooght was admitted as a surveyor by the Council of Holland; from about 1675 to 1696 he was active as an instructor of navigation in Amsterdam. He was a man of more learning than most practitioners. Besides publishing some works on navigation, instruments and such like, he translated Euclid's *Elements* from Latin into Dutch (1695). He also contributed some laudatory poems to Dirk's books and a translation of a Latin poem by Jeremiah Horrox. [152] There exists an undated letter written to him by Dirk Rembrandtsz wherein different theories on planetary orbits are discussed. It appears that Vooght advocated an elliptical motion of the planets, following Boulliau rather than Kepler. Boulliau construed the irregularity in the planets' movement by means of an *punctum equans* instead of Kepler's second law. This procedure led to more or less the same results, but the calculations it entailed were easier.

Vooght had defended this view in an earlier (lost) letter to Dirk; it may well have been written in reaction to Dirk's views as expressed in the book of 1677. In fact, Dirk's response contains a reference to this work. Dirk was clearly irritated by Vooght's appeal to learned authority: 'Why should this be taken from Boulliau? You know I do not have the book and do not understand the language. So, your statement that I have not scrutinised it with due attention, what else is it but a mocking reproach?' Dirk agreed that the motion of the planets could seem to be elliptical, 'but I join Descartes who proves with good reasons that their course is in a circular vortex, but such that their course is more rapid at perigee than at apogee. So I calculate it by means of a *punctum equans* and say this gives the same result as an ellipse, to which you agree. So, I try to prove these irregular courses with good natural reasons. But if you prefer to remain ignorant along with Boulliau, it is all right with me.' [153] Clearly in remembrance of this discussion, Dirk wrote a few years later about

---

[151] Van Nierop (1677)a 79.

[152] Van Nierop (1977)a 31-32.

[153] *MML* 5 (1757), letter no. 28, p. 337.

elliptical orbits as fictitious inventions, introduced because of a lack of comprehension of Cartesian celestial physics.[154]

As Vooght's example shows, Dirk's ideas were not self-evident among practitioners. However, they quickly came to set the tune. The straightforwardness of the calculations, based upon a simple physical model, must have appealed to most instructors, who generally were people of humble origin like Dirk himself. It should be added that the accuracy of the methods used was deemed significant. Vooght's sources were inaccessible to most practitioners, who had little schooling in languages. Vooght himself never wrote a handbook of astronomy, although he probably taught his ideas in his courses. So, several writers on navigation followed Dirk's footsteps.

Very soon, Dirk Rembrandtsz' 'Dutch astronomy' had a competitor. In 1659, a year after the second edition of van Nierop's book appeared, Abraham de Graaf's 'Astronomy' was published.[155] De Graaf, a Mennonite like Van Nierop, was a teacher of navigation and mathematics in Amsterdam. He seems to have received at least part of his education with Van Nierop.[156] By 1659, however, he was aligned with another practitioner, Claes Jansz Gietermaker. Quite naturally, he preferred to use a book of his own in his courses, rather than the one by Van Nierop, who was a professional rival. On the other hand, Van Nierop was rather critical of De Graaf's astronomy, which according to him was better titled 'An elucidation and short explanation of the *Dutch Astronomy*'.[157] Indeed, De Graaf's astronomy appears largely inspired by its predecessor. Like Van Nierop's book, De Graaf's is mainly a manual for the calculation of stellar positions, eclipses and such like, and is not really a book on cosmology. The tables it contains were copied from Van Nierop's book, as De Graaf himself admits. It starts with an exposition of the world according to the ideas of Descartes (referring explicitly to the *Principia philosophiae*). The arrangement of the work is thus very similar to Van Nierop's. De Graaf even inserted a plate of the solar system which derives its basic idea from Van Nierop: a Copernican system superadded to Descartes' scheme (from *Principia philosophiae*) of the universe.

None of this means, of course, that he would not have been sincere in his profession of Cartesianism. De Graaf wrote an extensive preface, wherein he commented upon his astronomical studies. Putting mathematics aside for the moment and turning to the study of astronomy, 'I found to my amazement that

---

[154] Van Nierop (1680), *Aen den leser*.

[155] De Graaf (1659).

[156] See the letter from Jan Christiaansen to Van Nierop, 20 Jan. 1659, in: MML 4 (1756), 91-96, letter no. 9.

[157] Van Nierop (1661)a 64.

nearly everything here was in truth, or at least very probably, different from how it appeared to the eye. Observing the motion or course [of the stars], it did not correspond in any part with that of sight. Looking into the causes, I found them to be that the sun, and not the earth stood still. Scrutinising why this should be so, I was nearly dumbfounded by the probability of the reasons given. So, putting aside antiquity, which was founded upon the probable reasons of the eye, which unfaithfulness I well knew, I sided with reason alone and its agreement with observations.' As de Graaf explains (referring to Lansbergen for further details), the many irregularities in the course of the planets are dissolved in a heliocentric system. But his arguments are not just mathematical. 'Adding to this the physics of R. des Cartes, which appeared to me to be true because of its compelling arguments, I not only found arguments why the above-said should be so: above that, all the aforesaid was solidly confirmed by it. What is more, it gives natural causes ['*natuurlyke reden*'] of all the circumstances of changes in their course, established by continual observation.'

Like Van Nierop, De Graaf held that the vortices move faster towards the centre than towards the periphery. He used the same model of planetary motion as Van Nierop, based on circular orbits, whereas the motion of the planets is construed by means of a *punctum equans*. Although De Graaf is less explicit than van Nierop on the subject, he defends the view that such irregularity is caused by the variation in the movement of the celestial matter. He specifically rejects the use of epicycles, 'as absolutely in disagreement with the Divine Creation'.[158]

Even if De Graaf did not outright copy Van Nierop, he had clearly followed his courses with attention. It is quite understandable that Van Nierop, or his pupils, thought it improper that De Graaf should publish all this under his own name. Initially, Van Nierop reacted rather condescendingly, and abstained from using abusive language. But De Graef's alleged 'plagiarism' remained an issue in the professional quarrels of the following years, and provoked some harsh words.[159] These attacks did not succeed in silencing De Graaf, who became one of the most prominent mathematical practitioners of the second half of the seventeenth century. He published a large number of books, covering most segments of the mathematical sciences; not just practical mathematics, but also important segments of theoretical mathematics.

A later book of a similar kind is the *Astronomia* from 1702 by Simon van der Moolen, another mathematics teacher in Amsterdam. He, too, was a convinced Cartesio-Copernican. As he explains, the hypothesis of a fixed earth

---

158 De Graaf (1659) 11, 19-20, 17.

159 Van Nierop (1661)a 64-69; (1677)b 18-24. de Graaf (1663) 49-51, 54-59. See on the latter pamphlet Crone (1962) 45-46 and Wolthuis in Langendijk (s.a.) 37-39.

was put forward by Ptolemy and followed by all astronomers until Copernicus. Copernicus was the first to state that it moves, 'but thereafter still many remained of the ancient opinion. Finally, however, the turning of the earth has been argued forcefully by Renatus des Cartes in his Principles of Physics, so that presently one finds few or no people of reasonable sense, who doubt it. So, too, this my book has been built upon this principle.' [160]

This book, too, is mainly a manual for computation. The tables are new and partly recalculated on the basis of newer data. But here again we find an introductory chapter on cosmology. Van der Moolen's view of the universe is clearly Cartesian: an infinite space filled with vortices. Here, too, the vortices move fastest close to the centre. Hence, as in the books of Van Nierop and De Graaf, the planets move fastest in their perigee. The planets' orbits are supposed to be eccentric circles. Still, Van der Moolen introduces some new elements in his cosmology. He ventures (be it very briefly, as his is not a book on physics) a theory of the cause of the daily rotation of the earth. As he sees it, the side which is turned towards the sun is warmed up. It evaporates and so becomes lighter. The planet then turns its heavier (backward) side to the sun, and everything begins anew. He also puts forward a theory on the distances between the planets and the sun. As he sees it, the planets take a place where their centrifugal force equals the force exercised by the solar vortex. Planets made from heavy material are therefore closer to the sun than those made from lighter stuff are. All this is a typical example of physical speculation in the later phase of Cartesian philosophy. [161]

A final example of a mathematical book opening with a summary of Cartesian physics is the *Cosmographia* (1698) by Pieter Maasz Smit. [162] He had been a student of Dirk Rembrandtsz. In 1682 he corresponded with him from Blokzijl in the province of Overijssel, but on the title page of his book he announced himself as instructor of mathematics, living in Amsterdam. The book is mainly a manual on the construction of globes, although the introduction deals with the motions of the planets and the Cartesian theory of vortices. Descartes is mentioned by name when Smit explains that the fixed stars have their own vortices, just like the sun. On the whole, however, his physical introduction is much shorter than those of the earlier three. Moreover, its function in the context of the book is not quite clear. It may well have been some kind of a standard element in the courses of Dirk Rembrandtsz and his 'school'.

Not all practitioners were students of Dirk Rembrandtsz. Arent Rogge-

---

160 Van der Moolen (1702) 165.
161 Van der Moolen (1702) 1-7 (cosmology), 31 (planetary orbits).
162 Smit (1698). A later (undated) edition is in the Boerhaave Museum, in Leiden.

veen, from Middelburg (the capital of the province of Zealand), was originally a pilot. Later he was active as a surveyor, map maker and teacher of navigation. He ended his career as examiner of the pilots in the service of the Middelburg department of the Dutch East-India Company. In 1665 he published a pamphlet on the comet which caused a lot of commotion in Europe in that year. In this work, Roggeveen presupposes the Cartesian theory of the universe, expressly referring to Descartes' *Principia philosophiae*. He uses the theory of vortices to account for the comet's velocity, more or less adapting Descartes' theory of the tides. He also reformulates Stevin's theory of a 'magnetic rest' of the earth's axis in terms of vortices: a special vortex was required to keep the earth's axis in position. As in the case of Dirk Rembrandtsz, Roggeveen's 'Cartesianism' encompasses or adapts older elements. Unlike most academic Cartesians of the time, he regards the comet as a sign of God's anger.[163]

Part of the pamphlet is devoted to the Copernican question. Here, too, one finds a mixture of old and new theories. According to Roggeveen, the course of the comet clearly proved the motion of the earth. From a motionless earth, one could not understand how the comet moved that fast contrary to the order of the degrees. The latter expression was clearly taken from Simon Stevin. Roggeveen did not deny that in principle God Almighty could well have created such an irregular motion. But in fact, God had created the world according to a fixed order which we can understand. According to Roggeveen, the comet thus solved the dispute between 'some theologians and astronomers' on the motion of the earth; just like the voyages of discovery had refuted the idea that the earth was flat instead of spherical.[164]

Mathematical practitioners who do not profess Cartesianism, do not present themselves as Copernicans either. An example of this is the navigational instructor Dirk Makreel, whose 'Illuminating leading-star of navigation' was a rather successful book in its genre. The introduction is concerned with the status of mathematics and the relation between theory and practice. However, there is no part on cosmology. Makreel is aware of the Copernican system: 'As it appears to us, all heavenly lights are turning on the world's axis in 24 hours...', but he does not commit himself.[165] That is not to say that Makreel was not interested in cosmological questions. Quite the reverse is true, as it appears from a pamphlet he published in 1681 on the comet of the preceding year. However, here too he evades the question of the world

[163] Roggeveen (1665) 23 (Descartes), 28-29 (velocity), 36-38 (earth's axis), dedication (sign of God's anger). On Roggeveen see: Davids (1985) 163, 294-295, 400.
[164] Roggeveen (1665) 30-34.
[165] Makreel (1671) 22: *Op ons ghesicht| doen alle lichten des Hemels...*

system: 'We shall here suppose a fixed earth and a moving sun, and thereafter (for the time being) regulate our arguments.'[166] Van Nierop's (and his followers') active propagation of Copernicanism was definitely founded in his acceptance of Descartes' physical principles. So, clearly, the correlation between Cartesianism and professed Copernicanism held true outside academia as well.

Popular discussions on Copernicanism

By the second half of the seventeenth century, a large part of the population was aware, be it only dimly, of the new horizons opened up by the developments in contemporary science and philosophy. Copernican ideas must have been widely known. In particular the correspondence of Dirk Rembrandtsz, who remained very much a man of the people, offers us a glimpse of the eagerness with which the Copernican issue was followed and discussed by both the educated and the uneducated. True, many letters are devoted to other fields of mathematics – gnomonics, chronology, cartography and geometry. Still, they do testify to a lively interest in cosmography. A certain Jan Hendrikken, otherwise unknown, preparing to leave for the East Indies, wrote to Dirk asking for an explanation of a passage in the astronomical work of Philips Lansbergen.[167] Another correspondent is a certain Balthasar Adriaansz, living in the village of Graft, addressed as a 'lens grinder'. This probably indicates a hobby rather than a trade. Anyhow, he probably belonged to the artisan class. In the eighteenth century, his descendants appear to have made a living by making harpoons. (And probably by some other trades as well – economic specialisation had not gone that far in the Holland countryside.) Dirk gave Balthasar instructions for the observation of the phases of Venus; instructions which he claimed to have sent to other lens grinders and amateurs.[168] The phases of Venus were a powerful argument against the Ptolemaic system.

One of the men fascinated by the new astronomy was Jacob van Veen (1604-1674). He came from a distinguished family in Alkmaar, and was a member of the town government and dike-reeve of the Heerhugowaard.[169]

---

[166] Makreel (1681).

[167] 8 Nov. 1660, *MML* 4 (1756) 527-528 (letter no. 20).

[168] DRvN to Baltsar Adriaansz, 13 Jan. 1678. *MML* 8 (1760) 299-300. The letter was recuperated from a certain Baltsar Smit, harpoon maker at Graft. An English translation is printed in Vermij (1996)b 68.

[169] This information was kindly provided by Mrs Wijnekus from the Noord-Kennemerland regional records office.

Initially, he seems to have been hesitant about the Copernican system. In 1658, he sent Van Nierop a question in verse form, asking him how it was – if one put the sun in the centre and had the earth turn around it – that the earth did not deviate to the north or south. Dirk answered with a little poem which stressed the smallness of the universe for God, and for the rest referred Van Veen to his 'Dutch astronomy'.[170] A later correspondence between Van Veen and Van Nierop, in 1665, occurred after the appearance of a comet. Van Veen held the view that two comets had appeared successively, while Dirk thought it was one and the same.[171]

Van Veen's enthusiasm carried him somewhat further, however, than observing comets and discussing matters with mathematicians. In 1670, he wrote a short treatise on cosmology, *De loopende werelt* ('The moving world'). It had the express purpose of demonstrating that the earth is moving, not the sun. To this purpose, Van Veen explained the courses of all planets, illustrated with large drawings. *De loopende werelt* was never printed. It is not clear whether it was ever intended to be printed. The fact that it has a dedication (to the governor of Dutch Ceylon) underscores the point that it was to be made public in some way. But as things turned out, it circulated in manuscript. Several manuscripts are known today. They are all neatly written exhibition pieces. In many cases, the illustrations have been coloured. For the title pages, even gold paste has been used.

Interestingly, the manuscript was added to over the course of time. One copy, preserved in the Amsterdam Maritime Museum, is undated. This version seems to be closest to the original as written by Van Veen around 1670 (the date of the dedication). A second one, from the estate of Petrus van Musschenbroek and now preserved at Leiden University, is largely identical to the first. It is dated 1670 and the title page bears the author's name. However, at the end of the text, a second scribe has added a few more pages; some of these concern astronomical problems, while others refute religiously-inspired objections against Copernicanism. These additions appear not to be by Van Veen, but by some seventeenth-century owner of the manuscript. A third copy, now in the Boerhaave Museum, has had a substantial part added to it, so that its comprehensiveness has been roughly doubled. The title has been changed to *Astronomische raadtselen* ('Astronomical riddles') and the title page bears the date 1673. It starts with a new introduction, followed by descriptions of the various celestial bodies: comets, the stars and the individual

---

170 Van Veen to DRvN, no date; DRvN to van Veen, 1658 July 12. *MML*, 3 (1755) 209-210 (letters nos. 5, 6).
171 Van Veen to DRvN, 19 Feb. 1665; answer 3 March 1665. *MML* 6 (1758) 99-102, 126-130 (letters nos. 35, 36).

planets, including the sun and the earth. Only then follows the text of the former manuscripts, starting with the dedication. There is no evidence that the additional part is also by Van Veen; his name does not appear on the title page. Still, this appears most probable. In any case, it has clearly been copied by the same hand as the earlier copies, and in the same careful way. The fact that it was worthwhile making copies and even adding to them shows that there was a demand for such texts.

It deserves notice that Van Veen's arguments remain astronomical and that there is no reference to physical principles, Cartesian or otherwise. The part on comets in *Astronomische raadtselen* contains some physical speculation on their nature. The author ponders whether they are new creatures – born from the corruption of the air – or whether they have been around since the beginning. He concludes that these things are hidden from us. Van Veen was probably mainly inspired by Lansbergen. The title page of the former's manuscript bears a vignette from the latter's works.

However, the arguments of Dirk and other mathematical practitioners clearly reached a wider audience. An interesting example is the religious meetings held by Willem Deurhof, an Amsterdam amateur theologian and philosopher (by trade, he was a manufacturer of trunks, cases and such like) who founded a sect with antinomian leanings. 'Deurhofianism' had its starting point in the philosophy of Descartes and was influenced by Spinoza. [172] It was generally regarded as utterly heretical by the Reformed, if not as atheism. Between July 1708 and February 1717, Deurhof held at his home a series of 394 sermons on the Book of Job. Many years after his death, his followers decided to publish them. Because of resistance from the Reformed Church, the edition was incomplete. It covers only nine of the chapters of Job, out of a total of 42. From the parts that were published, however, it appears that Deurhof did not just discuss spiritual truths, but at times also went into physical matters (the Book of Job provides ample opportunities to do this). He even performed some simple experiments, for instance on the constitution of the air. In the main, his world-view is rather Cartesian.

Two sermons are devoted to a rather elaborated discussion of the world system, based on Job 9:7. His aim here is to demonstrate that the earth's motion is not contradicted by Scripture. [173] But in another sermon, related to Job 9:9, Deurhof discusses also extensively the causes of the earth's rotation. He mentions the opinions of Descartes, Van Nierop, Van der Moolen and Hartsoeker, but in the end he defends an opinion of his own. In his view, the

[172] The literature on Deurhoff is very scant. For an introduction, see Fix (1990). See also Evenhuis III, 348-349.

[173] Deurhoff (1741) II, 130-172.

celestial vortex around the sun varies in density. It therefore flows over the earth in an asymmetrical way and affects one side more than the other. This causes the rotation of the earth. Deurhof rejects Descartes' view that the space between the stars is completely filled with vortices. The sun's vortex does not extend much beyond the orb of Saturn and has the shape of a disc rather than a globe. There is no reason why God would have made it larger, as a larger extension would be superfluous. [174]

Now, not all popular authors who discussed the motion of the earth regarded it favourably. The biblicist criticism of Copernicanism appears to have made some headway outside academia as well. Jan Verqualje is a case in point. In 1657 and 1658 he obtained letters patent from the States of Holland and the States-General, respectively, for the construction of a new kind of watermill. [175] In 1661, he explained that he had designed these mills to drain the water from the Amsterdam canals. He asserted that building these mills would have been a remunerative business, had not one of his employees deceived him and established a business of his own based on Verqualje's invention. [176] After his attempt in this field had thus come to nought, he turned to the problem of establishing longitude at sea. In 1659 he presented his solution to the United East Indian Company, the States-General and the Admiralty of Amsterdam. [177] The request called him a diamond cutter. He also published his invention in a book, which came out in 1661. Verqualje apparently was a man of much wit, but was not very successful in his undertakings; his ambitious attempts to explain the riddles of nature may partly have served to compensate for this. His speculations reveal that mixture of ingenuity and naiveté which characterise the zealous but unschooled amateur. Proudly, he announced that he had come up with his inventions not by reading books, but as a result of his experiences. In fact, his education appears to have been very limited, not only in mathematics and philosophy, but also in writing. His works are very confusing and often difficult to follow.

His book on the invention of 'finding longitude' dealt with some other subjects as well, including the question of the motion of the earth. Verqualje defends the traditional view that the earth is resting and the sun moving. His arguments are partly traditional, partly hard to follow. His main argument, however, appears to be purely religious: 'The scholars or other people may try to find, but their guesses may easily miss. I myself believe Holy Scripture; for

---

[174] Deurhoff (1741) II, 189-212.
[175] Doorman (1940) 225, 293 (nos. G.453 and H.76).
[176] Verqualje (1661), preface.
[177] Davids (1985) 130.

those writers were moved and commanded by the Holy Ghost when they wrote.' [178]

Another 'east and west finder', indeed one of the most notorious of the century, was Lieuwe Willemsz Graaf. Largely forgotten by now, his ideas were widely discussed among the navigators and practitioners of seventeenth-century Holland. His celebrity appears to have been largely based on the self-conscious way he announced (but never fully disclosed) his ideas, and the contempt with which he dismissed anybody else's idea. He made many enemies over the course of time and was at the centre of many controversies. In 1696, he published a small pamphlet: 'Short, general, and modest refutation of the writings, tempers, thoughts, hypotheses and understandings of many astronomers, who have tried for a considerable time (and still try today) to obscure the light of God's clear and true Word by the uncertain and human art, in particular in the rest of the sun and motion of the earth.' [179] The title sufficiently indicates the tenor of the work. In the same year (1696), Graaf also gave vent to his anti-Copernicanism in an almanac he had calculated. In it, he dismissed the Copernican theory as unproven and improbable, referring for further details to the work just mentioned. [180]

Graaf's book results from the larger project of finding east and west by an overall reformation of astronomy. As he explains, this project had been attacked by envious people, so he had decided to write a short refutation of their main thesis, the motion of the earth; 'the more, as it not so much concerns the art, as the wisdom and trustworthiness of God, and therefore can be done without disclosing any foundation of secret knowledge'. [181] The argument of the work is purely biblical. As for his opponents, Graaf simply states that they blindly follow Copernicus and Lansbergen, even if by now gross errors have been shown in their work. Graaf discusses the main biblical passages used to support the motion of the sun and the rest of the earth, and emphasises that it is fully unwarranted that some people try to explain these passages in a figurative way. In the end, he gave what he called a hint of the hidden knowledge he planned to disclose in a major work. He stated that he knew exactly the positions of the sun and the moon at the time of Joshua, and could prove that they were indeed above Gibeon and the valley of Ajalon, respectively, where Joshua ordered them to halt (Joshua 10:12). As he explained, Joshua did not order the halt because he needed daylight; had he needed light, he would simply have asked for a fiery column. Joshua wanted

[178] Verqualje (1661) 19-29, quote on p. 28.
[179] Graaf (1696).
[180] Salman (1999) 91.
[181] Graaf (1696) 4.

to suspend the impending new moon, as the obligation to celebrate it would force him to discontinue the battle.

A person who deserves some credit from historians of science is Hendrik Stevin, son of the mathematician Simon Stevin. Although not an active scientist himself, he devoted a lot of time to propagating the scientific reputation of his father, and published a fair part of the latter's manuscripts. In 1667 he published a work, 'Mathematical philosophical practice', containing some leftovers from earlier publications, along with some considerations of his own. One of these concerned the Copernican system. His father, as has been discussed above, had been an ardent supporter of Copernicanism. Generally speaking, Hendrik was the propagator of his father's ideas; but on this point, he refused to follow and decidedly rejected the motion of the earth. 'So, thou Nicolaus Copernicus! Thou Simon Stevin! Thou Philippus Lansbergius! Thou Renatus Descartes! Wherever you are: so, I say, the earth is not a planet, erring in the heavens. But you on this point have been erring and seducing on earth.' He left no doubt as to his main motive: the motion of the earth was contrary to Scripture. He spoke some harsh words against the scholars who regarded it as an established theory: 'Everybody speaking against it is reckoned by those scholars as someone who, because of his feeble judgement, cannot understand those mysteries, as a layman and as a cipher.' [182]

Having said that, he still felt urged to refute the Copernican system by natural reasons. He developed one single argument, taken from the motion of projectiles. In fact, his objection is not really stupid. The Copernicans asserted that all motion on earth is relative to the earth. The movement of projectiles with respect to the earth is the same whether or not the earth is moving. Hendrik accepted this as long as the earth was moved by a simple (rectilinear) translation. But he maintained that the argument did not hold if the earth had a rotation or a curvilinear movement. He surely had a point here. The argument is not developed mathematically, nor does he display any real insight into the role of inertia in those more complicated cases. But he was one of the very few philosophers who asked these questions in earnest. Most Copernicans (apart from Huygens) had never thought about them. Stevin was quite right to put the problem to them. [183]

Hendrik Stevin's work met with little response. As he recounts, he went to Leiden to consult the experts but they were apparently not very interested in his ideas. According to Van Nierop, 'the writings of this Stevin are considered

---

[182] H. Stevin (1667), general appendix, 46-57.

[183] Ibid. See also general appendix, 12-13, and the appendix to book v, 59-61. The figure to which Stevin is referring in the text is missing. The book's figures have been published separately (Hendric Stevin, 1663), but this volume does not contain it either.

here of no value at all, but rejected as mere caprices, without being remembered by anybody.'[184] Van Nierop included a refutation of them (together with a refutation of the earlier pamphlet by Cardinael) in his 1677 appendix to his 1661 book on the motion of the earth.[185] Still, in at least one case Hendrik Stevin's work had some impact. In 1681, Dirk Rembrandtsz van Nierop received a letter from Johannes Droomers, controller at the Royal Audit Office in Bruges, in the Spanish Netherlands. Today, he is remembered as a playwright. As Droomers explained, he had become a Copernican upon reading Lansbergen's books. But now, he had also read Dirk's works, among them the refutation of the anti-Copernican stance of Hendrik Stevin. He was not convinced by Dirk's refutation and instead felt that Stevin really had a point. (Droomers' interpretation of Stevin appears highly personal, but of course, he had not seen the original work). So, he had become a geocentrist again. This resulted in an exchange of some rather prolix letters between the two. The by now aged Dirk in the end found it difficult to follow Droomers' arguments. He asked the advice of a former pupil of his, Pieter Maasz Smit, then living in Blokzijl in Overijssel. Smit sent an extensive reply, based on Cartesian physics. Dirk thereupon urged him to write to Droomers himself as well. It is not known whether Smit did so.[186]

Apart from people simply arguing for or against the motion of the earth, there are those who frame their own cosmography. One of the more curious contributions to the discussion on Copernicanism came from Jacob Coccaeus. Coccaeus was an educated man. He had studied at Leiden, and had matriculated at the age of 18 in the faculty of arts as Jacobus Cochaeus Hanoviensis on 3 March 1633. He does not seem to have taken a degree. Still, he obtained the post of conrector of the Latin school at Haarlem. He resigned this position in the summer of 1662.[187] He appears to have died some ten years later, as in 1672 his library was auctioned off at Haarlem.[188] The following year, his *Epistolae duae posthumae circa nobilissima Cartesii meditationes de philosophia prima*

[184] Van Nierop to Droomers, [1682 Jan 30], *MML* 8 (1760) 25.
[185] Van Nierop (1677)b 7-11.
[186] Droomers to Van Nierop, 20 Oct. 1681, *MML* 7 (1759), 511-515; Van Nierop to Droomers, [30 Jan. 1682], *MML* 8 (1760), 24-26; D to vN, 21 March 1682, *MML* 8 (1760), 245-256; vN to D, *MML* 9 (1761), 41-46; D to vN, 25 April 1682, *MML* 9 (1761), 265-276; vN to D, n.d., *MML* 9 (1761), 412-418. Smit to Van Nierop, 25 July 1682 (st.vet.) and 12 Sept., *MML* 10 (1762), 84-94, 130-131. Van Nierop to Smit, n.d., *MML* 10 (1762), 164-166. I owe gratitude to Dr N. Geirnaert, keeper of the records of the town of Bruges, for providing me with additional information on Droomers.
[187] This information was most kindly recovered for me at the Haarlem municipal archives by G. Dorren (Amsterdam), who is preparing a thesis on the inhabitants of Haarlem in the seventeenth century.
[188] *Catalogus* (1672).

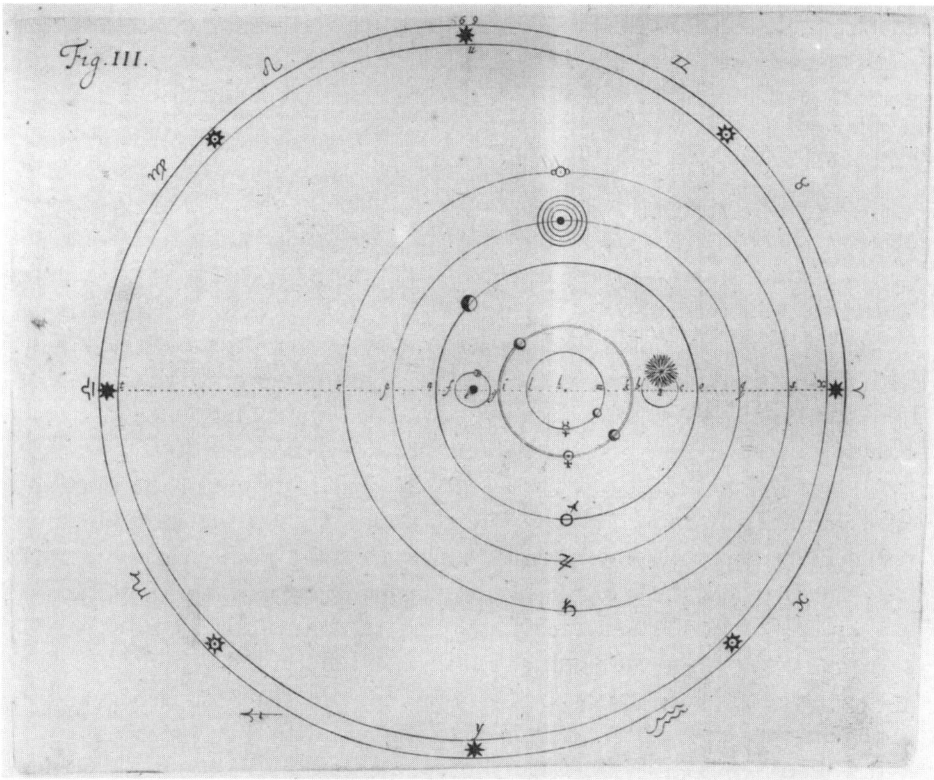

Figure 5: The system of the world according to Jacobus Coccaeus, *Epistola de mundis, quae circumferuntur, systematis*, 1660. The earth is immobile at the centre of the universe. The sun is moving around the point *a*; the point *b* is the centre of the orbits of the planets. (Leiden university library.)

was published in Brussels.[189] The connection with the southern Netherlands seems to indicate Catholic sympathies. Indeed, his library contained several edifying works of Catholic signature.[190] His retreat of 1662 may partly be due to his Catholicism, but this is just a guess.

A glance at Coccaeus' library gives the impression of stunning erudition. Mathematics and astronomy appear to have been his main field of interest. In astronomy, Coccaeus owned all the main works. The collection of philosophical works is impressive, too. Partly, these are the main works on Cartesian philosophy, but another part consists of works of Hermetic or Paracelsist

[189] Mentioned in Jöcher *Allgemeines Gelehrten-Lexicon*. I have not been able to find a copy.

[190] *Catalogus* (1672) nos. (folio) 59, 63, 75, 77. These are among the very few works in the vernacular.

character. Nearly all of these books are in Latin. The catalogue also mentions some instruments, such as a pair of globes, two telescopes, two microscopes, magnets, and a number of Latin, Arabic and Hebrew manuscripts. This was the library of a serious scholar and leaves one unprepared for the intellectual quixotry Coccaeus displayed in actual life.

In 1660, Coccaeus published in Amsterdam an *Epistola de mundi, quae circumferuntur, systematis et novo alio illis certiore dialogismum paradoxum complexa* ('Letter on the current systems of the world and another new system which is more certain than the former, comprehending a paradoxical dialogism').[191] The letter is in the form of a dialogue. The author recounts a discussion he has had with two fictitious characters, Caelus – a Copernican – and Polus, who upholds traditional astronomical views. Coccaeus refutes them both and advocates a completely new system. According to him, the earth is the centre of the sphere of the fixed stars. He accepts the daily rotation of the earth, but rejects an annual motion. The moon is moving around the earth, as in any system. The sun is moving in small circles around a point at some distance from the earth. So, the sun does not turn around the earth, nor does the earth turn around the sun. The planets, finally, are moving around a point halfway between the earth and the point around which the sun is turning. They do so in their usual order. Mercury and Venus are moving between the sun and the earth, and the other planets comprehend the two with their orbits.

It will not be necessary to follow the arguments of the author in detail, but a few points should be mentioned. The Copernicans, Coccaeus says, claim that their system is simpler than Ptolemy's, but in practice they, too, need all kinds of complicated constructions to save the appearances. It has come so far that some authors have imagined that the planets do not move in circles, but in ellipses.[192] It is really not worthwhile, in order to correct the old anomalies, to displace the earth from the centre, just to create new anomalies which are as intolerable as the former ones. Moreover, Coccaeus rather elaborates upon the phases of the planets. The phases of Venus and Mercury were regarded as decisive arguments that at least these two planets turned around the sun. However, Coccaeus flatly denies that such phases do in fact occur. To prove this, he refers to some observations of Venus by the Polish astronomer Hevelius, which are not in accordance with Copernican theory. Coccaeus concludes that Venus has light of its own, but only on one side; its other side is dark.[193]

It is striking that in the booklet there are few or no traces of Cartesian philosophy, which was to be so well represented in Coccaeus' library ten years

[191] Coccaeus (1660)a. I apologise for referring to its Dutch translation, Coccaeus (1660)b.
[192] Coccaeus (1660)b 24.
[193] Coccaeus (1660)b 22, 29-30.

later. There is no mention of vortices or even of celestial physics in general. Coccaeus briefly mentions Descartes, but only in order to refute him. He fully applauds it if people want to imitate Descartes. But those Cartesians are ridiculous, 'who, having rejected the yoke of the common philosophers, make themselves the slaves of this one man in such a way that they try to draw all wisdom from him only.' [194] One should realise, of course, that Coccaeus' had had his education before Cartesianism had entered the scene. His own system looks rather old-fashioned: the exclusion of all non-circular motions, the careful symmetry (the sun's orbit mirrors that of the moon), the universe closed by the sphere of the fixed stars. Moreover, Coccaeus calls his system the system of Moses. [195] He may well have drawn his main inspiration from his Hermetic readings. His colleague from Hoorn, Andreas Cellarius, who was equally well-read in alchemical and Hermetic literature, later suggested that the centre of the universe (which in Coccaeus' scheme would be the centre of planetary motion) was the place of neither the sun nor the earth, but the source of divine light. In Coccaeus' world system, this indeed would be possible. [196]

Coccaeus' work, funny as it may seem to us, was discussed by many people, ranging from the commoners in Haarlem to Christiaan Huygens. Dirk Rembrandtz, as the foremost proponent of Copernicanism in northern Holland, clearly felt that it was his task to put things right. At the same time, he seems to have felt somewhat uneasy about Coccaeus' argument from the phases of Venus and Mercury as observed by Hevelius, for these observations were indeed incompatible with the Copernican system. Accordingly, he turned to a greater authority for advice – Christiaan Huygens. The letter is lost, but there is a reply from Huygens, dated April 1660. Huygens approved of Dirk's intention to publish a refutation of Coccaeus' book, adding that it probably would not be very difficult to compose. As to Hevelius' observations of Mercury, he emphasised that they were clearly wrong, as they were not in agreement with the Copernican hypothesis. At about the same time, Huygens wrote to Boulliau about these observations by Hevelius, stating that initially he had not remarked their incorrectness, but that he had been alerted by Coccaeus' work. He called Coccaeus' system 'the most absurd one you have ever seen. His book on the subject has been printed and I am eager for you to read it, in order that you may laugh.' [197]

---

[194] Coccaeus (1660)b 33.

[195] Coccaeus (1660)b 37-40.

[196] On Cellarius see below, p 235–237.

[197] Huygens to Van Nierop, 27 April 1660, *OC*, III, 73-75 (no. 747); this letter is also printed in *MML* 4 (1756) 449-451 (letter no. 18). Huygens to Boulliau, 22 April 1660. *OC*, III, 66 (no. 743). See also *ibidem*, 95.

The matter was urged upon Dirk still more when, later that year, a Dutch translation of Coccaeus' book appeared. The translation is anonymous, but appears to have been made by Johan Kies van Wissen, a wealthy citizen of Haarlem and a member of a prominent Catholic family. In December 1660, Dirk Rembrandtsz received a letter on the issue from some inhabitants of Haarlem. They had been studying the mathematical arts for a long time, they wrote, and had finally accepted the Copernican system. But now, the book by Coccaeus, as well as some discussions with its translator Kies van Wissen, who apparently was zealously propagating the new system, confused them. They asked Dirk, as the local oracle on astronomy, what to think about it.[198]

Dirk took the matter to heart. A short refutation of Coccaeus' book was appended to his work on the motion of the earth, which happened to be in press. Later in 1661 he published a full volume on Coccaeus' system.[199] In writing this, he had the support of a young student at Leiden, Abraham Boddens. Boddens had scientific leanings, although it cannot be established whether he had any real talent. His career was ended prematurely by his death in 1661, at the age of 20. In his Leiden years, he had tried to gain access to the republic of letters by offering his services to the established men of learning. He had been introduced to Huygens in 1658, and since then had made himself useful by presenting him with books (an almanac by Van Nierop, among other things) and acting as a go-between in Huygens' contacts with Hevelius.[200] Boddens tried to make himself useful to Hevelius as well. He sent him the book by Coccaeus and asked his opinion of it, pointing out that Coccaeus argued his case by referring to Hevelius' observations of the phases of the planets. Hevelius sent him a long reply, wherein he explained that his observations were as accurate as could be, but not accurate enough to reject the whole system of Copernicus on this sole base. Boddens sent a translation of this letter to Van Nierop at the beginning of 1661.[201] When Van Nierop published his extensive refutation of Coccaeus, he eagerly used Boddens' material. He quoted extensively from the letter from Hevelius, as well as from some other writings, which a learned friend had provided him with. Maybe this friend was Boddens, but this cannot be proved.

---

[198] Jan Harmensz and Jacob van den Velt to DRvN, 22 Dec. 1660. *MML* 5 (1759) 38-39 (letter no. 21). An English translation of this letter in Vermij (1996)b 64-65.
[199] Van Nierop (1661)a 76-79; Van Nierop (1661)b.
[200] Cf. *OC*, II, nos. 501 (p.198), 504, 508, 644; III, 713, 714 (p.16). See also Vermij (1996)b 64.
[201] See for these letters: Vermij (1996)b 65-67.

Surpassing Descartes

At the universities, several stages of Cartesianism can be noted. Those who had had their education around 1650, when Cartesianism was novel, tended to emphasise the superiority of the theories of Descartes and closely follow in the footsteps of the master. Later generations developed a more detached view of his theories and tended to criticise or elaborate upon them. Outside academia, 'Cartesianism' was much more chaotic. Still, a similar development seems to have taken place.

One interesting point is that by the end of the century, people started elaborating upon Descartes' cosmological theories, in particular his theory of vortices. People framed theories about the cause of the earth's rotation, the distances between the planets and other points not dealt with by Descartes. We met something of the kind in Van der Moolen, and to a certain degree in Deurhof. The physician Cornelis Bontekoe, one of the main propagators of Cartesian theories in the Dutch Republic, explained that the system of earth, air and moon was kept in motion by the larger vortex of the sun. In his view, all celestial vortices geared into each other like cogwheels and thus were moved by each other. [202] All this was mere speculation. Such theorising renders it understandable that in the eighteenth century, experimental philosophers could call Cartesianism a bundle of mere chimeras.

Nevertheless, at the end of the century it was fairly common to have some dissatisfaction with Descartes' ideas. I am speaking here not of mere traditionalists, who were hostile to Descartes anyhow, but of people who regarded themselves as eager proponents of the new philosophy and the new physics. Philosophical dissenters had earlier stated that Descartes had done many useful things, but that people should not slavishly follow him. Coccaeus had said this, as noted before. Hendrik Stevin said something of the kind. [203] But by the end of the century, this became something of a commonplace. People came to feel that Descartes' theories had served their end in cleansing the Augean stables of scholastic philosophy, but had now become something of a stumbling block.

The general feeling was most eloquently worded by Janus Montanus, whose burlesque poem I quoted earlier. After Descartes has conquered the armies of Aristotle in a big battle, Montanus points out that his overlordship soon became as oppressive as that of Aristotle. He closes with the promise of a new hero who will come to vanquish Descartes in his turn, and lay the

---

[202] Bontekoe (1689) vol. I part II, 342-343. Something similar had been said by Roggeveen (1665) 37.

[203] H. Stevin (1667) preface; see also general appendix, 3-4.

foundations of a truly sound philosophy. Undoubtedly his hints concern some real person, but, regarding the overall character of his work, he may be poking fun at him rather than being serious.[204] Still, there can be little doubt that some people were all too ready to act as such a 'hero'.

Towards the end of the century people tried to emulate Descartes' ideas – and some did so with amazing self-assurance. Like Descartes before them, they pose as the prophets of science. So far, man has erred in darkness and has not been able to find the key to true knowledge; but they, they will discover the secrets of the universe. In practice, recognising the need of a philosophical alternative proved simpler than supplying one. More often than not, the would-be innovators remained completely under the spell of Descartes. Actually, none of their solutions has had any impact on the course of science or philosophy. They deserve to be mentioned here only because they offer an insight into the actuality of the cosmological debate at the time.

A philosopher who illustrates this tendency is Nicolaas Hartsoeker.[205] Hartsoeker was a scientist of international renown, even though he never taught at a university. Although for a great part of his life he lived outside the Netherlands and published most of his work in French, he deserves to be mentioned here. His *Principes de physique* appeared, in French, in 1696. A Dutch translation was published four years later by Jan ten Hoorn. In this work, Hartsoeker designed a new system of natural philosophy. Although he did so clearly because of dissatisfaction with the dominant Cartesian philosophy, his own system followed Cartesian lines. He started with the laws of motion, deduced the rules of percussion, and went on to the more complicated systems that actually do occur in nature. He dealt with gravity, the origin of the sun and the stars, the loadstone, and meteorological phenomena. On points of detail, he corrected Descartes' earlier assumptions, but the overall framework is clearly Cartesian. Apparently, his mind was not able to reach beyond Descartes' basic tenets. Certainly he had little interest in the new experimental philosophy or in Newtonianism. In fact, he was one of Newton's fiercest opponents on the continent.[206] His innovations follow rationalist rather than empirical lines.

Hartsoeker did not doubt the truth of the Copernican system for one moment. He did, however, add some new elements of his own. Hartsoeker was a physicist, not a mathematician. His most famous accomplishments were in the field of microscopy. Turning to cosmology, he dealt with the field without having much recourse to astronomical theory. He limited himself to speculations on the way the solar system had come into being and how it would

204 Montanus (1701) 51-52, 76-77. He may mean Dirk Santvoort, on whom below.
205 On Hartsoeker, see Wielema (1992).
206 Cf. Berkvens-Stevelinck (1975).

develop. He supposed that the motion of the earth was caused by the light of the sun. In principle, the earth was in equilibrium with the celestial matter, but the light of the sun pushed it backward and sideways. As there was nearly no resistance, this resulted in a eccentric orbit around the sun. The earth's motion around the sun caused a vortex of celestial matter, which then swept all the planets along in its movement around the sun. The diurnal rotation of the earth is caused by the motion of the moon, which causes a vortex of the celestial matter around the earth. The moon was formerly turning around the sun, but it has been caught by the earth by means of a mechanism that does not appear very clear. Study of the orbits of the planets will shed light on the local properties of the celestial matter. So, in the end, Hartsoeker's ambitions led him to little more (as far as cosmography is concerned) than the kind of speculations the eighteenth century would regard as Cartesian excesses.[207]

The desire to abandon Descartes was found not only among the learned. Probably the most curious would-be scientific figure of this period is Dirk Santvoort, an ironmonger from Amsterdam. He must have been rather well-to-do, but he does not seem to have read any language besides his native Dutch. In 1699, he published in Utrecht, where he was living at the time, a book called 'The cause of the motion of the world'. A much extended edition appeared in 1703 as 'The cause of motion and the principles of solid bodies', and a sequel appeared in 1707.[208] As he explained, until then philosophers had merely quarrelled over the question which of the universe's great bodies (i.e. the sun and the earth) was moving and which was at rest. They had not investigated the causes of such motion. This was exactly what he himself proposed to do. With the causes known, all remaining questions would be easy to solve.[209]

Rejecting both Aristotle and Descartes, Santvoort undertakes to develop a wholly new system of natural philosophy. The main source of inspiration for Santvoort's philosophy (if one may call it that) is clearly Cartesianism. Santvoort starts to exposit the general laws of motion which govern the universe. His propositions are indeed put forward in a similar way as Descartes' rules, but as to content, they are quite dissimilar. In Santvoort's view, the universe is governed by attractive and repulsive forces. A hot and a cold body attract each other, while hot bodies repulse each other, as do cold bodies.[210]

These principles serve as (among many other things) foundations for a kind

---

[207] Hartsoeker (1696) 125-144 (Chapter VII: *Du mouvement de la terre & des planetes*).

[208] Santvoort (1703), (1707). There also appeared a Latin translation: Theodorus Santvoort, *Dissertatio de causa motus et principiis solidorum corporum* (Utrecht 1704). There are some slight adaptations as compared to the Dutch text, which I follow here. On Santvoort: Vermij (2001).

[209] Santvoort (1703) 2.

[210] Santvoort (1703) 3-22.

of cosmogony. This part, too, is clearly inspired by Descartes' theories on the subject, published in *Principia philosophiae*. Santvoort starts from the hypothesis that once, there were no solid bodies in the universe. This hypothesis was counterfactual, as he asserted at a later occasion. The universe had existed from eternity, filled with worlds which came into being and disintegrated according to the mechanism which he was to explain. But in order to elucidate this mechanism more clearly, he supposed that there were no bodies at all to start with. Under those circumstances, the world was filled with 'a saline, thick, airy fluid'.[211] Hot saline particles coagulated and formed the seeds of the solid bodies. They attracted a layer of cold material; this cold surface then attracted a new layer of hot material; and so the bodies grew, until so much material had been used that the air in the universe became so thin that no further material could be extracted.[212]

The sun is one of these solid bodies. Because it became overheated, it caught fire. As hot bodies attract cold material (and vice versa), the sun started heading towards a large cold body: the earth. Under each other's influence, the two bodies started moving on their axes. By doing so, each created a vortex around itself. The combination of the two vortices caused air to amass between the two bodies, which drove them away from each other again. At the same time, the mass of air in between acted as a kind of lens, by which means even at a large distance the sun warmed the earth, and the earth cooled the sun. The larger the distance becomes, the larger this airy lens grows, so that by this means sun and earth 'remain so to speak interminably and infrangibly connected to each other'.[213] The sun and the earth have started moving around the airy lens. So, Copernicus, Ptolemy and Tycho Brahe were wrong. Neither the earth nor the sun is turning around the other: they are both moving around the central lens.[214]

As for the other planets, Santvoort does not pay much attention to them. He thinks they are mostly small, hot objects which have been trapped in the vortex around the bigger bodies. If they were cold, a planet like Venus would show phases like the moon – which, according to Santvoort, is not the case. Comets are in fact the outer planets of similar systems nearby, which occasionally become visible when they are closest to us.[215]

Santvoort draws his inspiration not only from Descartes. His work also betrays the influence of such authors as Hobbes, Lucretius and Spinoza.

---

211 Santvoort (1703) 261.

212 Santvoort (1703) 254-300. Santvoort (1707) postface.

213 Santvoort (1703) 333.

214 Santvoort (1703) 300-392.

215 Santvoort (1703) 392-408; 384 (phases of Venus).

The whole thing is a strange hotchpotch of ideas and arguments, but they have some common ground. As a whole, Santvoort's world is a decidedly materialistic one. This becomes particularly clear at some other places in the book, where Santvoort explains the principle of life and of human and animal intellect (no real difference, according to him) according to these same principles of attractive and repulsive forces. Indeed, Santvoort seems to have been rather hostile towards established Christianity. Another book, in which Santvoort professed his aversion to clericalism more openly, was banned in Utrecht in 1709. 'The cause of motion' escaped this fate, probably because it contained no direct references to the Bible or to theological dogmas.

Hardly any scholar ever paid attention to Santvoort's amazing theory. But it did draw some attention in the world of philosophical debating clubs in the Dutch Republic, where Santvoort himself probably was quite active. So it happened that Santvoort's book got an extensive rejoinder from a young Amsterdam office clerk, Hendrik Wyermars, who published in 1710 a book entitled 'Chaos imagined'.[216] He objected mainly to Santvoort's idea of an eternal universe of worlds arising and decaying. According to Wyermars, our world as we know it today has existed from eternity. Wyermars was clearly not very interested in empirical reality. Interest in those circles was often religiously oriented, although not necessarily in a way the Church would welcome. Wyermars drew his arguments mainly from what he considered a correct insight into the transcendent and immanent operating of God. In this metaphysical exercise, which is clearly influenced by Spinoza, he did not abstain from theological issues and spoke out rather frankly on the Bible. His heterodox ideas were considered so dangerous that he was sentenced to 15 years of imprisonment by the Amsterdam court of justice, an unusually severe punishment in the relatively tolerant atmosphere of the Dutch Republic.

Despite Wyermars' critical attitude to Santvoort's work, he still found much commendable in it. 'Especially his understanding of the motion of sun and earth is, in my view, better than anybody's I have read until now. The hypothesis of Copernicus is in my view too much in favour of the sun, and those of Ptolemy and Tycho Brahe in favour of the earth. This man very nicely puts them on an equal footing; as it is a foolish imagination that the one would have come into being because of the other.'[217] It was the philosophical consideration that nature worked uniformly and without preference which undermined the Aristotelian cosmos and contributed to the acceptance

[216] Wyermars (1710). On Wyermars, see the tendentious but still usefull article by Vandenbossche (1974).
[217] Wyermars (1710) 123.

of Copernicanism. But without further astronomical knowledge, the same consideration could be mobilised against the Copernican world as well.

The impact of the new astronomy: poetry

There was debate about the Copernican system among all layers of the population. But one might still doubt how widely the issue was known, and how it affected the general outlook of the common man. The opinions of the learned are relatively easy to assess. It is quite another question how far the general public was aware of the topic. People representative of the general public seldom put their thoughts on paper, and if they do, one may for that very reason doubt their representativeness. If we want to assess the opinions on Copernicanism among the general population, we have indirect evidence at best.

An interesting field of study concerns the way in which the system of the world, or the new science in general, is represented in art and literature.[218] This is a rather tricky business, however, if one's aim is to study how far such ideas were received as scientific theories. Modes of representation in art and literature are heavily dominated by conventions. Poets may refer to the system of Ptolemy because of its poetic associations, without bothering much about its scientific value. Still, the way they react may be revealing as to the significance of such theories to the non-specialist.

The three major Dutch poets of the seventeenth century are Joost van den Vondel, Pieter Cornelisz Hooft and Constantijn Huygens. All were born in the sixteenth century, but they can be said to have dominated the whole of the seventeenth century. In part, this was simply due to longevity: Vondel died in 1679 at the age of 92, and Huygens died in 1687 at the age of 91. All three were Renaissancers: they venerated ancient learning and tried, each in his own way, to incorporate the ideal of the humanist, the Renaissance man of letters. I agree that this makes this overview somewhat one-sided. It would have been interesting to know how a poet who grew up in the 1650s or later dealt with cosmography. The point is that I know no good instances of that.

Hooft came from the Dutch regent class. His father had been burgomaster of Amsterdam and he himself fulfilled several civic duties. Hooft was a sceptic in philosophy and a libertine in religion, an admirer of the Stoics and Epicureans among the ancients, and of Montaigne among the moderns. He was nearly exclusively interested in the world of man, in active, political life. Consequently, he paid little attention to cosmographical questions. Undoubtedly

[218] Work in this field includes: Korninger (1956), Tuzet (1965), Dauphiné (1983); see also Russell (1972) 224-227.

he was aware of contemporary discussions (he was related to Willem Jansz Blaeu), but his poetry and prose writings reflect very little of this. It has been suggested that a few isolated passages could be read as expressing sympathy for the Epicureans' and Bruno's idea of an infinite universe, but the evidence is not conclusive. Where Hooft uses cosmological images in his poetry, and he does so only seldom, he takes them from Ptolemaic astronomy.[219]

Vondel was quite a different man. His parents were Mennonites from the southern Netherlands, who had taken refuge in Holland. As such, Vondel could perform no public duties. He earned his living as a merchant, but took a passionate part in the various civil and religious disputes of his days. At the end of his life, he converted to Catholicism. Vondel embodies the 'baroque' in Dutch literature. He was particularly famous for his tragedies, many of which retold episodes from the Bible. Cosmic imagery recurs at several instances in his poetry. Vondel knew of Copernicus and his poems contain a few isolated references to him. But wherever he refers to the cosmos and the system of the world, his world-view appears Ptolemaic. He even used elements, like the heavenly spheres, which by then (the 1660s) were regarded even by advocates of Ptolemy as outdated. Quite evidently, Vondel is less interested in an exact physical description of the heavens than in the power of God which is behind it.[220]

For our purpose, Huygens is the most interesting person, and not just because he was the father of the scientist Christiaan Huygens. Constantijn Huygens was a man of many talents, active on the public stage as secretary to three successive stadholders, and well versed in the arts and sciences which became a man of the world. He wrote music as well as poetry. He entertained himself with optical experiments and befriended many men of science and letters, among them Descartes. In fact, he was one of Descartes' main protectors in the Dutch Republic. He seems to have felt a sincere admiration for the French philosopher and praised him several times in his poetry. If Hooft embodies the Renaissance man of letters and Vondel the baroque poet, then Huygens was the seventeenth-century *virtuoso*. His literary work is often of a didactic and somewhat intellectual character. Consequently, he not uncommonly speaks on scientific or scholarly topics.[221]

Huygens knew the Copernican system. He referred to it in his poetry on several occasions and, although he never spoke out openly, he apparently felt some sympathy for it. Still, when he came to speak on the cosmos as a whole, his images presuppose a Ptolemaic system. Some people take this as an ex-

[219] Verkuyl (1985).
[220] Boeles (1906) 1-33.
[221] Verkuyl (1987). Matthey (1973). See also Verkuyl (1965) and (1984).

pression of the difficulties a well-educated, seventeenth-century man had to overcome to accept the Copernican system. In this view, Huygens' use of geocentric images was not a deliberate stance, but rather quite unconscious. All seventeenth-century learning, after all, was based on Aristotelian philosophy. Speaking on cosmology, it was rather hard to find alternative terms, even if one disagreed with some technical details. As long as there was no clear alternative to Aristotelianism, its basic assumptions were likely to turn up even among adversaries.

There may be some truth in this, but then again, Huygens was not so much looking for scientific correctness as for poetic force. The Ptolemaic world had much more to offer in poetic imagery than the as yet unsystematic new cosmology. The old cosmos of order and harmony was closely linked with ideas on the divine and on the world generally. If we find Vondel or Huygens using outdated cosmic images, this does not so much indicate that it was hard for them to think of the universe in more modern terms, but that those ancient images still served some poetic purpose, regardless of how they were thought of in actual life. In a similar way, the occurrence of pagan deities in seventeenth-century poetry says little about the religious beliefs of the poets involved. Hooft frequently referred to omens in his literary work, whereas in reality he was very sceptical about their existence.

At the same time, it must be admitted that the new world-view apparently failed to serve such a poetic purpose. The poets were not sufficiently inspired by the new discoveries and theories to mould them into powerful poetic images (whether this explains the lack of contributions to the theme by younger generations is difficult to say). What these older poets might occasionally express was their bewilderment at the new ideas – which does not necessarily mean, of course, that they rejected them. This can be illustrated by a long consolatory poem Constantijn Huygens wrote in 1647 (*Ooghen-troost*; 'Eye's consolation') for a friend who had recently lost the use of one of her eyes. Huygens also presented Descartes with a copy. [222] Among other things, Huygens reviewed various classes of people who act as though they were blind, even if their eyes are quite healthy. One of these classes comprised scholars, people 'who think they see everything'. An adequate translation of Huygens' verse regrettably is beyond my ability. I have tried to be accurate as regards content. As to form, I only more or less kept to the length of the original lines, simply in order to remind you that, originally, it was poetry.

---

[222] *AT*, V, 648-649.

The scholars are all blind.
They have their books for eyes. They stumble as a child.
They feel they stand secure in all they do or say.
Five shout: That is the way, ten rise to disagree.
And truth is only one. [223]

Of course, the vanity of knowledge is a traditional theme. The reference to bookish learning was commonplace even among the new philosophers. What is less usual is that Huygens takes his examples exclusively from natural philosophy. After the introductory passage just quoted, he reviews several learned disputes, accompanied by the sober refrain 'And truth is only one'. Among these disputes are traditional bones of contention – such as perpetual motion or the finding of east and west – as well as more recent ones, such as the circulation of the blood or the general principles of things:

Here, things are heaps of globules that are swarming,
And thus cause moist and dry, created sun and stars.
There, this is a waking dream, and one should keep to say
That every creature acts according to its nature. [224]

The dispute on the system of the world figures rather prominently: it takes the first place. The dispute on the way the eye functions, which in the context of the poem as a whole might be deemed more relevant, is only second. Huygens states, put into plain prose: 'The heavens are not allowed to go round. By now, it is the turn of the earth. Quickly, tables will turn again. There the earth is standing still; the next moment, she again is like a piece of meat on the spit, roasted by the sun. And truth is only one.' [225] This sounds rather sceptical. But something more should be said about it, as these lines are only preserved in Huygens' manuscript. All printed versions are much shorter and read: 'Here, the sun is turning all round. There, it is the work of the earth, after a new and firmer find.' [226] This sounds as though Huygens, thinking it over, found the dispute on the system of the world not so completely hopeless as he had first suggested.

---

[223] Const. Huygens (1894) 114-115 (lines 875-879): *De letter-luij zijn blind,| En sien maer door haer Boeck: Sij struijcklen als een kind,| En meenen vast te gaen in all haer doen en seggen:| Vijf roepen, dat's de wegh, tien konnen't wederleggen,| En waarheid is maer een.*

[224] Ibid., 116 (lines 901-904): *Hier zijn de dinghen stoff van bollekens, die werren,| En maken voght en droogh, en schiepen sonn en sterren.| Daer is't een wackre droom en soo genoegh geseidt,| Dat ieder schepsel werckt naer syn verborgentheid.*

[225] Ibid., 115 (lines 881-884): *Den Hemel magh niet draeijen,| 'Tis nu de werelds beurt. flus sal de kans weer swaeijen;| Daer staet de wereld stil, strax werdt se weer Gebraed,| En light gelyk aen 'tspit daer 't sonne-vier om gaet.| En waerheit is maer een.*

[226] Ibid., 115 (footnote): *Hier loopt de Son in't ront,| Daer is't de Werelts werck by nieuwer fixer vont .*

On the whole, however, Huygens is not dealing with the various disputes from the comfortable point of view of a man who knows it is all vanity. His repeated sigh 'And truth is only one' strongly suggests that he really took to heart the prevailing uncertainty in natural philosophy. How far his case is representative is difficult to say. After all, Huygens was deeply committed to the values of late Renaissance culture, as it existed when he was a young man in the first half of the seventeenth century. Still, he may serve as a spokesman for those who may have been quite interested in the new science, but were left rather bewildered and disoriented once they came to realise its full consequences.

The testimony of world maps

Art and literature offer only very limited possibilities for assessing the general view on Copernicanism. Artists generally have preoccupations other than representing the knowledge of their times. However, this is not true in all cases. There is at least one field of applied art which strives, or pretends, to give an accurate and reliable picture of the world: map-making. In the seventeenth century, this art blossomed in the Netherlands, and particularly in Amsterdam. Part of the map production of course served purely practical ends, i.e. for travelling and navigation. But the large wall maps, the precious globes and the voluminous atlases were never intended to be taken out to sea. They were purchased by rich merchants and burghers, to satisfy their curiosity and to serve as status symbols which displayed both their owner's wealth and intellectual standing.

World maps, in particular the large wall maps, offered more than just an image of the earth. They were crammed with explanatory notes and images of the world's large cities, reigning monarchs, allegories of the four elements or the four parts of the world, episodes from biblical history and, in some cases, cosmographical emblems. Such decoration may be abundant especially on maps which show the earth in two hemispheres, a design which leaves a lot of empty space to be filled in some way or another. Historians of cartography tend to neglect such border decorations and concentrate on the map itself. It may be surmised, however, that for the original customers, a beautiful and impressive decoration was as important as an accurate reproduction of the earth. Most of them, after all, were not geographical experts. They bought the maps for intellectual orientation and to impress their friends.

In the seventeenth century, the Dutch Republic, more in particular Amsterdam, was the world's largest production centre of maps, globes and atlases. Although not all of this production was sold at home, rich Dutch citizens were among the map makers' most important customers, and they probably called the tune for the fashion in this field. Map makers suffered from high production costs and had to sell their wares. Moreover, they were locked in fierce competition with each other and could not afford to neglect the wishes and whims of their potential customers. Any development on the demand side forced a reaction from the producers. A close study of these maps, therefore, may teach us something about the changes in the vision of the world, not really of the population at large, but at least of the dominant elite. In the following paragraphs, therefore, I shall have a closer look at the cosmographical imagery on world maps.[227]

Before 1648, Dutch world maps seldom display any cosmographical imagery. The only example is the world map by Willem Blaeu from 1619 (re-edited in 1645-1646), which has been mentioned already. However, around 1650 there was a real boom in border designs displaying various cosmographical emblems, often in a rather prominent way. During the course of the second half of the seventeenth century, such cosmographical imagery maintains its place in map decoration, although really new designs are rare. In most cases, only elements from earlier maps are copied and modified. The most successful cosmographical emblems, copied many times since, were introduced around 1650 by two map makers, Joan Blaeu (the son and successor of Willem Blaeu) and Nicolaas Visscher.

Blaeu probably set the tone for the new trend. In or around 1648 he published a splendid new wall map, which was later reprinted several times.[228] Its border decorations are particularly lavish. Among its most prominent characteristics is a new scheme of the Copernican system. On his father's map from 1619, the Copernican system had taken a rather modest place. On this new map, it occupies almost all the space left free in the upper centre between the hemispheres. The planets are represented as the mythological figures from whom they take their name. Most of the images are taken from a border decoration on a map from 1606 by Joan's father, Willem, representing allegories of the seven planets.[229] The earth – which is not a planet according to the traditional view, and hence is not represented on the map of 1606 – is reproduced as a simple globe, accompanied by the allegory of the moon. The sun, which in 1606 was represented by Helios on his chariot, is now represented by an enthroned figure with a sceptre, apparently Apollo. Probably, the original figure was too large to position within the orbit of Mercury. Also, perhaps a

---

227 See also Vermij (1997).

228 Shirley (1983) no. 371. NB Shirley reproduces the third edition of this map, the reprint by De Wit from 1665/1666. As the original map is inaccessible, I base my description on it.

229 The sources of the images on the 1606 map are discussed by Schilder, *Monumenta cartographica Neerlandica* IV (*Losse kaarten...*) 173-175.

moving chariot was not thought to be a fitting emblem in a heliocentric system.

Apart from this Copernican system, the two competing systems of Ptolemy and Tycho Brahe figure on the map, though in a far less prominent way. They are shown as small schemes, captioned 'Hypothesis ptolemaica' and 'Hypothesis Tychonica'. These schemes are at the bottom centre, flanking a small map which shows the earth as it was known to the ancients.[230] Apparently, the resulting association with obsolete knowledge was purposive. Joan, like his father, was an adamant supporter of Copernicanism, so this stance need not surprise us.

Before long, others followed Blaeu's example. Nicolaas Visscher introduced some cosmographical emblems on two of his maps, which he issued about 1650. One of these is a large wall map, a new edition of a map by Jodocus Hondius from between 1611 and 1618. Visscher appears to have adapted Hondius' border decorations. Visscher's map is only known from editions from 1660 and 1669, though the first impression must date from before 1652.[231] Although one cannot be completely sure that no alterations have been made between the first edition and 1660, the similarities of the emblems to those on Visscher's next map from about the same time (to be discussed below) seem to indicate that these elements were original.

The space between the bottom of each hemisphere is filled by two celestial hemispheres with some additional images: portraits of Tycho Brahe and Cornelis Houtman, as explorers of the northern and the southern skies, respectively, and a table of stellar magnitudes. (Visscher had added the names of Tycho and Houtman to a pair of celestial hemispheres on an earlier map, but not their portraits.[232]) As on the map by Blaeu, the celestial hemispheres are flanked by two small representations of world systems. The one is called '*Systema mundi juxti mentem Claudii Ptolemaei Alexand*' ('The system of the world according to the opinion of Claudius Ptolemaios of Alexandria'), the other '*Genuinum Mundi Systema secundum N. Copernici hypothesin*' ('The true system of the world according to the hypothesis of N. Copernicus'). Neither system is reproduced more prominently than the other, but in his captions Visscher leaves no doubt that he prefers the Copernican view.

The same schemes, with identical captions, are also to be found on another of Visscher's world maps. This one is part of a series of maps designed for insertion into the luxury Dutch Bibles published by the Amsterdam house of

---

230 This element appears inspired by a similar map, from 1590, by Abraham Ortelius (Shirley (1983) no. 176).

231 Shirley (1983) no. 416. On the dating, see Schilder (1978) 25.

232 Shirley (1983) no. 294.

Ravesteyn. The series consists further of maps of the Holy Land, the voyages of St Paul, and similar topics. Visscher's map is the first world map purposely produced as a Bible map. The map is presently dated around 1650. It is found in Bible editions up to 1670.[233] The cosmographical schemes are placed between the hemispheres, the Copernican scheme above, the Ptolemaic below.

The emblems introduced by Visscher are among the most reproduced representations of the world system of the seventeenth and eighteenth centuries. Visscher's schemes are characterised by the indication of the solar rays which illuminate the earth and the moon, thus illustrating the cause of the moon's phases. A rather curious element in the Ptolemaic system is the small epicycle of the moon, superposed upon the moon's orbit around the earth. Ptolemy used an epicycle to describe the motion of the moon, but it is not clear why it should be reproduced in the scheme, as all other epicycles in the Ptolemaic model are ignored.

The illustrations introduced by Blaeu and Visscher were picked up enthusiastically by their colleagues. Apart from re-editions or copies of the maps themselves, we also find the cosmographical symbols reused in various combinations, particularly on the large wall maps. The earliest example is a map by Hugo Allard from about 1650 which will be discussed below. A map by Frederik de Wit published before 1663 is largely a copy of Blaeu's 1648 wall map, but with Blaeu's smaller world systems replaced by the schemes by Visscher (with the captions omitted).[234] Subsequently, Justus Danckerts put those emblems in exactly the same way on a map of about 1670.[235] Danckerts' map was reprinted in its entirety by Etienne Roger in 1713.[236] Joachim Ottens republished in 1690 another world map from Frederik de Wit, now lost, which must have dated from about 1670. The border decorations are rather similar to those of de Wit's 1663 map, but the smaller cosmographical schemas are those by Blaeu, not Visscher.[237]

Gerard Valck, finally, published a wall map in about 1690, which reproduced in its decorations the small systems of Ptolemy and Tycho Brahe from Blaeu's map, but not the large Copernican configuration. Copernicus' system was thus not represented on the map. It is not certain whether or not

---

233 Shirley (1983) 401. Poortman and Augusteijn (1995) 142. I have studied the map in Bibles edited in 1657 and 1660, Leiden University Library, 1169 A 2 and 578 A 1 (I owe gratitude to Mr Berkhemer for providing me with the shell numbers). Contrary to Shirley's assessment, the Bibles are Dutch, not Latin.

234 Shirley (1983) no. 422. On De Wit: van Eeghen (1990).

235 Shirley (1983) no. 454.

236 Schilder (1977) 13-14. Zie also Schilder (1990) 55-59, 60-61, where it is stated (not wholly correctly) that the decorations are largely based on Blaeu's world map.

237 Shirley (1983) no. 453.

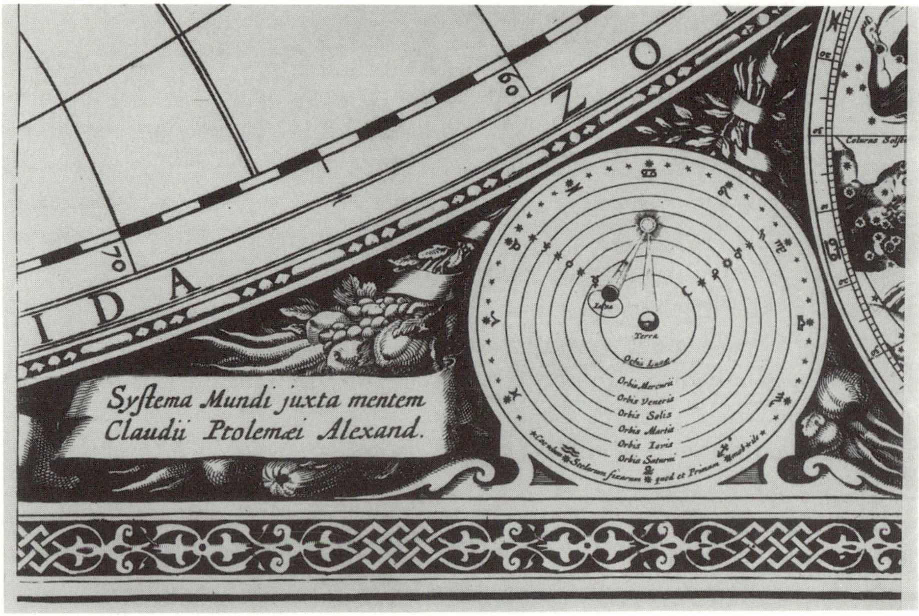

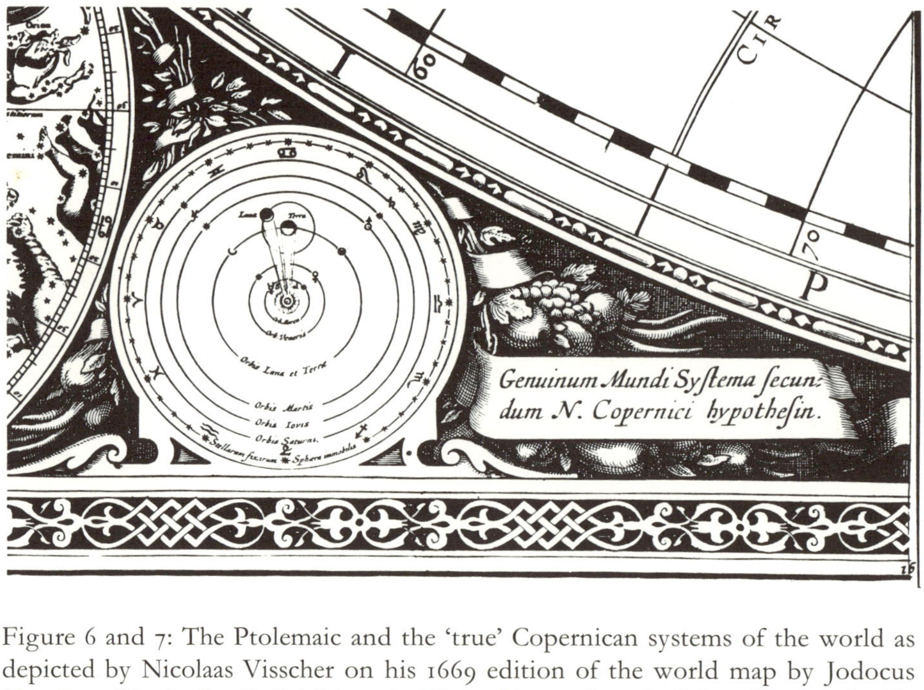

Figure 6 and 7: The Ptolemaic and the 'true' Copernican systems of the world as depicted by Nicolaas Visscher on his 1669 edition of the world map by Jodocus Hondius. (Facsimile: G. Schilder ed., *The world map of 1669 by Jodocus Hondius the elder & Nicolaas Visscher* (Amsterdam 1978); photograph Leiden University Library).

this was done on purpose. Anyhow, the map was aimed not at the Dutch but the English market, as the title is in English. A second version was published after 1700 by the English publisher George Wildey.[238]

Apart from world maps, we also find cosmographical schemes on celestial maps. Here again, the example is set by the schemes on the large world maps. A map by Ludovicus Vlasblom, published by Johannes van Keulen in Amsterdam in about 1675, reproduces Visscher's Copernican scheme and Blaeu's Ptolemaic one.[239] A map by Andreas van Luchtenburg, published in Rotterdam around 1688, offers Visscher's Copernican scheme, as well as a scheme of the lengthening and shortening of the days, taken from Dirk Rembrandtsz van Nierop's 'Dutch astronomy'. Its caption is: 'Natural proof of the sun's rest and the earth's motion.' Luchtenburg's celestial map was reissued by the publisher J. Robijn between 1688 and 1701.[240]

As for Bible maps, the designs by Visscher remained foremost. Visscher himself issued a new series of Bible maps in about 1660, this time for the Bibles of Elzevier. It included a new world map, which offered the world systems and the captions exactly as on the earlier map. It remained in use until the 1780s.[241] Visscher's earlier map was anonymously copied in about 1660.[242] Then the publishing house of Keur at Dordrecht ordered a set of maps for its Bibles. Their world maps were produced by Bastiaan Stoopendaal in 1682 and 1688, by Daniel Stoopendaal in 1714, and by Jan de Jager in 1748. All of them copy Visscher's Copernican and Ptolemaic schemes. However, they omit the captions. The user has to know for himself.[243]

Blaeu and Visscher had consciously introduced cosmographical emblems on their world maps and, it may be assumed, consciously shown their preference for Copernicus. How far this is true for their many imitators is less clear. The boom in cosmographical symbols after 1650 rather suggests that map producers had discovered that such symbols were in demand, and therefore provided them – with diligence, but not necessarily with understanding. Nor with much originality, for they simply copied some of their predecessors' successful emblems. All wall maps after 1650 are based on the old models.

---

[238] Shirley (1983) no. 544.

[239] Warner (1979) 256-257. Leiden, univesity library, collection Bodel Nijenhuis 169 no. 2.

[240] Warner (1979) 166-167. I also used a photograph of the copy of this map in Paris, Bibliothèque Nationale, Ge.D. 12.671, kindly provided me by Dr Peter van der Krogt (Utrecht).

[241] Shirley (1983) no. 431; Poortman and Augusteijn (1995) 160. I have studied the map in a 1661 Bible from Haarlem (UB Amsterdam, OF 65-14) and a 1664 Ravesteyn-Bible (Amsterdam, University Library 405 A 5). Shell marks were obligingly provided by Mr Berkhemer.

[242] Shirley (1983) no. 414.

[243] Poortman and Augusteijn (1995) 195, 204, 196, 205.

The most interesting case of such borrowing is provided by the wall map of Hugo Allard, which I shall discuss in some detail.

Allard's map,[244] which was published somewhere around 1650, is very rich in ornamental imagery. Several of these images are cosmographical. Most prominent is the large Copernican configuration in the upper centre, which is copied straight from the 1648 wall map by Blaeu. In the two upper corners of the map are two celestial hemispheres, flanked by two smaller world systems, and in the two corners below two other astronomical schemes, with explanatory captions in Dutch.

The two small world systems are those introduced by Visscher, viz. his Ptolemaic and Copernican configurations. This seems an odd choice. Blaeu had had the Copernican system accompanied by the two rival systems. Visscher had opposed the old Ptolemaic system to the new (and true) Copernican one. On Allard's map, we find two Copernican schemes and one Ptolemaic, which seems less well-considered. The haphazard composition of the map becomes really apparent, however, when we consider the captions. In the case of the small Copernican scheme, Allard has largely copied Visscher's caption as well: '*genuinum mundi systema secundum N. Copernicum*'. The Ptolemaic scheme, however, is provided with a caption taken from Blaeu's map: '*hypothesis tychonica*' ('Tychonian hypothesis'). This is a flagrant error. There cannot be any doubt that the world system depicted is the Ptolemaic and not the Tychonic one.

As for the two astronomical schemes at the bottom, they are not very original either. One is taken from the works of Philips Lansbergen. It represents the earth in its course around the sun and the changing length of the day. Lansbergen used it as an emblem in several of his books. It appears to have been rather popular as an emblem, as it was also used in works by Jacob Lansbergen, Willem Blaeu and Jacob van Veen. The other scheme explains the phases of the moon. It is taken from Gassendi's popular introduction to astronomy, *Institutio astronomica*, the first edition of which appeared in 1647. Image and caption are not well harmonised. The text refers to lettering in the figure, which is not to be found there.

Allard clearly aims at superficial splendour rather than scientific accuracy. He indiscriminately borrows elements which fit his purpose, and integrates them into an aesthetically attractive but scientifically inaccurate image. The cosmographical elements are there to impress rather than inform. Still, Allard seems to have guessed what it was the market wanted, for his map became a commercial success. In 1660, it was reprinted twice, once by Allard himself

244 Shirley (1983) no. 392. I also used a photograph of this map, provided by the Royal Library at Copenhagen.

and once by Pieter van den Keere.[245] These new issues contain no changes which concern us; the erroneous caption has not been corrected. Another reprint is known from about 1720.[246] There is also a world map by Jan Houwens from about 1660, which reproduces a map by Pieter van den Keere from 1611, but copies the entire border decorations from Allard's map. In this case, too, the erroneous ascription of Ptolemy's system to Tycho Brahe has not been corrected.[247]

Allard's example makes it clear how the average map producer worked. From the boom in world systems on maps around 1650, it can be assumed that they scented the public's interest in the subject. Even so, it is striking that no new symbols were created. Apparently, the emblems by Blaeu and Visscher had gained a strong position, so that one did not risk replacing familiar schemes.

Atlas maps and separate maps of the same format show a slightly different development. (As atlas maps were sold separately as well, and separate maps could be included in atlases at some stage, there is no strict separation between the two genres.) Starting in 1650, these maps follow the trend of displaying cosmological imagery, but they copy the emblems on the wall maps only to a very limited degree. None of them reproduces the schemes of Blaeu and only one reproduces those of Visscher (a map by Frederik de Wit, which was included in Doncker's sea atlas from 1660 onwards[248]). Nor did they copy each other in the same way as the other maps. As they were much cheaper, innovations were less dangerous for the producers. Thus, new designs kept on appearing during the second half of the century. I shall review the most important of them.

About 1650, there appeared two different world maps with (but for some unimportant details) similar border decoration. As both date from about the same time, it is not possible to say for sure which map was original and which a copy. There even may have been some kind of collaboration. One of the maps, which was based on a map by Geelkercken and de Meine from 1610, was published by Hugo Allard, whose large wall map was discussed above.[249] The other map was published by Jacob Colom.[250] Both were used in atlases. The border decoration of both maps was largely based on a map by Hondius

---

245 Schilder and Weelu (1980) 24, where one finds also a reproduction of a later edition of Allard's map, with a dedication to the English King Charles II (1660-1685).
246 Informations on reprints by Shirley (1983), no. 392.
247 Shirley (1983) no. 412.
248 Shirley (1983) no. 421; the captions are included.
249 Shirley (1983) no. 378.
250 Shirley (1983) no. 381.

Figure 8: Large wall map by Hugo Allard, about 1650, carrying various cosmographical diagrams. The large Copernican system top centre, with mythological figures representing the planets, is borrowed from the wall map by Blaeu. The smaller world-systems are those from Visscher's maps (compare figure 6-7). The astronomical diagrams below are borrowed from textbooks by Lansbergen and Gassendi. (Copenhagen, Royal Library.)

from 1630. However, some new elements had been added, all of which concerned the discussion on the world system. Two of the four portraits of famous geographers in the corners have been replaced: Mercator and Hondius have ceded their places to Copernicus and Tycho Brahe (the other two are Ptolemy and Julius Caesar. Ptolemy, of course, is now regarded as an astronomer rather than a geographer). At the bottom centre is a picture of two scholars who apparently are discussing the system of the world, as one is holding a geocentric armillary sphere, and the other a heliocentric one. This illustration is borrowed from the frontispiece of Lansbergen's astronomical tables from 1632, where the figures are identified as the ancient astronomers Hipparchos of Rhodes and Aristarchos of Samos. Another element from the same frontispiece is reproduced at the top centre: an image of the earth moving in the zodiac, with the sun in the centre. The latter element was also reproduced on De Wit's map in Doncker's sea atlas [251] and, ironically, on a Greek map printed at Padua in 1700 by presbyter Chrysanthos, patriarch of Jerusalem. [252] One wonders whether he realised that it was a Copernican emblem.

In 1662, Joan Blaeu published a new world map in his *Atlas maior*. This atlas appeared in the years 1662-1667 in several languages. The world map was also included in an atlas by Van Keulen in 1681-1685. The upper border is decorated with a panorama of 'celestial figures seated amidst clouds'. [253] In fact, these are allegories of the planets in their respective orbits. As on Blaeu's 1648 wall map, the planets are represented by classical deities. Central is an enthroned and sceptred Apollo, who clearly symbolises the sun. What is curious about the design is that the hemispheres of the map at the same time represent the earth in this planetary configuration. The map itself is, so to speak, integrated into the border decoration. In the upper centre, the moon sticks her head between the two hemispheres. The whole design was copied on a map by Dirck Davidsz, which is included in a sea atlas by Lootsman from 1676-1677, as well as in an English version from 1677 and in atlases from 1689 and 1694. [254]

Curiously, in the textual part of the atlas, Blaeu is much more cautious about the world system. He explains the system of Copernicus and that of

---

[251] Shirley (1983) no. 421.

[252] Shirley (1983) no. 598.

[253] Shirley (1983) no. 428 (quote on p. 450). NB Shirley erroneously does not reproduce Blaeu's map at this point, but the map by Dirck Davidsz., no. 477. A reproduction of the map by Blaeu can be found on the flyleaves of J. Blaeu, *De grote atlas van de wereld in de 17de eeuw*, J. Goss ed. (Lisse 1992).

[254] Shirley (1983) no. 477.

Ptolemy (but not the Tychonic one), and says that the Copernican system 'is followed by the best and learnedst mathematicians of this time.' Yet, he does not want to say which is true. Geocentric and heliocentric systems are equivalent. In the following chapter on spherics, he presupposes a fixed earth, as this is easier to understand for those inexperienced in astronomy. [255]

Two other maps will be briefly touched upon. Both of them were engraved by Jacob Harrewijn. One was published by Jacob Allard in 1685, [256] the other – entitled *Nieuwe wereld* ('New world') – was published by Doncker and Robijn in 1687. [257] Both maps were sold as single sheets, probably for connoisseurs. Their imagery is rather complicated and perhaps for that reason, they have found no imitators. The 1685 map shows again a panorama of allegories of the sun and planets in the upper border. Although the sun is represented by a figure in a chariot, the configuration is clearly heliocentric. A figure representing the earth is depicted in a large zodiac, which has the sun for its centre. [258] The 'new world' of 1687 does not display a system of the world, but shows the portraits of Copernicus, Ptolemy, Tycho and Dirk Rembrandtsz van Nierop. The whole seems to symbolise the advance in knowledge. Both ancient and modern scientists are honoured, but the moderns deserve most sympathy.

Cosmographical imagery does not seem to have become as dominant on atlas maps as on the large wall maps. Probably, on wall maps display – and hence fashion – was more important. On the other hand, small maps required less investments to produce and were probably easier to sell, so their makers could afford to take some risk. Perhaps this is why the makers of wall maps did not abandon the familiar emblems once these had become fashionable, while the makers of smaller maps kept on introducing new designs. Still, these smaller maps do confirm the overall trend. By 1650, map makers were adorning their maps with cosmographical imagery.

At first sight, the way cosmological imagery was used on world maps may seem somewhat capricious. However, the whole history becomes perfectly understandable if seen as the result of the map makers' reaction to a new interest in cosmographical questions among their clientele. Now, can we go further and draw conclusions about the views and preferences of this clien-

---

[255] J. Blaeu, *Grooten Atlas oft wereldbeschryving...* I (Amsterdam 1664), introduction, Chapter I (no page numbers). I owe this reference to the Hooykaas papers.

[256] Shirley (1983) no. 525. Leiden, university library, 002-15-009/014.

[257] Shirley (1983) no. 534.

[258] At about the same time, some other map makers also portrait a gathering of classical deities in the sky: Johannes Ram (Amsterdam 1683: Shirley no. 519) and Justus Danckerts (Amsterdam 1685: Shirley no. 529). It is difficult, however, to interpret these as planetary configurations.

tele? All the original designs appear to exhibit a preference for the Copernican system, whereas until about 1690 no examples are known of maps propagating a geocentric system. One may wonder whether the many imitators really made a conscious choice, considering their at times evident lack of understanding. Still, even if they did not know much about cosmography (and probably, many of these practitioners knew some basic principles), they had to know about the exigencies of the market.

It is striking, anyhow, that foreign map-makers preferred a geocentric system when making their world maps. This is attested by the cosmographical imagery on English maps from 1625 and 1627, and on French maps from 1636, 1647, 1688, 1670 and ca. 1700.[259] The difference is especially striking as French cartographers often borrowed from their Dutch colleagues. Two French maps, from 1666 and 1680, reproduce Visscher's Ptolemaic world system; in both cases, Visscher's corresponding Copernican system is simply ignored.[260] We noticed above that the only example of a Dutch map which displays geocentric but no heliocentric schemes – Valck's 1690 map – was aimed at a foreign market, the English one. Another map quite certainly aimed at a non-Dutch audience was published in Amsterdam in 1695 on the presses of the Armenian archbishop, Wuscan Theodorus Warthabeth. All the text is in Armenian. Still, the world system (top centre) is easily recognisable and decidedly Ptolemaic.[261] Foreign maps from this period showing a preference for Copernicus are rare. I know of only two: a French map by Jean Boisseau from 1645, which reproduces only a Copernican scheme,[262] and an English map from about 1687.[263] One should add that map production in England – as in most other European countries – was quantitatively insignificant compared to the production in the Dutch Republic. Only France offers enough material for comparative purposes.

Regarding these national differences, the preference for the Copernican system is not really fortuitous. Probably, map purchasers were interested in Copernicanism rather than in the traditional view. This does not necessarily mean that they were convinced Copernicans. Their interest may have been simple curiosity, because Copernicanism was new and topical. Anyhow, they were not really averse to it. As to international comparison, the relative free-

[259] Shirley (1983) nos. 313, 317, 340, 370, 449, 538, 613.
[260] Shirley (1983) nos. 441 (Piere du Val, 1666) and 496 (Avuray de Garel, ca. 1680).
[261] Shirley (1983) no. 575. See also C. Koeman, 'A world map in Armenian printed at Amsterdam in 1695, in: *Imago mundi* 21 (1967) 113-114.
[262] Shirley (1983) no. 362, which offers a single scheme: *Disposition des orbes ou spheres celestes selon l'hypothèse de Copernic*. Boisseau's earlier map from 1636 (Shirley no. 340) offers only a geocentric scheme, *Du nombre des Cieux et de leurs Mouvementz*.
[263] Shirley (1983) no. 535; see also no. 460.

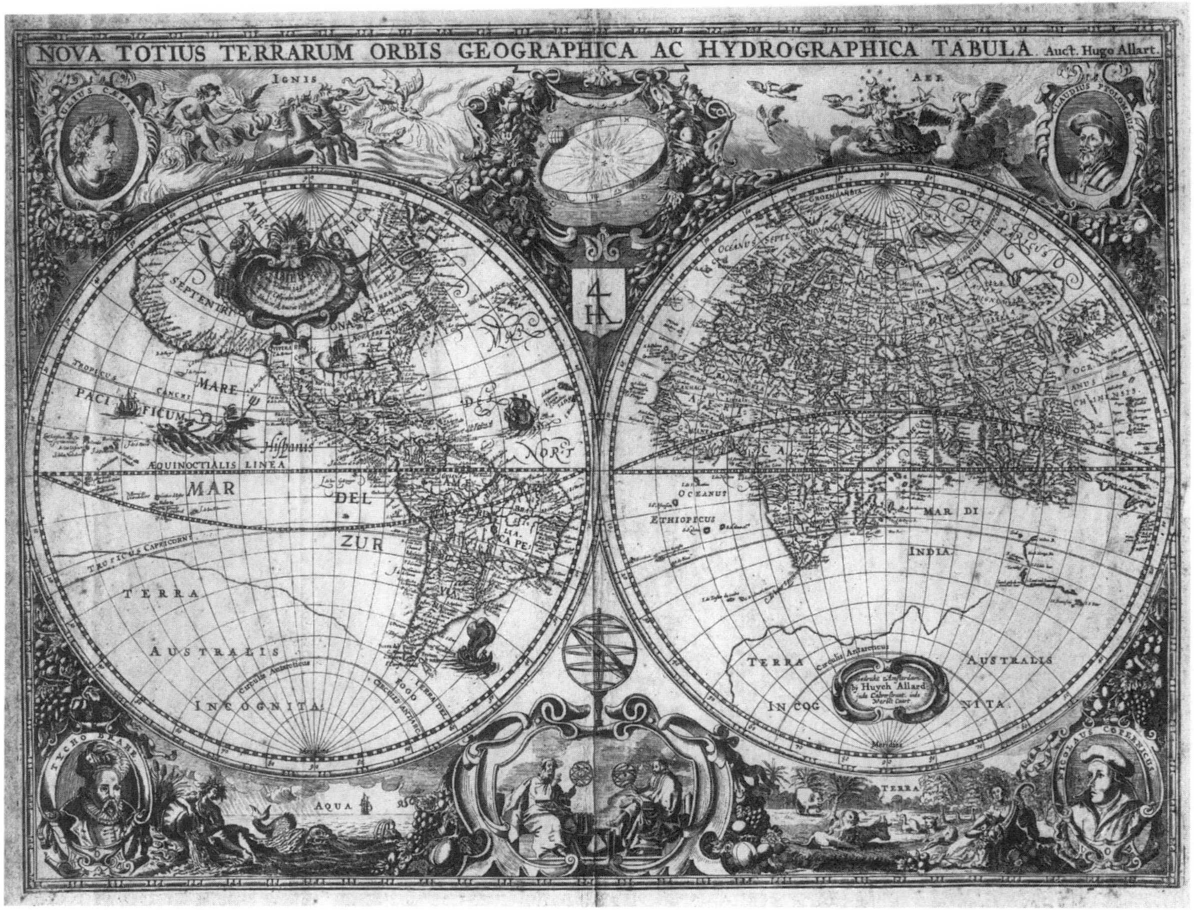

Figure 9: Small world map by Hugo Allard, about 1650. The zodiac above and the two seated figures below were taken from the title-page of Lansbergen's tables, compare figure 3. (Nürnberg, Germanisches nationalmuseum.)

dom of the press in the Dutch Republic should not be overlooked. Copernicanism could be freely propagated in France, but it is difficult to believe that an image of the 'true Copernican system' would have been tolerated in French Bibles.

What is striking, finally, is that the style of the decorations is baroque, with many allegorical figures and so on. Such a style seems more apt for representing a biblical, hierarchical and meaningful cosmos than Descartes' universe of matter and motion. The new world system did not automatically bring a new intellectual outlook in other parts of life. This seems to tally with what we remarked above in the section on poetry, that is, the new world-view had found no expression in powerful poetic images or a convincing artistic style. Only around 1690 would new artistic styles develop in map decoration. By then, however, the great age of Dutch cartography was largely over. Later maps are designed by foreigners, such as Doppelmaier in Nürnberg.

The most famous storehouse of cosmographical images is Andreas Cellarius' *Harmonia macrocosmica*, which was first published in 1661. Cellarius was a German from the Palatinate, who had settled in the United Provinces and in the end became rector of the Latin school in Hoorn.[264] In 1645, his work on military architecture was published by Johannes Janssonius, one of the two Amsterdam publishers who controlled the market of luxurious, multi-volume atlases (the other was Blaeu). Shortly after 1645, Janssonius expressed his idea of complementing his series of atlases, which showed the surface of the earth, with a volume aimed at displaying the whole universe. *Harmonia macrocosmica* was clearly aimed at the same kind of clientele, i.e. rich burghers with some intellectual leanings, as had the earlier atlases or world maps. It was a large, luxuriously illustrated volume. Its core comprises 29 cosmographical plates, which borrow extensively from existing imagery. The illustrations clearly were more important than the text.

Cellarius deliberately does not take a stance on the issue of the true system of the world. He simply exhibits the various views and rehearses the arguments put forward to support them. The Copernican, Tychonian and Ptolemaic systems are all depicted in various ways, in plain systems and scenographically. There is also a picture of the *Aratea* (drawn after Jacob de Gheyn's earlier engraving). There is no picture of the Cartesian universe, although it is discussed in the commentary on the Copernican planisphere.[265] After these cosmographical discussions follows an astronomical part, with plates depicting the vastness and distances of the celestial bodies according

---

[264] Cellarius (1661). For biographical details, see van Gent (2000).
[265] Cellarius (1661), main part, 26-27.

to various theories, armillary spheres, a world map, celestial maps and such like. In these parts, everything is depicted from the standpoint of an immobile earth.[266]

In the original version, the pictorial part was preceded by an introduction over 100 pages long in which Cellarius, somewhat in the tradition of Mercator's *Atlas*, gives a description of the entire universe. He starts by describing the Creation following the seven days, whereupon he describes in detail the heavens, the elements, the mineral, vegetative and animal kingdoms, and the history of the mathematical sciences. In passing, he also discusses such topics as the immortality of the soul and Descartes' theory of the passions (which he rejects). What is striking are the many references to Paracelsus and similar authors. He even gives a list of alchemistical axioms, theorems and precepts of alchemy. Cellarius' world-view is clearly in the Hermetic tradition. Biblical arguments are to be found, too. According to Cellarius, the fluidity of the heavens follows from biblical passages as well as from modern observations. Anyhow, at no place does he take a position in the Copernican debate. Rather, he emphasises the pettiness of human knowledge. We do not know the cosmic causes and reasons. Only motions, orbits and proportions have been investigated, and even there our knowledge remains dubious and uncertain.[267]

Although this profession of ignorance was probably sincere, it does not mean that Cellarius did not have some private opinions. He defended his refusal to take sides in the debate on the true system of the world with the argument: 'What if someone else were to come who would not state that the sun or the earth was at the centre of the whole universe, but would uphold that all astral bodies including the earth were moving around the centre, and would place in this centre the Divine light, which is reigning and conserving all these bodies?...'[268] One is curiously reminded of the ideas of Cellarius' colleague from Haarlem, Coccaeus, who was equally well-read in Hermetic literature, and who indeed devised a system whereby both the sun and the earth were at an equal distance from the centre of planetary motion.

The work proved quite a success. There were two re-editions, one after 1680 and one in 1708. This success was due to the illustrations rather than the text: the later editions contain only the illustrations; the entire text has

[266] Cf. the *praefatio*. All plates are enumerated, with small reproductions, in van Gent (2000) 19-24.

[267] Cellarius (1661), *praeloquium*, 10 (ignorance; see also p. 1 of the main work), 11 (fluidity of the heavens), 28-29 (passions), 45-47 (alchemical precepts). The various world systems are still discussed on pp. 64-67. NB The main part, with the plates, is paginated separately from the *praeloquium*.

[268] Cellarius (1661), *praeloquium*, 10.

been left out. One should add that the plates were not really up to contemporary standards. Christiaan Huygens made some harsh criticisms.[269] The book was just a showpiece, useless for astronomers. However, it apparently was just what the public wanted. Its reception confirms the trend illustrated by world maps: interest in cosmography was fashionable rather than learned. People were curious about, impressed by or liked to show off their knowledge of cosmology, but – apart from some rare exceptions – did not bother with the details. Nor did most people feel uneasy about the new learning. They were aware of the new developments in science, but whether they really knew what was going on is another question.

[269] Van Gent (2000) 14-15.

Part IV. Biblical authority and Christian freedom: Copernicanism and Bible interpretation

11. Copernicanism in the theological discussion: preliminaries to a problem

The Bible in Reformed theology

The Copernican world system was not only contradicted by Aristotelian philosophy, but it also ran counter to the world-view as it was taken for granted in the Bible. Because the Bible was regarded as God's infallible word, this presented a serious problem. The problem how to accommodate the Copernican theory to the standard interpretation of the Bible was recognised early, among others by Osiander in his preface to Copernicus' book. A large number of texts could be adduced to demonstrate the stability of the earth or the motion of the sun, for instance: '...the earth abideth for ever. The sun also ariseth, and the sun goeth down, and hasteth to his place where he arose' (Eccl. 1: 4-5). God 'laid the foundations of the earth, that it should not be removed for ever' (Psalm 104:5). The sun 'is as a bridegroom coming out of his chamber, and rejoiceth as a strong man to run a race. His going forth is from the end of heaven, and his circuit unto the ends of it' (Psalm 19: 5-6). The list could easily be extended. Theologians were very good at discovering texts which could be used to support their view. However, the text most often alluded to in discussions on the Copernican system was not a description of the order of Creation, but the history of the miracle God performed in response to Joshua's prayer when the latter was leading the Israelites in their victory against their enemies the Amorites:

> 'Then spake Joshua to the Lord in the day when the Lord delivered up the Amorites before the children of Israel, and he said in the sight of Israel, Sun, stand thou still upon Gibeon; and thou, Moon, in the valley of Ajalon. And the Sun stood still, and the moon stayed, until the people had avenged themselves upon their enemies. Is not this written in the book of Jasher? So the sun stood still in the midst of heaven, and hasted not to go down about a whole day. And there was no day like that before it or after it, that the Lord hearkened unto the voice of man: for the Lord fought for Israel.' (Joshua 10:12-14.)

As it seems, the resistance against the Copernican system was brought about not only by exegetical difficulties, but also by a general feeling that the Co-

Figure 10: Joshua ordering the sun to halt, as depicted on a stone tablet at an Amsterdam mansion. The inscription says: 'Who obeys God, is obeyed by the elements'. After an engraving by F.W. Zürcher in J. van Lennep and J. ter Gouw, *De uithangteekens*, II (Amsterdam 1868, repr. The Hague 1974) p. 61.

pernicans ignored God's hand in nature. In a like way, the miracle of 2 Kings 20: 9-11 and Isaiah 39: 8 was adduced in the discussion. Here, King Hezekiah asks the prophet Isaiah for a divine sign. Upon Isaiah's prayer, God 'brought the shadow ten degrees backward, by which it had gone down in the dial of Ahaz.' In Isaiah's text: 'Behold, I will bring again the shadow of the degrees, which is gone down in the sundial of Ahaz, ten degrees backward. So the sun returned ten degrees, by which degrees it was gone down.' It seems a queer argument, as the motion of the shadow would be equal in any cosmological system. Referring to such miracles rather attests to a more general fear that the Copernico-Cartesians, by making all depend on the laws of nature, left no room for God's presence in the universe.

The history of the debate within Roman Catholicism on the compatibility of heliocentrism with the Bible is well known. Less is known about the reaction within the Protestant Church. The problem with Protestant theology is that Protestantism is not a unity, and that even the various Churches into which it is split are less centralised than the Church of Rome. It is therefore more difficult to draw a coherent picture. The decentralised character of the Protestant Churches was in itself an important element of the debate, as it was much harder to resolve, or at least to end, the issue with an authoritative decree.

Anyhow, the ecclesiastical background put its own constraints on the discussion. Important to our purpose is the attitude of the Reformed to Scripture. After all, the question was whether the Copernican world system was compatible with the Bible. It was therefore principally an exegetical issue. As Scripture is more important in the Protestant than in the Catholic tradition, one might surmise that the argument from the Bible against Copernicanism would have much more force within the Protestant tradition than within Catholicism. Still, the exegetical principles within Protestantism left the theologians considerable freedom.

What were these exegetical principles? First of all, the Reformed acknowledged only the literal (or historical) sense of the Bible. Catholic exegesis knew, beyond the literal sense, of various types of spiritual meaning: anagogical, allegorical and tropological.[1] Reformed theologians rejected all hidden meanings, and returned to the plain text. Biblical exegesis came down to establishing the literal sense of the text. For our purpose, the distinction is not very important. Among the Catholics, too, the question of Copernicanism had only to do with literal meaning. Literal meaning could be either simple or figurative; in the latter case, the language used was metaphorical or in any other way figurative.

Secondly, the Reformed maintained that Scripture could only be explained by Scripture itself. That is, if there were any differences in laying down what exactly was meant by a Scriptural sentence, the only legitimate way of investigating the question was to turn to other Scriptural places for elucidation. If the Bible spoke of 'God's hand', one could only conclude that this expression was meant figuratively because of other texts which attested that God is Spirit. Likewise, if two texts seemed contradictory, one should solve the problem by turning to yet other texts.[2] This is a striking difference from Catholicism. Catholics were bound in their exegesis by the interpretation of the fathers. Moreover, the Church itself, instituted in its Supreme Pontiff at Rome, was the ultimate judge in all disputes of interpretation. So, despite their attachment to the literal meaning of the Bible, the Reformed were more free in their dealings with the text than post-Tridentine Catholics were. Neither the interpretation of the fathers, nor the decrees of Rome (exactly the grounds upon which Galileo's attempt to reconcile his theories to the Bible had foundered) mattered very much.

Over the course of time, this attitude underwent many modifications. Calvin's own dealing with the sacred texts shows a sophistication which may

[1] Cf. Blackwell (1991) 33-34.
[2] Laplanche (1989) 122.

derive partly from his humanist education. He showed a good deal of common sense and refused to follow the letter of the text against all odds. For instance, he asserted that people who conclude from the mentioning of 'the waters above the earth' in Genesis 1: 7 that there is a real ocean in the skies, 'too servilely tie themselves to the letter of the text'. In cases like these, the Bible accommodates to the understanding of the vulgar. Therefore, 'it would be a preposterous cause, to reduce their [Moses' and the prophets'] phrases to the exact rules of philosophy'. In the case of the waters above the earth, he thought they must simply be clouds. [3] Calvin was not an advocate of the Copernican system. In so far as he took any notice of it at all, he regarded it as a crazy idea. [4] But the principles laid down by him and the other Protestant reformers enabled their followers to deal more freely than their Catholic counterparts with secular knowledge. The Bible was not a book which contained in a hidden way all possible knowledge; it simply taught the true faith. Theology was no longer the queen of science which systematised all possible knowledge; it was simply the knowledge of the biblical message. [5]

As, however, over the course of time the Protestant creeds became confessions which fixed boundaries, they set their theologies on a scholastic footing. And as the Bible was their sole legitimation, this put ever more demands on their exegesis. The Protestants increasingly emphasised that the Bible was inspired in its very syllables. It was not just a book wherein God taught the true faith: it was the alpha and omega of knowledge. [6] The founding fathers of the Dutch Reformed Church had had their education at the Calvinist academies of Heidelberg or Geneva. Still, they soon went much further than Calvin himself. One of the leading theologians was Franciscus Gomarus, a professor at Leiden and Groningen famous for his leading role in the rejection of Arminianism. Gomarus rejected allegorisms and kept to a single (or, sometimes, a double) literal interpretation of the text. By now, however, the text was supposed to solve all kinds of current theological controversies unknown to the biblical authors themselves. By syllogistic arguments, the text was forced to extend far beyond its plain meaning. [7]

---

[3] I follow the discussion of Calvin's views in Hooykaas (1984) 176-178; the quotations are from p. 177.

[4] Quite a lot has been written on the attitude of the reformers on the Copernican question, in fact much more than the subject deserves. There is no need to refer to this literature here. Let me just mention Scholder (1966) 57-65.

[5] Cf. Deason (1989). Deason goes too far, in my view, in suggesting that it was the scholasticism dominant in Roman Catholicism which forced Galileo to frame a new synthesis of science and religion, which considerably complicated his attempt at accommodation.

[6] See Laplanche (1989) on the various ideas on inspiration among Reformed theologians.

[7] Laplanche (1986) 71-72.

The king of Reformed neo-scholasticism in the Dutch Republic was Gisbertus Voetius, whom we met earlier as the main adversary of Descartes in the Dutch Republic. Voetius wrote many theological works, but not a large synthesis. He preferred to discuss isolated topics, which he analysed with the help of scholastic instruments. He was the true master of the disputation, a genre which enabled him to comment on any current problem. Voetius' influence reached further than academic theology. He became the virtual leader of a movement generally called the *nadere reformatie*, or Further Reformation. The Further Reformation had many traits in common with the Puritan movement in England. As a movement, it started largely in Zealand, which had strong connections with England. When Voetius was appointed to the chair of theology at Utrecht, however, he quickly became the major spokesman and propagator of the movement. The Further Reformation aimed at the total reformation of Christian life and Christian society: the whole of private and public life should be brought under the law of Christ. This entailed the furthering of private piety – the Further Reformation had a strong Pietistic bent – and the uncompromising rejection of anything which could be seen as opposing, or even distracting from, Christ's law. [8]

Still, the anti-scholastic ideals of the Reformation were not completely forgotten. First of all, there was more theology in the Netherlands than just the Reformed theology. The earlier tradition of anti-scholastic theology was continued particularly by Mennonites and Remonstrants. Both groups generally did not look to the Bible for the solution of physical problems. Of course, certain individuals might reject Copernicanism at least partly because of its incompatibility with the Bible. We saw an example in the Mennonite navigational instructor Sybrand Hansz Cardinael. But it was not an issue in their theology. The Remonstrants, despite their roots in Reformed theology (their founding father, Jacob Arminius, used the scholastic method himself), developed a rather undogmatic spirituality and were relatively critical of dogmatism and scholasticism. One of the most important Remonstrant exegetes was Hugo Grotius. To him Joshua 10:12 was not suitable for deciding the Copernican question. In the wake of various Jewish authorities, he considered it probable that the text was not referring to any real miracle or actual prolongation of daylight at all, but was just a poetic, hyperbolic expression. Still, he did not exclude the possibility of a miracle. God, of course, would have been able to stop the course of the sun or, alternatively, display an optically generated image of it above the horizon after sunset. So, what happened physically is left totally undecided. Grotius' explanation of other texts, too, leaves little

[8] Cf. van Lieburg (1989).

room for their being applied to physical discussions.[9] Grotius' exegesis was no isolated case. Especially his suggestion that the seeming standstill of the sun was really an optical phenomenon was repeated by several other interpreters outside the Reformed Church, such as the Mennonite Dirk Rembrandtsz van Nierop, the philosopher Spinoza,[10] the Remonstrant Jean Le Clerc and the sectarian leader Willem Deurhof. During a religious meeting at his house, the last mentioned even demonstrated the principle by means of a globe, a candle and a mirror.[11]

But in the Reformed Churches, too, there arose opposition against the new neo-scholasticism. Nor was there a lack of people who opposed Voetius' programme for the reformation of society. Voetius' main opponent in many issues was the professor of theology at Groningen, Samuel des Marets (Maresius). The animosity between the two men was notorious. It was a kind of private war and as such widely regretted as unworthy among Christians. A more principled opposition arose at Leiden. The dominating person at the theological faculty there was Abraham van der Heyden (Heidanus). His background was very different from that of Voetius. Family ties connected him with the political elite of Holland and he himself owned considerable wealth. In his youth, he had travelled and studied extensively, and as a consequence, Heidanus was a humanistically minded man with wide-ranging interests. Unlike Voetius, he had remained aloof during the Arminian troubles. After some years as a minister, he had been appointed professor of theology at Leiden in 1648.[12]

Still, these people needed a kind of intellectual programme to set against Voetius' pretensions. The most vigorous anti-scholastic movement within the Dutch Reformed Church became, in time, Cartesianism. Voetius had rejected Cartesian philosophy, largely because it explained the world without having recourse to biblical revelation. Now, the question was raised whether it might not be his exegetical principles which were wrong. The debate was an old one. As stated, many theologians had pointed out that the Bible taught not physics but the true faith. However, this debate took on a novel character as Voetius' opponents argued their case not from theological principles, but from Cartesian philosophy. And it was exactly Copernicanism which forced the issue upon the theologians.

---

[9] Grotius (1732) 106; cf. also 258-259 (on Eccl. 1) and 225 (on Psalm 19). Cf. Gebhardt in Spinoza (1925) V, 17-18, on Jewish and other authorities for this view.

[10] Spinoza (1925) II, 36.

[11] Deurhoff (1741) II, 164-169. Equally, the miracle of Ahaz' sundial was demonstrated by means of a candle, a mirror and a dial.

[12] See on him Cramer (1889).

Early theological positions on Copernicanism

By 1650, very little had been written in the Dutch Republic on the problem of Copernicanism and the Bible. Any theologian who stumbled upon the issue could evade it by applying the common argument introduced by Osiander, i.e. that the Copernican theory was a mere astronomical hypothesis which need not correspond to physical truth. Sibrandus Lubbertus, a prominent and very staunch defender of orthodoxy, appears to have answered in this vein when confronted with the outburst of his friend Ubbo Emmius against Stevin.[13] (Emmius himself, by the way, was not a theologian.) In 1638 the leading theologian, André Rivet, preacher at the court of the stadholder, when asked by Mersenne, denied that the opinion of the earth's motion should be deemed heretical.[14]

However, the international discussion on the subject was clearly known. Most arguments were available from the Latin translation of Galileo's *Dialogo*, the *Systema cosmicum*, from 1635. An extract from Kepler's introduction to *Astronomia nova* – wherein he spoke on the compatibility of the new astronomy with the Bible – and a translation of Foscarini's 'Letter on the motion of the earth' of 1615 – which had been banned by the Roman Church in 1616 – were appended to this work. Unlike these earlier books, the *Systema cosmicum* was easily available (although it was brought out in Strasbourg, the publisher was Elsevier at Leiden) and drew considerable attention. A second edition appeared in 1641 in Lyon, and a third in Leiden in 1699. Moreover, in 1636, a year after it had brought out the *Systema cosmicum*, Elsevier published (also in Strasbourg) Bernegger's Latin translation of Galileo's Letter to the Grand Duchess Christina with, in an appendix, an excerpt from Zúñiga's commentary on Job. All classic examples of the thesis that Bible exegesis should be accommodated to science were therewith readily available to the educated Dutch public. The general tenor of these writings is that the texts in the Bible which seem to speak of a moving sun and an unmoved earth should not be taken literally. The Holy Ghost had accommodated Himself to the understanding of the receiver. (Galileo's view is partially more complicated, but this does not concern us here.)[15]

---

[13] Cf. Emmius' reply, 1608 Oct 20, in Emmius (1911-1923) II, 55. (Lubbertus' letter has not been preserved.)

[14] See the letters by Mersenne to Rivet, 20 Nov. and 20 Dec. 1638, in Mersenne (1932-1986) VIII, 222, 239.

[15] On Kepler's accommodation of Copernicanism and the Bible, see Scholder (1966) 66-71; Karpp (1970); on Zuniga, Brotóns (1995). The literature on Galileo is overwhelming. An excellent work is Blackwell (1991), which also contains important sections on Foscarini. See also Westman (1986) 90-95.

The one authoritative author within the Dutch Republic to have paid systematic attention to the matter was Lansbergen, who had devoted five pages of his 'Considerations' to the refutation of the arguments from Holy Scripture against the Copernican system.[16] (He was joined by his son Jacob, who on this point also took up his father's defence against Froidmont.[17]) Lansbergen, too, uses the accommodation argument. He undoubtedly knew Kepler's arguments from *Astronomia nova*, but he appears largely independent. He offers two basic arguments. The first is that astronomy or mathematics are sciences independent of biblical knowledge and have their own foundations. Even people inspired by the Holy Ghost derived their secular knowledge from the gentiles: Moses was taught by the Egyptians and Daniel by the Chaldeans. Trying to obtain such knowledge from Scripture easily leads to error. Lansbergen, following Copernicus himself, points to the example of Lactantius, who on biblical grounds denied the sphericity of the earth, and of St Augustine, who denied the existence of antipodes. He offers some other examples as well, where it should be evident that one has to follow the learning of the mathematicians rather than the letter of the Bible. One is the 'molten sea' of King Salomon from 1 Kings 7: 23 and 2 Chronicles 4: 2. The Bible says it was 'ten cubits from the one brim to the other: it was round all about... and a line of thirty cubits did compass it round about.' These clearly are rough numbers, not an exact representation of the truth. Lansbergen took the example from his own *Cyclometria*, from 1616. In its dedication to Maurice of Nassau, he had given a short history of the attempts to measure the circle. As the oldest recorded attempt, he had referred to the account of King Salomon's molten sea, which showed how undeveloped this branch of mathematics had been in biblical times.[18]

Lansbergen's second basic argument turns up when he reviews the various texts commonly invoked to refute the Copernican system.[19] As he explains, in these instances the Holy Ghost does not speak according to the truth of the matter, but according to its appearance. Here, too, Lansbergen offered some examples to argue that this is the Holy Ghost's common way of speaking. One is taken from Psalm 19: 5-6, where it is said that the sun 'is as a bridegroom coming out of his chamber, and rejoiceth as a strong man to run a race'. The example may not have been wholly convincing, as the text was

[16] Lansbergen (1650) 17-22.

[17] For some extensive quotes from Jacob's work, see Dunin Borkowski (1933-1936) IV, 521-522.

[18] Lansbergen (1650) 18-19. Lansbergen (1616), dedication: '*Erat itaque tum temporis Cyclometria quaedam in usu, rudis scz. illa, quae diametri & peripheriae ratione ponebat triplam, hoc est, ut* X *ad* XXX.'

[19] Lansbergen mentions Ecclesiastes 1:5, Psalm 19: 6-7, Matthew 5: 45, Joshua 10: 12 and 2 Kings 20: 11 against the rest of the sun; Psalm 24: 2 and Psalm 104: 5 against the motion of the earth.

among the stock texts of the opponents of Copernicanism. (As such, it is also discussed by Kepler.) Lansbergen's other example is from Psalm 24: 2, where it is said that God 'hath founded it [the earth] upon the seas, and established it upon the floods'.

Lansbergen's short commentary remained a *locus classicus* in the Copernican debate in the Dutch Republic. Several of the examples introduced by him were to have a long and notorious career. Psalm 19: 5-6 (the sun compared to a bridegroom) would become one of the most discussed texts in the Dutch debate on Copernicanism and the Bible. Lansbergen's discussion of King Salomon's basin had even more impact. This presented a serious problem to the champions of a strictly literal interpretation; the text would remain a challenge for exegetes in the centuries to come. In the second half of the seventeenth century it would also, and this was not intended by Lansbergen, enter the *Tractatus theologico-politicus* by Spinoza and become a stock text of Deists and libertines who sought to undermine the very authority of the Bible. [20]

An entire treatise on the problem of Copernicanism and the Bible in the Dutch Republic was printed in 1651 by the Utrecht printer Johannes van Waesbergen. However, this was not the work of a Dutchman, nor was it contemporaneous. As Hooykaas has demonstrated, it concerned a work by Copernicus' pupil Joachim Rheticus, and at the time was about a century old. It defends Copernicanism in the usual way, i.e. by asserting that the Bible is not meant to teach physics or astronomy. The work does not appear to have had much influence. For our purpose, the publication is of interest as it attests to a growing local concern of the problem of Copernicanism and Bible interpretation. [21]

The one other major authority on the issue, but on the anti-Copernican side, was Gisbert Voetius. We have already dealt with Voetius' polemic with Batelier. At this point, we may go somewhat deeper into his arguments. The main reason for Voetius to spend so many pages on the issue of the earth's motion was Batelier's exegetical principles. For Voetius, Batelier's view of the Bible is downright blasphemous. If the Bible is not fully trustworthy in every detail, there are only two possibilities: either the Holy Ghost *cannot* speak the truth, or He does not *want* to. Both possibilities are absurd. Whoever takes such liberties in exegesis will go from bad to worse. The argument that the Bible accommodates to the understanding of the 'common man' would imply that the Patriarchs, the Prophets and all the ancient people were too stupid to understand the Copernican system, even when dictated by the Holy Ghost

---

[20] Spinoza (1925) II, 36. On its later use, see Vermij (1991) 82-83.
[21] Hooykaas (1984).

Himself. And then, should one surmise that the Bible only directs itself to ignorant contemporaries, and not to intelligent followers of Copernicus in our own time? Voetius concludes: 'It cannot be admitted that we undermine the certainty and divine authority of Scripture in order to support clever human conjectures, or what they call probable hypotheses, regarding celestial matters.' [22]

Voetius was a key figure in the formation of a new generation of ministers and theologians. One might assume that he imposed his view on the Copernican issue upon his students, even if he did not bother to tackle the matter in print any further. This assumption is difficult to assess, however, since most of his students left no writings, least of all on physics or astronomy. Still, we might offer an example of what appears to be Voetius' influence in this respect. One of his favourite students had been a Gualtherus de Bruyn, who had answered to his four disputations on atheism in 1639. In 1640, this De Bruyn chose to obtain a doctorate in philosophy. His doctoral disputation, tripartite as was usual in Utrecht, dealt with evil, unwilled and voluntary deeds, and the motion of the stars. In the last-mentioned part, he defended the traditional view that the stars were fixed to solid orbs. De Bruyn gave some consideration to the fact that the Bible does not mention these orbs, but concluded that his view need not contradict Scripture. Finally, he proposed an argument against the Copernicans. The Copernicans (he probably was thinking of Lansbergen) regarded the immense velocity which the heaven should have according to the Ptolemaic system as an impossibility. But according to De Bruyn, their own principles forced them to allow for such great velocities in another respect. Christ, Elia and Henoch had been taken from this earth to the empyrean, the third heaven, as would happen with the blessed on the Day of Judgement. Now, given the immense extensiveness of the first and the second heaven which the Copernicans assume, one might calculate that they had crossed these at a velocity of about the same order of magnitude as that of the stars. [23]

Another writer who shows Voetius' influence is Martin Schoock, who had been his second in the conflict at Utrecht in 1641-1643. Schoock had defended three disputations on scepticism, and during the conflict Voetius urged that he rework them into a full book. [24] However, it was published only ten years later, in 1652, and then only the first part of it (the second part never appeared). It is a comprehensive and erudite work. Its starting point are the

---

[22] Voetius (1635) 283.

[23] De Bruyn, disp. Utrecht (pro gradu) 1640 Dec., *Problemata de motu siderum*, 1, 2, 7. See also the astronomical corollaries nos. 3 and 4.

[24] Voetius (1648-1669) 1, 1157. Duker (1897-1915) 11, 175.

opinions of the ancient sceptics (Sextus Empiricus, among others), but in fact it is Cartesian philosophy that is attacked. The work consists of four books. The fourth is devoted to a review of the various disciplines, among them astronomy. Schoock asserts, against the sceptics, the certainty of the discipline. [25] In a later chapter, he explains the difference between the true hypotheses of physics and the notional hypotheses of the mathematical disciplines. [26] Finally, there is a chapter on the authority of Scripture in natural things. [27] Herein, Schoock argues, against Lansbergen and Foscarini, that the Bible does not speak according to the opinion of the people, but *ex rei veritate* ('according to the truth of the matter'). Schoock repeats Voetius' argument that the other opinion implies that the Holy Ghost is either unable or unwilling to speak the truth. The Bible's prime aim is indeed to show us the way to salvation, but that does not exclude other intentions. It accommodates to our understanding, 'but not so that it lies with the liars and errs with the erring'. [28] Schoock ends by giving five rules for a legitimate exegesis of Scripture concerning natural matters: one may not question articles of faith; one should account for the way of speech used by the Holy Ghost; one should not distort the meaning of the words, because that would seem more reasonable; truth may not conflict with truth; and the sentence taught should be more reasonable than the opposite.

Schoock shows some sophistication in dealing with the problem of Copernicanism. To Voetius and Revius, it will be remembered, there was no problem at all. The Bible was clear. Philosophical or astronomical considerations were simply not worth looking at. Schoock, however, felt urged to discuss the philosophical aspect as well. But then, of course, he was a philosopher himself.

Cartesianism enters the fray

Schoock wrote his treatise mainly as a refutation of Cartesian philosophy. His dealing with Copernicanism still does not take a very prominent place in the book. But as at this very time Cartesianism became the subject of a serious discussion, the problem of Copernicanism and the Bible became a spearhead in the debate. Defenders of the motion of the earth found new arguments in

---

[25] Schoock (1652) 367-379 (book IV Ch. 11).
[26] Schoock (1652) 381-398 (book IV Ch. 18). Page numbers are rather confused; between 379 and 381 are a hundred other pages.
[27] Schoock (1652) 399-426 (book IV Ch. 19).
[28] Schoock (1652) 406.

Cartesian philosophy. In reaction, anti-Copernican theologians came to feel the need for a more elaborate defence of their view.

Daniel Lipstorp, whose *Copernicus redivivus* we discussed earlier, was not a theologian. Still, he devoted some pages to biblical exegesis in his book.[29] Lipstorp's avowed aim was to propagate Cartesianism, and as such his book is the first instance of a 'Copernican' reading of the Bible in the service of Cartesian philosophy. Still, the relevant chapter is not very philosophical and largely a compilation of earlier opinions. Nevertheless, Lipstorp takes a clear stance. It is based on the idea, made popular by Galileo, of the 'two books', virtually two divine revelations, the book of nature and the book of Scripture. The Bible has not been given to man in order that he may learn mathematics or knowledge of nature. This should be drawn from the book of nature. Of course, there is only one truth. The theologians are quite right in asserting that what is true in theology should be true in philosophy as well. But on the other hand, what is false in philosophy cannot be turned into a truth in theology.[30] Lipstorp then explains that there are four kinds of arguments which people draw from Holy Scripture in order to refute Copernicanism: sentences which state or presuppose the rest of the earth, sentences which mention the motion of the sun, sentences which confront earth and heaven as centre against circumference, and sentences which speak of the earth below and the heavens above. As Lipstorp sees it, all these cases can be solved by supposing that the Bible demonstrates things as they appear to us, not as they really are. In natural things, one should follow one's senses and necessary demonstrations, not the Bible. 'For it would be a sin so to cut off the variety of arguments, conceded to us by Divine goodness, and therewith abuse the authority of Holy Writ, that we would dare to oppose manifest experiences and necessary demonstrations.'[31]

At about the same time, there was a certain theologian for whom combating the theories of Copernicus became a major task: the Leiden minister Jacob du Bois. Du Bois served as a Reformed minister in several villages in northern Holland from 1631 onwards. In this period, he engaged in polemics with the Mennonites, who were numerous in the region. In 1646 Du Bois became Reformed minister at Leiden, a position he retained until his death in 1661.[32] He was interested in natural sciences and appears to have been an early, if critical, reader of Descartes' *Principia philosophiae*. In 1648, he wrote to Descartes on the latter's theory of the elliptical form of the moon's orbit around the earth. It is

---

[29] Lipstorp (1653) 150-160.
[30] Lipstorp (1653) 151.
[31] Lipstorp (1653) 156.
[32] Biographical data in BWPGN, I, 476.

not known what exactly his objections were; the matter is only referred to in some later pamphlets. Descartes answered him in a letter of 8 April 1648. Du Bois sent an extensive reply on 28 April, whereupon Descartes sent back a small note with the request to let the matter rest.[33]

In 1650, Du Bois published a major work: a chronology of world history – from Creation to Crucifixion – in the vernacular. As such, the book supplemented the chronology by Paullus Merula, who had dealt only with recent times. In itself, the book is not very important. Du Bois consciously ignored most that had been put forward in chronological studies, as he chose to limit himself to biblical history and sacred authors. One should not mix the sacred and the secular, he explained. In Du Bois' career, the book is not insignificant, however. His dealings with chronology forced him to become acquainted with astronomical theory. He used Julian years (and the Julian period, which had been introduced by Scaliger). In order to find the correct places of months and days, he had to calculate the dates of the new moon. 'This entailed no small labour, but I have gladly spent it to obtain my further purpose, to wit: to bring God's Word in accordance with nature, by astronomical calculations. And all my labour has been sweet, as I found this agreement.' In order to be able to carry out these calculations, Du Bois had the mathematician Samuel Karl Kechel give him lessons.[34]

Kechel, who was of Bohemian descent, was curator of the astronomical instruments at Leiden University. He also gave some courses. Later, he assisted another Leiden theologian when he drew the figures of the temple from the vision of Ezekhiel for Cocceius' 1668 commentary on this book. Kechel appears to have been a well-known figure in his time. He published little, however, and today is known mainly for some observations of comets and meteorological phenomena. His opinion on Copernicanism is unknown.[35] One may ask how deep the tuition he offered Du Bois was, but one easily understands that among his fellow theologians, Du Bois was henceforth considered an astronomical expert.[36]

Du Bois' work on chronology also reveals its author's attitude to the Bible. Du Bois was forced to admit that in the Book of Chronicles some copyist's errors had crept into the text, as the literal text amounted to chronological

---

[33] [Hudde] (1656)a 29. [du Bois] (1656)b 6.

[34] Du Bois (1650), introduction. Kechel is mentioned in the preface.

[35] On him: van Maanen (1987) 225; de Waard in NNBW, II, 651. Applebaum and Hatch (1983) mention a *Systema novum* by him, of which I have not found a trace elsewhere. On his part in Cocceius' book: van Asselt (1996) 189.

[36] The respondent of a theological disputation by Cocceius, *De Patre, Filio & Spiritu Sancto* (Leiden, 30 Jan. 1655), Sibrandus Vomelius, in his dedication called Du Bois '*Mathematico insigni*'.

impossibilities. He admitted this only reluctantly, 'But what shall we do, when necessity forces us? And what shall we complain, if the correction follows from God's Word itself?' He was ready to accept any better solution, 'but meanwhile, I ask to be kept for the one who was brought to that against my will by force of reasons; always retaining that one should not bring oneself to saying such things, unless compelled by force of reasons.' [37]

After publishing his chronology, Du Bois applied his astronomical knowledge to the problem of the motion of the earth. In 1653, he published a *Dialogus theologico-astronomicus* entirely devoted to the problem. At the very outset, he stated that the problem is actually a non-problem, as the Bible testifies in a perfectly clear way that the sun moves and the earth stands still. However, advocates of Copernicanism claim that they can prove their point in a compelling, mathematical way. As most people have insufficient knowledge of astronomy to judge this claim on its own merits, they may become confused, as truth in astronomy would seem to contradict truth in theology (which is actually impossible, of course). A serious inquiry into their claims – theological as well as astronomical and philosophical – is therefore necessary. [38]

The work itself is written as a dialogue between Asterictus, who posits the Copernican arguments (for these, Du Bois used as sources the works of Philips and Jacob Lansbergen, the relevant passages in the *Systema cosmicum*, and the six disputations held at Rostock by Daniel Lipstorp), and Eudoxus, who refutes them. After some introductory points, the first part of the book is devoted to the arguments from biblical exegesis. [39] The arguments why the Bible should not be taken literally when speaking about the sun and the moon are rejected as utterly wanting. Du Bois emphasises that the Bible should be taken literally 'unless clear and sufficient reason leads and forces us to the contrary'. [40] In some cases, indeed, the Bible speaks in a figurative way. If firm evidence convinces us that the Bible cannot be taken literally (as in the case of the molten sea of King Salomon, where the Bible evidently uses rounded figures), we can concede without problem. However, if the arguments are only probable, as they are in astronomy, our respect for the Scriptures should gain the upper hand. 'We have to submit and give ourselves as captives to divine authority, and not depart from the proper sense of the text because of some human *it seems*: for, if this were accepted, one could have no truth, not even in matters of faith, from Holy Scriptures.' [41]

---

[37] Du Bois (1650), introduction.

[38] Du Bois (1653) preface.

[39] Du Bois (1953) 8-35.

[40] Du Bois (1653) 21 (*'nisi ratio clara & efficax nos ducat & cogat in contrarium'*).

[41] Du Bois (1953) 32.

Thereupon, Du Bois discusses a large number of astronomical arguments advanced by the Copernicans to prove their point.[42] These arguments are mainly taken from Lansbergen. It is striking that at this stage Cartesianism seems to play a minor role. The point is, however, that Du Bois' argument is rather scholastic. Not only are all arguments transformed into syllogisms before being discussed, but Du Bois also departs from set topics (or *quaestiones*) and simply collects ready-made arguments from known authorities. Descartes, however, does not offer such arguments, as nowhere does he argue the motion of the earth (instead, he takes it for granted and describes how it is brought about by the laws of nature). Hence, Descartes' works were of little use to Du Bois. However, this does not mean that Du Bois was not aware of Descartes' work or thought it irrelevant. In the preface, he states that the theory of a moving earth is much propagated by a 'sect of new philosophers'.[43] The rise of Cartesian philosophy may well have been Du Bois' main motive to tackle the subject in the first place. However, at this stage he is not aiming at specific Cartesians or their works.

As one may expect, Eudoxus finds all arguments put forward by Asterictus utterly inconclusive. This section concludes with a discussion of Tycho Brahe's geoheliocentric system, which appears to have Du Bois' preference.[44] In the end, of course, Asterictus is completely won over to Eudoxus' side.[45]

So, by 1653, Copernicanism had become something of an issue in the Dutch Reformed Church. It appears that it was mainly the rise of Cartesianism which made it a matter of theological concern. However, it was rather a marginal issue and so far there had been no indication that it would become of any real significance. But within three years, the situation changed radically. By 1656 Copernicanism was the centre of a huge debate which ended up virtually splitting the Dutch Reformed Church in two. How this came about and what made it possible is the subject of the following chapters. Actually, it was not a matter of a single debate: several controversies were waged at the same time. For clarity's sake, I shall not keep to chronological order, but discuss these parallel debates one by one. One should keep in mind, however, that in practice, these polemics were part of one large conflict, and contemporaries were well aware of this. The nature of this conflict will become clear in the course of time.

---

[42] Du Bois (1653) 35-65.

[43] Du Bois (1653) preface. The preface is dated 1 March 1653. The dedication of Lipstorp's *Copernicus redivivus* is dated 1 April 1953, so Du Bois cannot be referring to this work. He could have known, however, Wittichius' two disputations held at Duisburg in 1652, on which below.

[44] Du Bois (1653) 60-65. The defence of Tycho's system was later repeated in du Bois (1655) 302, even with a scheme of Tycho's system.

[45] Du Bois (1653) 65-66.

12. Copernicanism as a theological problem: the Wittich affair

A theologian championing Copernicanism

So far, the theologians had been quite successful in keeping Cartesianism at bay. Even if they could not prevent its winning over many philosophers, they could insist that Cartesianism concerned philosophy only. As far as the philosophers were kept within the limits of their proper discipline and not allowed to address questions which the theologians considered legitimately theirs, their efforts could be ignored and no real harm was done. It soon appeared, however, that the spirit of Cartesianism would not confine itself to these well-defined limits. Cartesianism entered the very field of theology itself. It has been argued that modern critical theology is primarily the result of the Cartesian method. That is, the method of systematic doubt and the new standard of mathematical certainty, when applied to the Bible. [46] Such claims may seem probable, but are difficult to prove. That Cartesianism had a profound impact on theological thinking is nevertheless certain. The discussion on the system of the world played a crucial part in this. Under the influence of Cartesianism, the debate on Copernicanism and the Bible got at the heart of theological debate in the middle of the 1650s.

The matter was brought to a head by Christophorus Wittichius. Earlier, we discussed his 'mathematical' view of Copernicanism. Wittichius openly connected Copernicanism and Cartesianism. It was no longer just the motion of the sun that was at stake in the exegesis of such texts as Joshua 10: 12, but the whole of Cartesian philosophy. But what gave Wittichius' arguments special significance was that this was not a philosopher or an astronomer but a theologian upholding Cartesian philosophy and accommodating it to the Bible.

How far theologians so far had accepted or rejected Copernicanism is difficult to say. Probably not all ministers agreed with Du Bois' or Voetius' stance. Most ministers, of course, had only a very hazy idea of astronomy, and readily agreed with any colleague who defended the dignity of the Bible.

---

[46] Popkin (1982). More nuanced: Scholder (1966) 131-170.

But others, for example, people like Heidanus, were more open to secular knowledge. So far they had not come into the open. Wittichius changed this, and in this way became the main instigator of the debate on Copernicanism within the Dutch Reformed Church. Therefore, at this point we may as well have a closer look at the way he theologically legitimised his defence of Copernicanism, and at his colleagues' reaction.

Wittichius was born in 1625 in Brieg (now called Brzeg) in Silesia to a Lutheran family.[47] He initially studied law, but then joined the Reformed Church and switched to theology. He studied at Groningen under Schoock and Maresius and then at Leiden, where he made friends with several fellow students who were later to become prominent Cocceian theologians. In 1651, he became professor of theology at Herborn, the reformed theological seminary of the German county of Nassau. At Herborn, Cartesianism had stirred up very much the same kind of disputes as it had earlier at Utrecht and Leiden. It was upheld by the professor of philosophy Johannes Clauberg, who was to become one of the philosophically most important followers of Descartes. His most outspoken opponent had been the professor of history, Cyriacus Lenz (Lentulus).[48] Wittichius and Clauberg had been friends since their student days[49] and on entering the seminary, Wittichius without hesitation took sides with Clauberg.

In 1652, Wittichius and Clauberg were appointed to the chairs of theology and philosophy, respectively, of the newly founded Duisburg University. This Reformed university, in the Duchy of Cleves, was situated close to the Dutch border and was tied to the Dutch rather than the German universities. It very soon became a hotbed of Cartesianism. At Duisburg, Clauberg and Wittichius continued their dispute with Lentulus. Soon after his appointment, Wittichius had two disputations defended. The first, 'on the abuse of Holy Scripture in philosophical matters', investigated, still according to the title, '1. Whether the Scriptures are the real principle of physics. 2. Whether the Bible, when it speaks of natural things, follows always the accurate truth, or rather more often the sense and opinion of the common man.' The second disputation, which formed a whole with the first, 'deals with the disposition and order of the whole universe and its principal bodies, and defends the opinion of Descartes on the true rest and true motion of the earth.' These works met

---

[47] On Wittich: *BLGNP*, II, 461-463; see also Bizer (1958) 340-347.

[48] Trevisani (1992) 25-27. See also 69-114 on Clauberg's ideas. Lenz' political writings are dealt with by Dunin Borkowski (1933-1936) II, 114-115, 434-435.

[49] Clauberg wrote a poem in praise of Wittich, *ob gratam in studiis* συμμαχιαν, in the latter's disputation *De aqua*, defended at Groningen under Schoock in May 1646. (FS)

with a response seldom encountered by academic disputations. In 1653, the two disputations were republished in Amsterdam.[50]

In his introduction, Wittichius stated his intentions in a rather polemical way. He attributes the current attacks on Cartesianism to manipulations of the Devil, the enemy of truth. Wittichius aimed to explain the motion of the earth in a clearer way than had so far been done, and also wanted to discuss the exegetical problems. In the first of the two dissertations, then, Wittichius argued that the Bible cannot be used as a source of knowledge in natural philosophy. The only foundation of philosophy is the light of reason. He extensively discussed the work of authors who had maintained the opposite, both sixteenth-century proponents of a *physica mozaïca* (Danaeus, Aslacus) and more recent authors, such as Du Bois, Revius and Schoock. In Wittichius' opinion, in matters of practice or morals the Bible does not speak according to the truth of the matter, but according to the opinion of the people. To defend this view, he musters an impressive array of theological authorities. Most of them are from the Reformed tradition, such as Calvin or Rivet. But he also refers to such writers as Joannes Chrysostomus and Maimonides, and even, be it incidentally, to non-theological authors, such as Francis Bacon.

So far, the argument rests completely on an appeal to authority. In modern eyes, this may seem unimpressive, but at the time it must have seemed an erudite and important piece of work. However, Wittichius added some considerations of his own. In his third chapter, he undertook to demonstrate his thesis in a large number of places where the Bible, speaking on practice or morals, apparently spoke 'according to the opinion of the people in deviation from the truth' ('*secundum opinionem hominum à veritate recedentem*'). Especially this third chapter would arouse the indignation of the Voetians.

In his second dissertation, Wittichius explained the Copernican/Cartesian world system. As explained earlier, Wittichius here offered as proof of the Copernican system a summary of Descartes' *Principia philosophiae*, put in a geometrical order. As it seems, Wittichius thought this argument conclusive. Two final chapters discuss the objections of his adversaries. The first deals with objections from the Bible, first in a general way and then from specific biblical sentences. The second chapter answers philosophical objections.

Detractors and rejoinders

Wittichius' work put the motion of the earth high on the theological agenda.

---

[50] I quote from this re-edition, Wittichius (1653); I did not see the original version. On these dissertations and Wittichius' subsequent works, see also van Bunge (1999) 311-313; Verbeek (1992) 74.

Up until then, most theologians had ignored the matter. After Wittichius, they had to take a position for or against his work. The two disputations elicited various rejoinders. It is not surprising that Jacob du Bois felt he needed to react. In 1655 he published a 300-page volume entitled *Veritas et authoritas sacra* ('Sacred truth and authority in natural and astronomical matters asserted and vindicated. Against the two dissertations (...) by Christophorus Wittichius (...). Added a refutation of the argument for the motion of the earth, taken by the same author from Descartes' Principles.')[51] In his earlier *Dialogus*, Du Bois had discussed Copernicanism after the major astronomical authors without speaking of Cartesianism itself. In this new work, Cartesianism is the main target. In an extensive preface, Du Bois discusses the philosophy of Descartes in general. His arguments form a heterogeneous collection, most of which had been used earlier in the general discussion on Cartesianism, which had been going on since the 1640s. So, he argues that Descartes rejects the traditional proof of God's existence; that there are examples of people who have fallen into atheism because of his philosophy; that Cartesianism contradicts Holy Scripture; and that Descartes upsets the established system of learning and leaves it all in disorder.

It would be going too far to follow Du Bois in all his arguments against Wittichius. Everything Wittichius puts forward is extensively discussed and rejected. Du Bois' general ideas have become quite clear already from the discussion above of his *Dialogus* (and we will review yet other works by him in the following), and in fact, Du Bois adds little to it here. His main argument is the authority of the Bible. One should not detract from the literal sense unless compelled by necessity. The Cartesians simply do not want to subjugate to divine authority. However, as they do not dare reject it openly, they use the excuse that the Bible speaks according to the opinion of the people rather than the truth of the matter. This opinion undermines the Church. The Bible is the sole weapon against heretics and against Satan himself. The opinion that the earth moves around the sun may be ancient in astronomy, but Wittichius is supposed to write not as an astronomer, but as a theologian whose task it is to teach young students the way to explain the Scriptures.[52] Wittichius is not a disciple of Christ, but of Descartes, that pupil of the Jesuits who drew his cogitations not from the clear sacred sources but from an impure pit, never becoming so wise that he abjured Catholic religion. 'He seems to swear on his [Descartes'] words, rather than on those of the prophets or of Christ himself.'[53]

---

[51] Du Bois (1655).
[52] Du Bois (1655) 117-122.
[53] Du Bois (1655) 213.

Whereas Du Bois was mainly refuting Wittichius' first dissertation, Jacob Revius, who was also from Leiden, took it upon himself to answer the second one. In fact, Revius appears to have regarded it as directed mainly against his own *Statera*. His book bore the title *Anti-Wittichius*.[54] It was particularly concerned with Wittichius' views on the infinity of the world, which formed the basis of his argument for the new cosmology. Revius defended the view – which, according to him, was supported by the Bible – that the world is limited and comprehended by the third heaven, the habitat of the blessed, but his concern is to demolish rather than to construct. He proceeds by quoting a passage from Wittichius and commenting on it, before going to the next passage. This does not make very good reading. Very often, he seems to be splitting hairs.

Another book was published by a young theologian who had not been much noticed until then, but who after Voetius' death became professor of theology at Utrecht: Petrus van Mastricht. Van Mastricht was born in Cologne and spent the first part of his career in various churches and at various universities in Germany, but his theological studies in the United Provinces brought him under the influence of Voetius. His *Vindiciae veritatis* ('Vindication of the truth and authority of Holy Scripture in philosophical matters against the dissertations by Chr. Wittichius') was published in 1655, as were his earlier-mentioned books. In a long dedication to the Duisburg town government, van Mastricht claims that Wittichius' assessment that the devil has a hand in the attacks upon Cartesianism puts the ball back in the other court. Cartesian philosophy, according to Van Mastricht, attempts to bring people first to scepticism, and by that route to atheism. First of all, Descartes attacks the book of nature. He rejects the received arguments to prove God from nature, and replaces them with his own nonsensical speculations. However, this is only a first step. As long as the Bible remains unmolested, Satan will not achieve his goal. So, the second object of Cartesianism is to attack the Bible itself. This cannot be done openly, so it is done by guile and deceit. Some philosophical comment is produced, squarely opposing the Bible and strengthened with false arguments against the truth. Now, it is maintained that the Bible, where it runs counter to this false opinion (the heliocentric structure of the universe), accommodates to the erring opinion of the people and need not be taken so seriously. Thus, the authority and infallibility of the Bible are rejected, and one can easily guess what will remain in the end of Christianity itself. The only problem left is that Descartes, as a Papist, has

[54] Revius (1655).

but little authority among the Reformed. However, by now the Devil has found someone else to spread this error in the Reformed Churches.[55]

In the book itself, Van Mastricht reformulates the problem posed by Wittichius in the form of two questions. In the first place, can the Holy Ghost tell untruths, simply in order to avoid scandal among the people? Of course not, answers Van Mastricht, as this would give rise to consequences for belief which are unacceptable. Secondly, is it allowed for a philosopher to decide whether pronouncements of the Holy Ghost are true or not? Again, certainly not. This would mean putting reason above the word of God in the way of the Papists and the Socinians.[56] These general principles are then put into practice in the exegesis of the relevant biblical sentences, whereby Wittichius' explanations are refuted.[57] Finally, Van Mastricht emphasises that even if one cannot deduce a complete course of physics from the Bible (a fact generally admitted, according to him), the Bible does speak about physical matters incidentally, and aims at explaining nature where the faithful need such explanation. The question itself, i.e. whether the sun or the earth is moving, is not discussed by Van Mastricht. As he sees it, this question had been sufficiently elucidated by Du Bois and in the Utrecht disputations which we are to discuss now.[58]

Even before Du Bois published his book, another attack on Wittichius had been launched from the theological faculty at Utrecht. On 31 May and 24 June 1654, the student Arnoldus Niepoort defended a disputation in two parts 'on the authority and truth of Holy Scriptures in philosophical matters'. Also on 24 June, Johannes Beusechum defended a disputation 'on the infallible truthfulness of the natural things, revealed throughout Holy Scriptures'. Then 4 October of the same year finally saw the defence by Henricus Troy of a 'philosophical-theological investigation, whether N.N. has proved, by way of some theses and hypotheses, so firm the rest of the sun and the twofold motion of the earth, that Holy Scripture, stating the opposite, should be taken as speaking according to the erroneous opinion of the people'.[59]

President at these disputations was the professor Andreas Essenius, but the respondents themselves are indicated as the authors. There seems to be no

---

[55] Van Mastricht (1655), dedication.
[56] Van Mastricht (1655) 13-26.
[57] Van Mastricht (1655) 70-188.
[58] Van Mastricht (1656) 222-258.
[59] For the disputations of Beusechum and Troy: disp. Utrecht (Essenius), 24 June and 4 Oct. 1654. I have not seen those by Niepoort. My description is based on the re-edition in Niepoort, Beusechum & Troy (1656), which contains the two earlier disputations by Niepoort on pp. 1-17 and 17-40 respectively. The re-edition of Beusechum's disputation is on pp. 40-66, and of Troy's on pp. 66-81.

reason to doubt this assertion (as Wittichius was later to do). It was not uncommon for professors to use the names of their students in order to attack each other more freely, but a disputation under one's presidency does not seem the right way to remain unnoticed in the background. Moreover, the four disputations show marked differences in style and content. This, of course, does not negate the fact that it was a concerted action (as the disputants admitted themselves), which must have had the active support of Voetius and Essenius.

The disputations attacked Wittichius severely, but they did not transgress the limits. They even distinguish themselves favourably from the attacks by Revius, Du Bois or Van Mastricht. They attack Wittichius' arguments rather than his person, and in no way is there an attempt at character assassination. Perhaps these young men really believed that arguments counted for something in this debate. To prevent scandal and contain the dispute within the academic sphere, the disputants do not even mention Wittichius' name (quite unlike the writings by Du Bois, Revius and Van Mastricht of the next year, who all mention it in their very titles). They only speak of their '*Opponens*'. Of course, every theologian in the United Provinces must have known what it was all about. For the sake of clearness, in discussing these writings I shall mention Wittichius by name.

Niepoort opened with a reminder of the fact that Christian apologists had always emphasised the excellence of the Bible over the Talmud or the Koran exactly in the field of secular learning. Whereas the Talmud and the Koran contain but philosophical nonsense, the Bible offers a treasure of true knowledge on the whole world. Wittichius now invalidates this argument.[60] Niepoort also argues that whereas Wittichius accuses others of being subject to prejudice, he himself only proceeds the way he does in order to save his own philosophical prejudices. Among these prejudices the idea that the Bible does not teach natural things figures prominently.[61]

As regards exegesis – the core problem – Niepoort proceeds carefully. He extensively discusses the ways of speaking in the Bible. Many expressions are indeed figurative. However, the Bible does not use these without reason. It may use them to indicate matters above our understanding; or, alternatively, in case of clear metaphors which nobody, not even the common people, will take literally. One should always follow the principle that Scripture can only be explained by Scripture itself.[62] The larger part of the second disputation is devoted to the discussion of specific texts and arguments. Moreover, if Witti-

60 Niepoort, Beusechum and Troy (1656) 1 (th. 2.).
61 Niepoort, Beusuchum and Troy (1656) 9-10 (th. 6).
62 Niepoort, Beusechum and Troy (1656) 2-7, 17-21.

chius were right in stating that the Bible follows the erroneous opinion of the common people, it is hard to see why this would concern natural knowledge only, and not theological and moral knowledge as well. After all, texts on natural things concern also articles of faith, as they teach virtue and obedience.[63]

Beusechum's disputation was principally devoted to a defence of Mosaic physics. He argued that Wittichius drew a caricature of it by pretending that it strove to deduce the whole of physics from the Bible. Nobody denies that the Bible ignores certain matters. But anything that *is* stated in the Bible, is true. In a similar vein, the fact that the Bible's principal aim is instruction into the faith does not invalidate the claim that it is infallible in other fields, too. If the Bible occasionally were to speak according to the erroneous opinion of the people, one would no longer be able to deduce any certain knowledge from it. Indeed, 'then one should (following that new invention) have to put all in doubt.'[64]

The whole discussion so far is purely theological. The authority of the Bible is discussed on the basis of theological premisses. There is not a word about astronomy. The relation between theology and philosophy is not touched upon, as the possibility that anything outside theology might be relevant simply does not occur to the disputants. Only in his last thesis does Beusechum state that all reason comes from God. If reason seems to stand against reason, one should suspend judgement. One should not try to mend affairs by means of a tortuous biblical exegesis.[65]

However, in the last disputation – the one by Troy – philosophy is the main subject. Troy investigates Wittichius' 'mathematical proof' of the motion of the earth. He successively reviews all Wittichius' theses, and then demolishes them. One hardly could expect otherwise. Wittichius' argument is purely Cartesian, whereas Troy grounds himself on Aristotelian philosophy. In several instances, he refers to his teacher, the philosophy professor Daniel Voet (Gisbertus' son). In the end, he remarks sharply that to Wittichius, Descartes seems to have written the Bible.[66]

These four disputations were answered by Wittichius in 1655 in a rejoinder, 'Theological consideration of the style of the Scripture, which it adhibits when speaking of natural things'. This work, too, was originally defended at Duisburg as a disputation. The next year (1656), Wittichius published it as a separate work in Leiden.[67] My discussion is based on the latter edition. In

---

[63] Niepoort, thesis 2 of the second disputation; see also th. 7 of the first one.

[64] Essenius/Beusechum, disp. Utrecht 24 June 1654, th. 3.

[65] Essenius/Beusechum, disp. Utrecht 24 June 1654, th. 9.

[66] Essenius/Troy, disp. Utrecht 4 Oct. 1654.

[67] Wittichius, disp. Duisburg 5 June 1655; Wittichius (1656).

the introduction, Wittichius clearly states that the work is a reaction to the four disputations, which he attributes to Essenius (without mentioning him by name). He writes that in principle his disputations could defend themselves and did not need a separate defence. However, he had decided to use the opportunity to elucidate his own point of view.

As a statement of Wittichius' views, the work is definitely successful. It is more readable than religious polemics used to be. The argument is clear and the whole book is only 82 small octavos (the original disputation ran to 13 quartos). Even so, however, quite a number of pages are devoted to a criticism of the way his adversaries have reproduced his ideas. As Wittichius does not hesitate to accuse them of calumny and suggests bad faith, it is doubtful whether in the end the book contributed very much to a better understanding.

Wittichius emphasises that the dispute is not about the authority of the Bible, as his adversaries suggest, but about the correct interpretation of the texts.[68] This leads him to an extensive treatment of the principles of exegesis. He introduces the concept of *notio vulgaris* (common notion) or *cognitio vulgaris* (common knowledge). Such common knowledge stands in opposition to philosophical knowledge. The Bible uses common notions, but that does not make its sayings untrue. His opponents, states Wittichius, put *vulgaris* (common) on a par with *falsa* (false). This is unjustified. A 'common notion' is defective, but it is not incorrect. It contains the truth *ad homines relatam* ('with respect to man').[69] Philosophical knowledge, on the contrary, 'transcends common knowledge, as it supplies its defects, leads to a knowledge of qualities and substances which is not confused, but distinct, and, basing itself on the common notions which are known in nature, corrects prejudices and errors. Philosophy investigates the interior of natural things and traces their hidden causes...'[70] The Bible does not offer this kind of philosophical instruction. It deals with things as they appear to us.

This treatment on a fundamental level makes the work more than just an occasional piece. Whether it was fit to convince his opponents is another matter. The general way it deals with the problem is not only the strength but also a weakness of the work. Wittichius says hardly a word on the question in hand, i.e. the motion of the earth. He seems to assume that acceptance of his premisses will automatically lead to acceptance of a heliocentric view, but this of course was too optimistic. One can subscribe to his views on

[68] According to Niepoort in Niepoort, Beusechum and Troy (1656) 106, this passage was not in the original disputation, but only in the Leiden edition.
[69] Wittichius (1656) 33.
[70] Wittichius (1656) 41-42.

nearly every point and still remain convinced that the Bible teaches that the sun moves. In fact, this had been stated before by Beusechum in one of his corollaries: 'Even if we grant our opponent that his fallacious statement (to wit, that the Scriptures at some places speak according to the erroneous opinion of the people) has been proved, he cannot prove in eternity that Holy Writ does so in Joshua 10: 12-13, Ecclesiastes 1: 5-6 and similar places.'[71]

This was in fact the main line of argument in the rejoinder from Niepoort, Beusechum and Troy. In the same year (1656), they republished their earlier disputations together with an extensive rejoinder wherein they answered Wittichius' arguments. They explain that initially, on seeing the Duisburg disputation, they had not deemed it worthwhile to react. But now that Wittichius had triumphantly republished this work in Leiden, they could not remain silent. The tone has become more acerbic by now. Wittichius is named openly, even in the title: 'Four theological disputations on the use of Holy Scripture in philosophical matters, against the dissertations by Christophorus Wittichius. Added a vindication of these disputations against the latter's Theological consideration on the style of Scripture'. The book is printed in the same format as Wittichius', apparently so that they could be bound together. It contained an official approbation by the theological faculty of Utrecht, signed by professors Voetius, Essenius and Nethenus.[72]

The new rejoinder, which is by far the largest part of the book, comprises a general introduction dealing with matters which concern all three authors, along with each authors' reactions. In the general part, the authors among other things report their attempts to meet Wittichius for a personal interview. According to them, their attempts were deliberately thwarted by Wittichius. They appear moreover particularly annoyed by Wittichius' repeated accusations that they have deliberately distorted his views; this theme occurs not only in the general part, but also in the separate contributions. 'We protest before God that we did not know that the author of the dissertation defended another opinion than the one we attributed to him in our disputations. In the following notes we shall prove that he defends exactly those opinions which we attributed him.'[73]

Of the individual answers, Niepoort's is the most comprehensive. The matter indeed concerns him most. He elaborates upon the differences between Wittichius' two books, as well as between the Duisburg and the Leiden ver-

---

[71] Essenius/Beusechum, disp. Utrecht 24 June 1654, annexa, 1.

[72] Niepoort, Beusechum and Troy (1656), preface.

[73] Niepoort, Beusechum and Troy (1656) 90. The '*Vindiciae pro disputationibus*' comprehends pp. 84-240. Of these, pp. 84-91 are a general introduction. The rejoinders of Niepoort, Beusechum and Troy are on pp. 91-186, 187-205 and 205-240, respectively.

sion of his book on the style of Scripture. It appears that Wittichius has become ever more cautious. In the earlier work, the contents of the Bible was at stake; in the later work, he only deals with ways of saying. In fact, he can largely subscribe to Wittichius' views in the latter book, but what does this have to do with the matter? The question is not 'whether the Holy Scriptures speak generally according to the opinion of the people; but whether it speaks contrary to truth according to the erroneous opinion of the people.'[74] Nobody denies that the Bible uses figurative expressions. The question is whether the Bible commits factual errors. 'Our dispute is on the matter which is signified by the expressions, not on formulae and tropical figures, unless such expressions, as rhetorical figures, denote a truth.'[75] Niepoort presumes that Wittichius' caution is just a tactical manoeuvre. After all, on the question of the earth's motion, he does not give an inch.

The contributions from Beusechum and Troy can be dealt with succinctly. Beusechum's objections add little to his earlier disputation – apart from the fact that he, too, has by now abandoned his polite manners. He values Wittichius' assertions of his respect for Holy Scripture as much, or rather as little, as similar assertions by Catholics or Socinians. Troy had not been attacked directly in Wittichius' book on the style of Scripture and accordingly had little to say. He undertook a refutation of Cartesianism in general on the occasion of the physical corollaries Wittichius had included in his Duisburg dissertation (but left out of the Leiden edition).

It was a few years before Wittichius' rejoinder appeared. He had an answer ready in 1657 and entrusted it to Johannes de Raei to have it printed in Holland, but for some reason publication was delayed.[76] When it finally appeared in 1659, it proved as long (some 800 pages) as the former had been short. Apparently, he had taken to heart his opponents' remark that his general theses did not prove anything about the exegesis of the Bible in particular, for the book is largely devoted to this issue. The very title indicates as much: 'The agreement of the infallible revealed truth of Holy Scriptures with the philosophical truth discovered by Renatus des Cartes.'[77]

Here, we can give only a very succinct idea of this book. Wittichius starts by repeating his earlier distinction between common and philosophical

[74] Niepoort, Beusechum and Troy (1656) 149.
[75] Niepoort, Beusechum and Troy (1656) 172.
[76] Wittichius to Clauberg, 1657 Nov 3/13: '*In Synodo sua Zutphaniae habita nescio quid iniqui contra sanam Philosophiam dium statuisse. Hucusque nondum de qui vehementer doleo defensionem meam hunc praelo committere, multis modus et a multis mea spe elusa, quo negligentia etiam quaedam, si liceat dicere, Dn de Raei omissit [?], cui eam curam tradito ipso tractatu commisseram, ita ut nunc datis ad ipsum literis eum inde necesse habeam repetere.*'
[77] Wittichius (1659).

knowledge, by now in a definitely Cartesian context. The Scriptures do not teach philosophical knowledge. This point is argued extensively, mainly by quotations from the Bible. The views of his opponents on this point are discussed and refuted. (In fact, such refutations occur on other points, too, throughout the book.) Thereupon, Wittichius explains the Cartesian theory of the universe, offering conclusive proof of its truth, as he regards it. This point has been discussed earlier and thus will be skipped here. Next, he argues that Cartesian philosophy is not detrimental to Holy Scripture, and repeats his views on the correct interpretation of the Bible. The remainder of the book is devoted to substantiating this claim, in the first place by means of a large number of examples wherein the Bible uses common language, subdivided into classes: dialectical ways of speech; expressions in moral and practical matters; figures of speech born from the obsolete opinion that the soul resides in the heart; *ad hominem* arguments; things the Bible denies not because they are untrue but because they are unknown to us (e.g. Hebrews 7: 3); expressions about God; and quotations in the New Testament from the Septuagint version of the Old Testament which the Reformed considered spurious. As an example of the latter, Wittichius triumphantly refers to Voetius, who (in explaining such a case in Lucas 3: 36), had explained that the Holy Ghost had done so 'in order not to disturb people'.[78]

Next, Wittichius musters an impressive array of theologians who appear to support his view with their authority.[79] After this, he starts refuting the contrary arguments of his opponents. He discusses 25 arguments in all; it would be going too far to summarise them here, especially as they contain little that is surprising. The final chapters are devoted to extensive discussions (this part comprehends well over 100 pages)[80] of the biblical sentences most often advanced by opponents of the Copernican cosmology, to wit Joshua 10: 12-13, Joshua 38: 8, Psalm 19: 5-7, Psalm 93: 1, Psalm 104: 5 and 19, Matthew 5: 45, Ecclesiastes 1: 4-5, and Genesis 1: 14.

Some other voices

Some authors, while joining the theological discussion, succeeded in remaining largely out of polemics. As by 1656 the compatibility/incompatibility of the motion of the earth with the Bible had become a hot issue, it was tackled by various authors, including some who did not regard themselves as one of the warring parties. Those who studied the issue without allowing themselves

---

[78] Wittichius (1659) 537-538.
[79] Wittichius (1659) 547-558.
[80] Wittichius (1659) 655-768.

to be drawn into local feuds were mostly foreigners. Johannes Herbinius was a student from Silesia. During his preparation for the Lutheran ministry he spent some time at the Reformed universities in the Dutch Republic, before finishing his course at Wittenberg. In June 1653 he matriculated at Leiden university as a student of theology. Later that year, he defended a disputation under the presidency of Adam Stuart.[81] According to the preface to his later work, at Leiden he debated the Copernican system with some of his countrymen. After long and animated discussions, he was nearly won over, until the problem of the interpretation of the Bible emerged. He studied the several attempts that had been made by Lansbergen, Wittichius and Lipstorp to accommodate the Scriptures to the new astronomy. However, none of these attempts pleased him; the Bible remained an insurmountable obstacle to the acceptance of the Copernican system. In the end, he took it upon himself to write a clear discussion of the issue. By this time, he had moved to Utrecht University, where he matriculated in July 1654. It was in Utrecht that his book finally appeared, in 1655. It is a small volume of over 300 pages, entitled 'Theological-philosophical investigation of the famous controversy on the motion of sun or earth, guided by the Holy Scriptures'.[82]

The work is divided into two books. The first discusses the relevant biblical passages; Herbinius often gives various versions of one and the same text. The second book deals with the matter in a general way. It discusses special questions and authorities. In a preface, Herbinius summarises his opinion in 12 points. To start with, he recognises that one cannot prove or disprove mathematical propositions with an appeal to Holy Scripture. Moreover, the Scriptures do not have the proper words to explain matters that are beyond our comprehension. Holy Scripture is not a judge of natural things; however, it is a witness. Its testimony should be taken after the meaning rather than the letter of the phrases. According to Herbinius, theology and astronomy are both incomplete. Neither can acquire true knowledge of natural things. Both disciplines can only have an opinion on the appearances, the one from the Bible, the other from mathematics. In practice, this restricts principally the astronomers in their theories. As Scripture does not speak of mathematical questions, they may freely frame hypotheses. However, they should not state that these hypotheses are true. In the seventeenth chapter, Herbinius discusses the question 'whether and how by the Copernican system God's honour is exalted and pious thoughts are evoked'. Herbinius' answer is that

[81] This disputation, *De causis, earum influxum realem asserens*, is mentioned in Walther (1862) 360, no. 119.
[82] Herbinius (1655).

'We cannot glorify God from the Copernican system, because the celebration of God's honour should be founded on indubitable truth'.[83]

This summary, however, does not do justice to the painstaking scrupulousness with which Herbinius tackles the problem. Especially in the second book he goes to a lot of trouble to give a judgement on all fundamental issues. Most of them concern matters of biblical exegesis. How does Scripture speak under various circumstances? In what situations does it accommodate to the understanding of the common man? The book is not really stirring, but it is a relief to find an author who tries to deal with the question in a nuanced and impartial way. This may partly be due to the fact that Herbinius, as an outsider, was not moved by existing tensions in the Dutch Reformed Church. Wittichius himself admitted that Herbinius was more polite than his other adversaries.[84] The two remained in contact. In 1688, Wittichius publicly announced that Herbinius had changed his mind. To that end, he quoted from a complementary letter Herbinius had written to him in 1664, in which he stated that he was content with Wittichius' answers and refrained from further disputing the issue.[85]

In 1656, a certain Ludovicus de Beaufort published in Leiden a small volume, *Cosmopoea divina*.[86] De Beaufort was born in Paris but had lived for a long time in Holland. He matriculated at Leiden university as student of theology in June 1647, at the age of 29, but then took his doctorate in medicine on 20 September 1649. In 1659, he was still regarded as a member of the Leiden academic community.[87] His book is dedicated to the Leiden professor of mathematics, Jacob Golius. It might well be a reaction to the discussion on Cartesianism in the Dutch Republic, but its character is quite different from the works discussed so far. In the dedication, de Beaufort asserts that the Bible is the highest touchstone of truth. Where the Bible remains silent, one should suspend one's judgement. His purpose is not to frame a physical theory, 'but to recount the divine construction of this world, as it is described by Moses, and to accommodate as far as possible all phenomena to its sense.'[88]

The work itself is in the form of a hexahemeron. It is a curious mixture of several elements – the Bible, Aristotle and Cartesian physics. De Beaufort fol-

---

[83] Herbinius (1655) 300.
[84] Wittichius (1656) 27-28.
[85] Wittichius (1688), preface to the second edition. Herbinius' letter is from 10 December 1664. Whether it is warranted to conclude from this letter, as Wittichius did, that Herbinius *veritati cessit*, is in fact not wholly clear.
[86] De Beaufort (1656).
[87] BGLU. De Beaufort's doctoral dissertation is at Saint Petersburg, according to Walther (1862) 358, no. 108 (*De febri in genere*). In 1666, he renewed his matriculation at Leiden.
[88] De Beaufort (1656) dedication.

lows the biblical text and digresses on the exact physical explanation. Under the heading 'the works of the first day' he exposes the principles of physics. He explains the Aristotelian concepts of matter and form, gives the Aristotelian definition of motion, and than goes on to discuss the laws of motion, which are mainly taken from Cartesian philosophy. On the possibility of a vacuum, he suspends his judgement, as it does not repugn either Christian faith – revealed truth – or natural reason. Curiously, de Beaufort is not explicit on the question whether the sun or the earth is moved, but it seems clear that he takes a geocentric world for granted. In cosmology, he discerns between subtle, heavenly and elementary matter, each with its own region. The region of celestial matter is what the Bible calls the firmament. Still, here again his world-view is supplemented with Cartesian elements. The celestial matter moves in huge vortices, at the centre of which are the stars. De Beaufort tries to accommodate celestial influence, sympathies and antipathies to mechanical physics.

Finally, in 1658, a short tract on the issue appeared, written by an old acquaintance of ours, the Utrecht professor Henricus Regius. It concerns a small Latin pamphlet, 'Conciliation of the places of Holy Scripture with the daily and annual rotation of the earth'.[89] Although Regius was not a theologian, his pamphlet, in Latin, tackled the same theological issues as the works discussed so far. It appears to be directed at a limited academic audience. The work is edited by a certain Carolus Fabricius. In the preface, he explains that some people had heard Regius explaining the biblical texts, which were often adduced against Copernicanism. They had liked his discourse so much that they had encouraged him to publish it. In the text, Regius explains that he has answered in his physics the objections against Copernicanism drawn from natural reason. What remain are the objections drawn from Holy Scripture. These have been answered in several ways, but not all of them are sufficient. Regius rejects the view that motion is only relative, so that the sun can really be said to move. He also rejects the idea that the Holy Ghost adapts itself to the opinion of the erring people: this would entail turning the Holy Ghost into a liar. Regius' solution is: the Holy Ghost uses the common people's habit of speaking, or the words used by the people, according to the appear-

---

[89] Regius (1658). Cf. de Vrijer (1917) 207-208. Curiously, The catalogue of the publisher Van Zyll mentions the work under the Dutch titles: *Ejusdem bevredinge der H. Schriftuer Vanden omloop des Aerdkloots 4°*, and under a different heading: *Regius van de bevreedinge der H. Schrifture vande loop der Sonne 4°*. The price is in the former case 0-4– (four *stuiver*), in the second 4-0– (four guilders, an apparent printing error). See *Catalogus* (1667). I have found no further traces of this Dutch version.

ance and the verisimilitude.[90] So, if one sails away from the land, one says that the land recedes, not because one thinks that the land moves, but because it is the common way of saying it. One remarks that the varieties and similarities between the various exegetical principles are sometimes rather subtle for an outsider. Regius concluded his short treatise with the wish that those who did not agree with him might at least discuss the matter in a friendly way.

[90] Regius (1658) 6: *...ex communi vulgi loquendo more, sive verbis a vulgo usitatis, secundum apparentiam & verisimilitudinem.*

13. Copernicanism as a political problem: the Velthuysen affair

In the debates dealt with so far, Copernicanism had been treated as a purely theological and academic issue. It was discussed in disputations and Latin books. Of course, discussions do not exist just in printed form, and in reality the issue will also have been addressed outside the context of learned dispute. At some places, it may even have reached the pulpit. This, of course, is difficult to determine. But that the issue was considered not just an academic one becomes clear from a Dutch pamphlet published in 1655: 'Proof that the opinion of those who teach the rest of the sun and the motion of the earth is not contrary to God's Word'. Its author, the Cartesian scholar Lambert van Velthuysen, later claimed that he had written it on behalf of someone who did not read Latin and whose curiosity had been raised 'by the preaching of the ministers against the motion of the earth'.[91] With this pamphlet, the debate entered a new phase. Velthuysen not only sought (and reached) a wider audience, but he also brought some new elements to the fore, which may have lurked silently in the background in the earlier discussions, but had not been ventured publicly so far. In order to understand his position as well as the vehement resistance it encountered from the Voetians, we first need to explain some of the wider connotations of Voetianism.

Voetianism as a political programme

Strictly speaking, the Further Reformation did not affect the field of politics. But, like English Puritanism, it had very pertinent ideas regarding the structure of a Christian society.[92] Dancing, public swearing, pawnshops and theatrical performances should be forbidden – that is, forbidden not just to members of the Reformed Church, which went without saying, but to anybody. Catholicism should be effectively banned, as should practically all other dis-

---

91 Velthuysen (1657) 23.
92 Israel (1995) 690-699.

senting groups. Political office should be held only by those sincerely professing the Reformed faith – 'sincerely', that is, according to the Voetians. The Further Reformation advocated the idea of the Republic as a 'Second Israel' with all its consequences.

Voetius' uncompromising attempts to reform society can be understood only against the background of common Calvinist ideas on the relation between Church and state. The Dutch Reformed Church had not been instituted as a formal state Church. It was a product of revolution, and suspicious of secular authority from its very start. The regents of Holland had welcomed the Reformed Church as an ally in the struggle against Spain and had given it a privileged position. This resulted in a rather independent position. This was in striking contrast to the situation in, for instance, Germany, where the national Churches had been instaurated by the governments and were fully dependent on them. Lutheran theologians developed various concepts to legitimate the preponderance of the state and the idea of a state Church. The government was acknowledged as the *praecipuum membrum ecclesiae* ('the foremost member of the Church' – that is, of the visible, institutionalised Church). Still later, they distinguished between an internal and an external power in the Church. The Church itself had primarily a spiritual task, and the government's task comprised all external ecclesiastical affairs.[93]

Such a separation of the spiritual and the secular sphere went back to humanist notions which had some popularity in the Dutch Republic, too, and which seemed compatible with Calvinism. But in the end, the Church's relation with the government was mainly dependent on local circumstances, and not determined by theological principles. The French Reformed Church came to take on a decidedly monarchical stance during the course of the seventeenth century, as they regarded their king as their protector in a hostile Catholic environment. Consequently, the theologians of Saumur advocated the *droit divin*. However, in the early Dutch Reformed Church, things had taken a different turn. The idea that the Church had only a spiritual task, leaving external affairs to the secular arm, had been adopted by the Arminians. (This, too, was a tactical rather than a theological stance.) As a consequence, the contra-Remonstrants took the opposite position and maintained that the Church was completely autonomous, not just in spiritual matters, but also in those concerning organisation and public worship. As the contra-Remonstrants carried the day, their idea of Church government became dominant in the Dutch Reformed Church. So, whereas the German Churches had, in their relations with the government and in ecclesiastical law generally, to

[93] Heckel (1962) (a reprint of an article from 1938).

stand on the *factum*, the law as it was in force, the Dutch Reformed Church, on the contrary, acknowledged only the *ius Dei*, the Divine law. [94] Each tendency to give the government more say in the affairs of the Church was identified with the Arminian heresy. Therewith, it had become not only a tactical but also a doctrinal error. [95]

This theocratic tendency was fully endorsed by Voetius. In principle, Voetius admitted that Church and state were two separate spheres with their own responsibilities. So, the state should not meddle with the appointment of ministers or the convocation of synods. As for the Church not meddling with the affairs of government, Voetius was less consistent. It was clear anyhow that it was the duty of the state to protect and support the Church. And, as Nobbs put it, none of the Dutch Calvinists 'was honest enough to admit that for them the two-kingdom theory was legitimate only if the ruler endorsed the will of the ministers.' [96] Theoretically, Voetius' position may have been sound. He was in no-one's service and not a member of any party. His ideas on the state and on Christian society were inspired by what he thought to be the law of God, not by any earthly aspiration. Still, in actual practice, Voetius' ideas were not politically neutral. In order to show this, we shall have to have a short look at the political situation in the Dutch Republic in this period.

The war against Spain, which had begun with the revolt in the sixteenth century, had finally ended with the peace treaty of Westphalia, in 1648. Although this did not mean that the Republic was now completely at peace – there were still struggles in the colonies, notably a conflict with the Portuguese over the possession of Brazil – the Republic's existence was no longer directly threatened. Without the pressure of war, long-hidden tensions emerged. The same had happened before, when the 12-year truce concluded in 1609 had given rise to the Arminian troubles. Now, immediately after the peace of 1648 there was a serious clash between the stadholder William II of Orange – a grandson of William the Silent – and the States of Holland. The actual dispute was over the continuation of the war with Spain and over army reductions. But it soon turned into a power struggle, whereby the Republic's very structure was at stake. The stadholder, representing the monarchistic element in the Dutch state, wanted to reduce the power of the town councils. William used military force and laid siege to the city of Amsterdam in 1650.

---

[94] Conring (1965) 185.

[95] On the Arminian theory, see Nobbs (1938) 25-107; Conring (1965) 30-43.

[96] Nobbs (1938) 254-255. See also uit den Bogaard (1955) 59-62. Voetius' ideas on the relation between church and state are discussed by J.Th. de Visser (1926) II, 391-403, those of some later theologians ibid. 403-423.

The crisis ended when in November of the same year William II unexpectedly died from smallpox, aged 24. After his death, his republican opponents quickly gained the upper hand. They convened an assembly, the *Groote vergadering* ('Great Assembly'), to settle the political structure of the Seven Provinces on a permanent footing. Originally, the republican state-form had not been instaurated by choice, but had been forced upon the provinces by circumstances. After so many decades, however, people had grown used to it. The Great Assembly confirmed the federative, republican state-form and the ascendancy of Holland within it. Moreover, the regents of Holland felt justified in appointing no new stadholder for the time being. This was partly due to their recent experience with William II, which made them rather cautious to put too much power in the hands of one man. But it also was due to a new self-consciousness. By 1651, the regents had become fond of their republican system of state. Republicanism was consciously formulated as an ideal – 'True freedom' was the name they gave it. As its leader emerged Johan de Witt, Grand Pensionary of Holland from 1653 onward. In the history of science, he is also known as a student of Van Schooten and an accomplished mathematician.

The settlement of 1651 was successful in keeping public peace, but some frictions remained. During his short stadholdership, William II had succeeded in building up an important following. His death left his party in disarray, but did not lead to its disintegration. Its head was now the stadholder of Friesland, a descendent of a junior branch of the Nassau family who had been regarded by many as William's evil genius. Moreover, William II left a son. For the time being, he was too young to play a role in politics, as he had been born a month after his father's death. But in this little boy were concentrated the hopes and aspirations of all those who for some reason or other were dissatisfied with the present regime.

One of the pillars of this so-called Orangist party lay within the Church. Ever since stadholder Maurice had successfully intervened on behalf of the strict Calvinists during the Arminian troubles, there had been a special bond between the house of Orange and the Dutch Reformed Church. In building up his party, William II had purposely and rigorously played on that. By adopting strict anti-Catholic measures in the areas under his jurisdiction and providing sustained support for the programme of the Further Reformation, he obliged the more intransigent wing of the Reformed Church. Thereupon, they willingly took his side in the struggle with the States of Holland. They decried the secular-minded regents of Holland as 'Arminians' or worse. After William's death and the instauration of the 'True freedom', a large number of the ministers resented the new regime. They had little confidence in the government of 'True freedom' and set their hopes on the restoration of the Prince of Orange. [97] There are strong indications that Voetius himself was among them. [98]

The mood of the most radical Calvinists is made clear by a pamphlet published in 1650, while William II was at the acme of his power. The pamphlet is ascribed to the minister Stermont from The Hague. The author, among other things, accuses the regents of Amsterdam of conspiring to suppress the true Reformed religion. He warns that the regents should better protect religion and leave the ministers in peace: 'For else, seeing that their priestly garb is scorched, and true Religion undermined (...), they might well awake from their sleep, and be roused to a holy zeal all over the country, so that they will publicly inform and caution the congregations against the fatal and offensive plans of many governments. (...) Mind, what so many hundreds of ministers as there presently are in the country could effect if they would join forces to protect their religion and the right of the Church, entrusted to them by God, and choose the righteous side of his Highness [William II].' [99]

To many regents, such language must have seemed downright seditious. Ministers should teach the people piety and obedience, not rouse them to insubordination. Dutch regents were wary of ecclesiastical pretensions anyhow. The events of William II's stadholderate made many of them suspicious of the true intentions of the Church. Voetius' uncompromising efforts to turn Dutch society into a truly Christian state were taken by many not just as a laudable, be it somewhat over-zealous expression of practical piety, but in a much more sinister way. In reaction, Dutch republicanism developed a strong anticlerical tendency. This anticlericalism was not so much religiously inspired, but rather of a political nature. The ministers should keep to their business and leave politics to the regents. [100]

This political anticlericalism clearly anticipated the general European trend of the eighteenth century, but it was not wholly without precedence. As for Holland itself, the regents could fall back on the Erasmian tradition, but they also were aware of developments overseas, of the political successes of English Puritanism. Probably, a minister like Stermont could feel encouraged in his threats by the English example. The secularising regents, however, could draw some lessons from the English as well, in their case from the opponents of Puritanism. The political writings of Thomas Hobbes were eagerly studied in Dutch republican circles. This seems rather surprising, as Hobbes was the

---

[97] Israel (1995) 595-609. uit den Boogaard (1955).
[98] Broeyer (1991) 181, 184.
[99] Quoted by uit den Boogaard (1955) 74-75.
[100] Schilling (1993).

theorist of monarchical absolutism. The Dutch, however, managed to read his work as a treatise on republican sovereignty. Common to Hobbes and the Dutch republicans was their distaste of ecclesiastical pretensions. [101]

Lambert van Velthuysen and the struggle with Voetianism at Utrecht

One of the main advocates of Hobbes' ideas in the Dutch Republic was the very Lambert van Velthuysen whose pamphlet we mentioned at the beginning of this chapter. Velthuysen was a learned man – a doctor of philosophy, of medicine and of law. He obtained his degree in philosophy at Utrecht in 1644 on a triple thesis, with a physical, an ethical and a mathematical part. The physical part – *De mundo* – is rather traditional and closely follows the Aristotelian model. Velthuysen acknowledges the division into a supralunar and a sublunar world, but argues against the existence of solid orbs. In the mathematical part, he tries to find the cause of the variation of the magnetic needle. The earth is a big magnet, and the needle is diverted to greater masses of land, or to where the soil contains greater masses of material apt for magnetism. Here, he appears rather interested in the new natural sciences. The whole thesis is dedicated to the professors of theology Voetius, Schotanus and De Maets, and the Walloon minister. Initially, Velthuysen intended to become a theologian. He accomplished a study in theology, but as he did not obtain a ministry, he shifted his attention to other fields of interest. He published a large number of books on a variety of topics, besides being active in the government of his native town, Utrecht. [102]

In Utrecht, Velthuysen could observe the Voetian political programme at work very closely. It was in Utrecht that tensions between the divergent tendencies in Church and society reached a climax. Voetius held sway in the church council and the theological faculty. His learnedness and his exemplary piety gained him many admirers, but on the other hand, he annoyed many people with his strictness. The issue at the centre of ecclesiastical-political debate in Utrecht was that of the secularised ecclesiastical benefices. With the arrival of the Reformation, the five Utrecht chapters had not been liquidated. They had been transformed into Protestant bodies, which exercised the political functions which formerly fell to the chapters. The successors to the medieval canons were generally drawn from the leading Utrecht families. They thus profited from the former chapters' substantial wealth.

To Voetius, this practice cried to Heaven. The chapters had obtained their

---

[101] Van Bunge (1999) 328-332.

[102] A good biography is lacking. For a very brief sketch, see Duker, III, 264-267.

wealth from donations for pious purposes. It was a grave sin to allow rich magistrates and notables to profit from them. He first ventured his opinion in 1642, much to the regret of the Utrecht nobility. At this stage already, the political struggle became mixed up with the philosophical debate. Descartes was befriended by several of prominent Utrecht nobles and appears to have been more or less used by them. In order to compromise Voetius they supported Descartes, whom they provided with some documentation on the Utrecht benefices. Descartes obligingly used this material to expose Voetius' political presumptions. However, the Utrecht city council which had to decide on the affair was not prepared to disavow their professor. As mentioned earlier, they simply imposed silence and Descartes was left empty-handed. His defeat was due not so much to unfamiliarity with Dutch ways of government, as to the fact that his Dutch friends had put him on the wrong track. [103]

The decision of the town council ended the open philosophical strife for the moment, but the political tensions remained. The town government asked Voetius to put his considerations concerning the Utrecht beneficies on paper, mainly as a way to keep him quiet, it seems. Voetius submitted an elaborate 'Theological advice' on the matter in 1645, but the regents temporised and nothing happened for a while. Then, in 1653, one of Voetius' friends decided to have the advice printed. It had the effect of a bombshell. As one Dutch church historian put it: 'Never before or since in the history of the Dutch Republic has a church council exercised such an open and principled criticism of the government's policy.' [104] Voetius dismissed established judicial practice with an appeal to the law of God. No Christian could take it on his conscience to take session in one of the Utrecht chapters, he declared. When in 1658 one of the canonries became vacant, the church council put heavy pressure on the claimants not to accept the benefice. No less then seven of them subsequently declined. [105]

The 'Theological advice' unleashed a fierce debate. Voetius himself published in 1656 a sequel, 'Cloud of witnesses', wherein he listed many authorities supporting his view. [106] It is difficult not to feel some sympathy for Voetius' opinion as such. But the way he brought it forward was not very tactful, to say the least. He refused any compromise, for which the town government probably would have been ready. He squarely opposed a legitimised practice and denounced the lawful government. Moreover, he had little patience with anybody who felt differently. In a 'register of the sins, which are common here

[103] Bos, in Descartes (1996) 18-19.
[104] Trimp (1987) 69. On the theological advice also Duker, II, 294-305.
[105] Duker, II, 323-324. Trimp (1987) 77-78.
[106] Duker, II, 316-319.

and elsewhere in the country', drafted by the church council in 1659 in preparation for a project of reformation, 'Arminian maxims tending to transfer the power of the church to the magistrates, and speaking against all reproaching of the magistrates', are put on a par with swearing, lying, drunkenness, fornication, etc. [107]

All this put many people on the alert. If there was one thing the regents did not like, it was ministers telling them what to do. In some of the subsequent pamphlets, these fears were clearly capitalised upon. An enormous stir was caused by an anonymous pamphlet which appeared in 1655, which claimed to present the text of a secret agreement between Voetius and his colleague Carolus de Maets, purportedly found in a book bought at the auction of De Maets' library. This agreement developed a plan for a gradual and peaceful take-over of government by the adherents of Voetius. The agreement may well have been a forgery, but if so, it was a very clever forgery which led the way to further attacks along the same lines. [108]

So, relations between Church and government were tense in the Dutch Republic in general and in Utrecht in particular. It is not amazing that Velthuysen, who was a principled republican and had written an apology for Thomas Hobbes' *De cive*, decided to teach Voetius a lesson. He did not attack him on the point of practical politics, however, but on that of the freedom of philosophy, in particular his anti-Copernicanism. These subjects were not unconnected in his eyes. As a scholar and a Cartesian, it must have particularly annoyed Velthuysen that Voetius used his authority to decry Cartesianism as impious.

In fact, Voetius and the Utrecht church council did meddle with philosophy. Shortly before, there had been an incident which might have served as an incentive to Velthuysen. In October 1654, the Utrecht church council took notice of a disputation *De trochlea*, recently defended at the university by 'a member of the Utrecht church' – to wit, professor De Bruyn, whose Cartesianism has been discussed above. It contained a corollary: 'the souls of animals are equally capable of immortality as those of humans.' [109] When asked for an explanation by the church council, De Bruyn answered that he had not intended to direct himself against Church or theology, and promised to abstain in future disputations from anything that might give offence. [110]

---

[107] Van Lieburg (1989) 23.

[108] Duker, II, 309-314.

[109] A dispute on this thesis when defended on another occasion by Regnerus a Mansfeld is discussed by Thijssen-Schoute (1954) 44-45.

[110] Resolution of the Utrecht church council, 9 Oct. 1654. State archives Utrecht, 222: church council Dutch Reformed Church, 6.

The affair did not remain at that, however. De Bruyn asked for the resolution in which the corollary was condemned to be cancelled. The church council several times postponed providing an answer, each alleging a different reason. De Bruyn had mentioned a disputation he intended to publish on the same subject. The church council thereupon decided to ask the curators of the university to prevent the said disputation, or at least that they might inspect it. Then, there appeared a little anonymous booklet (or manuscript), entitled (according to the resolutions) 'Further message of the opinion of Mr [Johannes de Bruyn] concerning the nature of the souls of animals', clearly written by a sympathiser of De Bruyn. De Bruyn himself disclaimed authorship. He had seen the book only after it had been published and had said nothing either in favour or against it. Among other things, the book claimed that the church council had condemned De Bruyn's opinion without having given him a fair hearing, or even given him a warning on the issue. The church council concluded that this could only have been inserted into the book by the connivance of De Bruyn. When asked whether the author did have this from him, De Bruyn was evasive. He agreed that he had been condemned without a hearing and that he had said so on various occasions. The church council dismissed the accusation. As they explained, De Bruyn had in no way been censured or condemned. They had just found his thesis offensive, and had asked him to abstain from similar theses in the future. Neither party wanted to take the issue any further. De Bruyn answered that he had not been able to conclude otherwise than that he had been condemned without a hearing, but that he gladly would accept the church council's explanation. The church council, on their side, seems eventually to have agreed to cancel the resolution.[111]

Velthuysen himself was a member of the Walloon congregation in Utrecht. The Walloon Churches had been founded by francophone refugees in the early years of the Revolt. By the 1650s, there was no longer any real need for francophone church services, but the Walloon Churches remained in existence as a fashionable variant of the Reformed Church. They were Reformed in every respect, but they had their own organisation, up to synodal level. As a member of the Walloon congregation, Velthuysen was not subject to the jurisdiction of the Dutch Reformed church council, wherein Voetius held sway.

---

[111] Resolutions Utrecht church council, 5, 19 & 26 Feb. and 12 March 1655. The original resolution was probably cancelled, as it is no longer there. When this happened is not clear. A letter from Velthuysen from Jan./Feb. 1656, printed as an appendix to Velthuysen (1656), still speaks of De Bruyn's problems with the church council.

Turning now to Velthuysen's pamphlet, Velthuysen states that the discussion on Copernicanism is just an extension of the debate on Cartesian philosophy. The latter has gained ever more the upper hand at Dutch universities. Now that its adversaries have failed to fog this 'rising light' with 'a cloud of detractors', they have changed their tactics. 'A large part of the philosophy of Descartes relies on the motion of the earth. Take away this foundation, the construction built upon it will collapse. They know that it is impossible to overturn these foundations by natural reason. So what is to be done? One should take the roof from the church, and make it God's cause.' Otherwise stated, the theological objections do not come forth from theology itself, but from guile. Velthuysen than takes it upon himself to combat this guile, for two reasons: to defend Christian freedom, and to protect Descartes' honour. 'So that nobody will be subjected to prejudice too easily; religion may keep its liberty; and learned speculations and opinions will not be subjugated and torquated under the laws of such people who estimate nobody but those who wear their colours and who will have everything sifted through their sieve.'

This is not just an abstract theological or philosophical discourse. Velthuysen makes it clear from the start that he is aiming not so much at a certain opinion, as at the pretensions of a certain group of people. Still, when it comes to the question itself, he reasons with great clarity and precision. His main task is a discussion of the Biblical texts commonly adduced against Copernicanism, but Velthuysen does so only after he has given some general rules for biblical exegesis. Velthuysen's main point is that one should discern between the intention of the text and the way this intention is expressed. Where the Bible 'teaches and dogmatises', one should take its sayings unconditionally as true.[112] However, one should not draw any conclusions from the expressions used. So, Velthuysen presupposes a theory of inspiration akin to those of the theologians of Saumur, whereby the Bible is inspired as regards its intention, but not in its ways of expression. As he made clear at a later stage of the debate, we should explain the Bible in the same way as we would explain the words of wise people, not stretching the meaning of the sentences beyond what is common among humans.[113] With this principle, he has little difficulty in showing that the standard texts have no bearing on the Copernican question.

Velthuysen's attack led to an extensive polemic. The charge was opened by the by now familiar Jacob du Bois. He published against Velthuysen a pamphlet 'Nakedness of the Cartesian philosophy'. Velthuysen replied, in 1656,

---

[112] Cf. Velthuysen (1656) 102.
[113] Velthuysen (1656) 106-107.

with a new and much amplified edition of his earlier pamphlet. Du Bois answered with 'Harmfulness of the Cartesian philosophy, or clear demonstration how harmful that philosophy is, both in the dissolution of God's Word, as in introducing new and harmful doctrines'. This pamphlet was published in Utrecht, as was the former. Moreover, it carried an approval of the three professors of the theological faculty of Utrecht, i.e. Voetius, Essenius and Nethenus. Du Bois had asked for the approval because, as he explained, 'in our classis [Leiden] the visitation of books is demanded more strictly than ever before, to that purpose that editions of anti-Cartesian books are suppressed. The visitators are all devoted and allied to the side of Heidanus.' Voetius and his colleagues were only too pleased to comply. [114]

Velthuysen rejoined with a 'Further proof that neither the theory of the sun's rest and the motion of the earth, nor the foundations of the philosophy of Renatus des Cartes, are contrary to God's Word'. After his own testimony, this pamphlet was attacked in public sermons by several Utrecht ministers. [115] Meanwhile, Du Bois had been joined by an anonymous pamphleteer who published in 1656 against Velthuysen, in Dutch, 'Short remarks on the unproved demonstration that the opinion of the rest of the sun and the motion of the earth is not contrary to God's Word, noted to suppress an upspringing root of bitterness against the Holy Writ'. [116] Velthuysen dismissed it as the work of some poor bungler, who apparently had sought to gain someone's favour with it; it clearly could not have been written by a Dutchman or a theologian. This was undoubtedly deliberate sarcasm. Du Bois asserted that Velthuysen knew only too well that this author was 'a prominent minister in the Reformed Church'. [117] In a second pamphlet, published later that year, this time in Latin, the author revealed himself as Caspar Streso, Reformed minister in The Hague. [118]

---

[114] The correspondence between Du Bois and the Utrecht faculty is printed in Cramer (1932) 388-392; see also 74-75. The quotation is from a letter from Du Bois dated 25 July 1656, ibid. 388. On p. 387 is reproduced a short notice with remarks on the pamphlet by Velthuysen, apparently for the purpose of a reply. There was no state censorship in the Dutch Republic, but according to the synod of Dordrecht, members of the Reformed church who wanted to publish on religious matters needed the approbation of the local classis, which appointed *visitatores* to investigate the writings.

[115] Velthuysen (1657) preface, [6].

[116] [Streso] 1656. The 'upspringing root of bitterness' is an allusion to Hebrews 12:15.

[117] Velthuysen (1656) 40-41. du Bois (1656) 35. 'Suetonius Tranquillus' (1656)d, 20, writes he knows the author well: 'it is an honest minister at The Hague, famous all over the country for his eminent learning, piety, peacefulness, and other good qualities.'

[118] Streso (1656)b. Cf. p. 13, where Streso reacts to Velthuysen's attack on his earlier animadversions, *quibus sermone Belgico represseram licentiosum istud genus torquendi & interpretandi ista sacra Dei testimonia, quae se quorundam Philosophorum opinioni de quiete caeli & motu terrae catervatim objiciunt*. The two pamphlets were by the same publisher.

It would be going too far to discuss each of these pamphlets in detail.[119] They are full of invectives and often repeat themselves. Besides the question of the motion of the earth, they discuss an ever-growing list of points from theology and Cartesian philosophy, e.g. whether animals have feeling, whether God can cheat, the nature of spirits, Descartes' proof of God's existence, the infinity of the world, and so on. We do not need to discuss all these, but some main points should be highlighted.

The fundamental issue was, of course, the interpretation of the Bible. Whereas Velthuysen wanted to discern where the Bible 'teaches and dogmatises' and where it simply uses common expressions for that purpose, Du Bois stated that Holy Writ 'teaches and dogmatises' in every letter and expression. In the heat of discussion, the antithesis was of course somewhat forced. Velthuysen pointed to Exodus 10: 21, the darkness over the land of Egypt 'which may be felt', as a clear example of figurative speak. Darkness, after all, is not something palpable, but just the absence of light. Du Bois did not give in: if the Bible says this darkness could be felt, it could really be felt. Apparently, it was caused by some heavy fog.[120]

Velthuysen was not impressed and pointed to the inconsistencies in Du Bois' own work. He drew attention to a passage in Du Bois' book on chronology, which was written before the controversy over Copernicanism. Speaking of the life of Abraham, Du Bois had to deal with a contradiction in the Bible. In Genesis it is said that Abraham's father was still alive when he left Ur in Chaldea, but Stephen, in his speech as given in Acts 7: 4, asserts that he already was dead at that moment. In his book, Du Bois had unhesitatingly followed the version of Genesis: 'It would be absurd to emend Moses, whose purpose it was to render the dates, by Stephen, who brings those words only in passing, without having an eye to the exact chronology'. Of course, this created a problem, viz. what to think of the words of Stephen. Either his words had been badly transmitted (in that case, we should follow the Samaritan version of the Bible), or he had simply been inaccurate and spoken 'after the opinion and the common idea of the Jews'. Du Bois chose the latter option. 'It is not strange to say that Stephen in those circumstances has spoken according to the common opinion of the Jews. Sure, he was moved by the Holy Ghost, so that he has not missed in the essence of the doctrine. But that does not contradict that in some circumstances of particular histories, he did not have all specific knowledge, or did not bother very much.' In a similar way, Du Bois said, the disciples had upheld, even after the sending down of

[119] The debate is investigated by van Bunge (1995) 49-54 and (1999) 324-327; McGahagan (1976) 281-289.

[120] Du Bois (1656)a, 16. Something similar had earlier been said by Kyperus, see above, p. 134.

the Holy Ghost, the incorrect opinion that heathen should be circumcised in order to become Christians.[121]

Velthuysen concluded from this passage, rather triumphantly, that his own opinion was in fact vindicated by Du Bois.[122] However, this was not completely justified. As Du Bois explained, he had tried to find a solution for a problem that arose from the Bible itself. But in the discussion at stake, it concerned plain and unambiguous biblical sentences. There was no reason to deny that these were taught expressly by the Holy Ghost.[123] Actually, Du Bois followed the Reformed rule that Scripture should be its own interpreter. 'Are there any reasons *in God's Word* why we should or could not take the places which speak of the sun's motion and the earth's rest [etc.] in their proper sense? Indeed, there are none...'[124] His grief was that Velthuysen applied external, philosophical principles to which Scripture should submit. Velthuysen, as Du Bois asserted, used reason as a rule of Scripture and refused to submit to Divine testimony. This came down to sheer Socinianism.[125]

Du Bois repeated his by now well-known thesis that one should not diverge from the literal sense of the Bible unless forced to do so by compelling reasons, that is, when moved by intra-textual reasons or when the literal sense would result in an absurdity. Here again Velthuysen turned the argument against him. How is it, he argued, that this man wants to accuse me of Socinianism? Me, someone who asserts that we should follow the Bible unconditionally whenever it teaches or dogmatises, even if it would seem absurd? Du Bois' own stance comes down to saying that we may diverge from the literal sense every time we feel that maintaining it would be absurd. This is really subjugating Scripture to reason. Velthuysen concluded that the feelings of Du Bois and the Socinians resembled each other like two drops of water.[126] Du Bois replied that he had drawn his exegetical principles from the works of Festus Hommius, a respected Reformed theologian, so they could not possibly be Socinian.[127] In his 'Further proof', Velthuysen thereupon extensively demonstrated the agreement of Du Bois' exegesis with that of the Socinians. On the other hand, his own exegetical principle – the distinction between 'teaching and dogmatising' versus 'not teaching and dogmatising' – was re-

---

[121] Du Bois (1650) 54-56.

[122] Velthuysen (1656) 42-43.

[123] Du Bois, *Schadelickheyt*, 18.

[124] Du Bois, *Schadelikheyt*, 18 (my italics).

[125] Du Bois (1655)b 53-54.

[126] Velthuysen (1656) 101.

[127] Du Bois (1656) 10.

commended by Voetius himself[128] (Voetius had applied the principle to other texts). Earlier, Velthuysen had stated: 'Two eggs are not more similar than my answer is to that of the theologians. But now that it is said by a Cartesian, it is a great heresy, a rupture of the authority of Holy Scripture, a devil's work, etc.'[129]

It seems quite clear that as regards their principles themselves, the combatants did not differ very much. They were both in the Reformed tradition. The problem was how those principles should be applied and, especially, to what purpose. Du Bois became angry not so much about Velthuysen's principles, but because of the conclusions he reached with them. Velthuysen liked to express things in a sharp and principled manner. After having accused Du Bois of Socinianism, he raised the question whether God could permit that man was misled by his natural reason. According to him, God's sanctity and truth could not suffer that man, when following God's testimony in nature, would eventually be deceived. But God could permit that man would temporarily err because of that testimony (as Aristotle and his followers had erred on the motion of the earth). In order to prove the latter point, he argued that the prophets themselves, even when moved by God's spirit, could lie.[130]

The exegetical issue was clearly linked with philosophical questions. Streso explicitly defends Aristotle's' theory of substantial forms, on the grounds that without them, the creatures can have neither soul nor reason.[131] But on the whole, philosophical arguments do not play an important part in the debate. However, there is another central issue in these pamphlets which soon outstrips even the exegetical question. That is the question of authority. A recurrent theme in the pamphlets is the relation between theology and philosophy. Du Bois and Streso uphold the old scholastic doctrine that theology should have the upper hand, and that philosophy should submit to the dominion of its mistress.[132] Freedom of philosophising has its limits where it contradicts the Bible, as commonly explained, and the accepted doctrine of the Reformed Church.[133]

Streso asserted that 'Scripture's literal sense should not be interpreted by anybody from a literal sense into a figurative meaning. If such an interpretation is necessary, it should be done by proper authority and order, restricting

[128] Velthuysen (1657) 35-36 (on Du Bois' 'Socinianism'), 38-42 (on his following the method recommended by Voetius and Nethenus). See also the preface.
[129] Velthuysen (1656) 32-33.
[130] Velthuysen (1656) 102-106.
[131] Streso (1656) 7.
[132] Du Bois (1655) 10; (1656) 6, 31. [Streso] (1656) 4.
[133] [Streso] (1656) 5-6. du Bois (1656) 1-6.

and opposing which is not allowed to everybody.' In the Reformed Church, the explanation of the Bible is subject 'to the ecclesiastical sentence of the official ministers and other leaders of the congregation, taken from Scripture, tried, approved, and admitted by charge and order of the Christian government, and accepted not just by the high and low government, but also by all sane and regular church members.' [134] He also asserted that the Reformed are bound in their exegesis by the authorised marginal annotations of the Bible. [135]

To Velthuysen, this came down to sheer popery. He commented, sarcastically: 'Sure, the purity of God's church is well preserved and the authority of Scripture is well defended when they have such proponents.' He vehemently defended his right to explain the Bible according to his own insight. Do his adversaries intend to be judges of their fellow citizens, in order to get into the government? One accepts doctrinal points because one judges them well-founded in God's Word, not because they have been imposed by some authority. The ministry has no right to censure church members for their unwillingness to accept some newly invented doctrinal points. Who teaches otherwise 'formally reintroduces popery.' [136] 'Those people who put our forebears to the gallows and to the stake, meant well too.' [137]

Velthuysen's aim was not so much to contribute to the theological discussion on the exegesis of such texts as Joshua 10: 12. He rejected the theological meddlesomeness as such. The conclusion of his first pamphlet makes this very clear: 'In vain so much has been done for liberty; so long fought against gallows and hangman's noose; thrown off the yoke of the world's mighties; if one has to suffer, here and elsewhere, all days from the pride, pity, scorn and persecution by just a few persons, who, referring to zeal and religion and with a stately aspect, are aiming at nothing else but an insupportable dominion; and are trying to have all classes of people and colleges under their whip.' It was particularly lamentable that those who deserved it least of all suffered most from such injuries: to wit, people like Descartes who spend their time investigating truth. 'Gave God that those to whom it becomes to resist such a spiritual tyranny, will not lend their hand to strengthen such an unlawful control. But alas! One feels that liberty is secured, when one is allowed to do what earlier one envied in another, and called tyranny then. Yet, true liberty consists therein that, living under a good government, everybody's honour, life and goods are secure from slanderers, murderers and thieves; that

[134] [Streso] (1656) 4.
[135] Streso, (1656)b 14.
[136] Velthuysen (1656) 44-48. See also 77, 122-123. See also Velthuysen (1657) 1-4, 7-9.
[137] Velthuysen (1656) 121.

virtue obtains its reward, and evil its punishment. And it is improper that one judges both according to the pronouncements of some impassioned man, instead of according to reason and equity.'[138] It is clear that Velthuysen was referring particularly to the situation in Utrecht. He openly admitted such on another occasion.[139]

Du Bois and Streso violently denounced his view. Du Bois rather predictably rejoined that this talk of Christian freedom reminded him of the Remonstrants, who also talked of freedom and tolerance while filling the various councils with their men, 'until they would have the majority of the votes and would play the comedy after their pleasure.'[140] The accusation of Arminianism made little impression on Velthuysen. Whereas his adversaries claimed that theologians should have the last word in philosophical matters, Velthuysen even did not want to give them full authority in theology itself. The foundations of faith are indisputable, but in minor points error is possible. 'So it is very unreasonable that the ministers get angry if one does not understand the matter as they do. That they address the congregation as masters, by way of commandment; rebuke and condemn them, even refuse them consolation; and pretend to do all this *in the name and by charge of God*.'[141] Velthuysen advocated a Church wherein all people subscribing to the fundamentals of Christian faith could be full members. He argued extensively for the case of 'moderation'. In his view, even the errors of Arminianism were not fundamental.[142] 'I refer to Holy Scripture and to ecclesiastical history whether not all schisms in God's church have arisen because of someone playing the master who, too obstinate to give in at any point, and having too much fancy in his own opinion, would not stand that any feeling contrary to his be tolerated in God's church.'[143]

Such ideas were anathema to the Voetians; also more moderate men had their reservations. Du Bois refers to professed Cartesians who disavowed Velthuysen's opinion in public, saying that Velthuysen had written these things not as a Cartesian, but as a Hobbist. By adding that Hobbes had been an adversary of Descartes, they denied any responsibility.[144] Velthuysen

---

[138] Velthuysen (1655) 30-31.

[139] Velthuysen (1657) 13. '*Die de gelegentheyt van de Kercke van Utrecht kennen, weten wel waer heen mijn woorden in mijn eerste andtwoordt streckten.* (...)'

[140] Du Bois (1656) 5.

[141] Velthuysen (1656) 77.

[142] Velthuysen (1656) 77-82.

[143] Velthuysen (1657) 14.

[144] Du Bois (1656) 73. See also 'Irenaeus Philalethius' (1656)a, 90: '*D Velthuysen heeft in desen deele sijn eigen gevoelen, en comt daar in wel met eenige andere gereformeerde Theologanten, maar, onses wetens, niet met eenige die voor Cartesianen uitgaen, over een.*'

was more radical than most opponents of Voetius. But he expresses clearly what people thought so irritating about his behaviour.

Related polemics in 1656

Velthuysen soon received support from several quarters. Especially the year 1656 saw a lot of controversy over the issue. We noted earlier that around this time, interest in Copernicanism at Utrecht University was at its height, too. The main reason for all this commotion was that at the time there was a major attempt by the Voetians to have Cartesianism banned from the Dutch Reformed Church, as we shall see in the following section. 1656 was an eventful year in other respects, too. 'This year 1656 is memorable above all other years so long as the world has existed' wrote a certain Jacob Vinck in a defence against the Utrecht church council, with which he happened to be in conflict.[145] Plague raged in the United Provinces and a general feeling of malaise prevailed. Vinck referred to a lunar eclipse in 1656 and other signs as indications of Doomsday. Around this time, there was even a slight increase in accusations of sorcery. This atmosphere gave extra weight to the ministers' denunciations of sinful behaviour.

Many people apparently were irritated by Du Bois' contemptuous dismissal of Cartesian philosophy. A satirical Dutch poem against Du Bois' first pamphlet appeared in 1656 under the title 'Cartesius renatus, or discovery of the envious and stubborn fanatic hidden under the guise of religious zeal, against the doctrine of Descartes'. The author seems to come from Gelderland, but prefers to remain anonymous: 'rebuking, criticising and slandering is in this country only allowed to ministers.' It was edited posthumously by a certain J.D.L.M.P.; it has been proposed that this should be read as 'Johannes [Antonius van] der Linden, Medical Professor', at Leiden. Its content is not very interesting, but it is significant that the controversy elicited this kind of reaction. The author comments on several passages in Du Bois' pamphlet and defends Descartes. As to the motion of the earth, he stresses that one should follow the intention of the Bible, not its letter.[146] Another instance of the opposition Du Bois met is offered by a complaint at the Leiden church council which he made in December 1656 against the professor of law, Van Thienen. Van Thienen had called him, 'in a public oration for several hundred people, a slanderer, who would have done a thief's and murderer's work, in

---

[145] '*Dit jaer 1656 is een besonder gedenck iaer boven alle iaeren so lang de werlt gestaen heeft.*' Manuscript in Amsterdam municipal archives, archives of the Dutch Reformed church at Amsterdam no. 231, p. 66. See on Vinck and an earlier version of this tract, van Lieburg (1989) 106-118.
[146] N.N., *Cartesius renatus* (1656).

slandering the candidate who would be promoted'. Du Bois therefore asked that the church council might call Van Thienen, being a member of the Reformed Church, to account for his unchristian behaviour. [147]

A minor controversy at the time concerned Cartesianism generally. Paulus Voet, a son of Gisbertus Voetius and one of the most adamant participants in his father's controversies, published in 1656 his *Theologia naturalis reformata*. In a chapter on 'the idea of imaginary space', he thought it necessary to refer to the disputes on Cartesianism at Utrecht University in the early 1640s, and even reproduced all the official documents. [148] An anonymous person (perhaps Voetius' enemy, Maresius) used the occasion to publish Descartes' side of the story as well. In a Latin pamphlet, he published the text of a hitherto unpublished apology by Descartes to the Utrecht government, from 1645. [149] However, not all publications attacking Voetius refer to Cartesianism. Another pamphlet was a collection of several doggerels sharply attacking the Voetian party. None of these mentioned the Cartesian question. One satirised the *Theologia naturalis reformata* (though probably because of the attacks against the Groningen professor Maresius it contained), and four others satirised Voetius' own 'Cloud of witnesses'; the main part is directed against a sermon of the Utrecht minister Lodenstein, a close collaborator of Voetius. [150]

Of greater importance was the support Velthuysen unexpectedly received from another side, equally in 1656. This support, too, took the form of an anonymous pamphlet. It added a new, mathematical element to the discussion, and gave rise to a second pamphlet war, which was waged parallel to the first one. The pamphlet appeared in both a Dutch and a Latin version: 'Mathematical proof of the stupidity of Jacob du Bois, reformed minister at Leiden, in his fight against the hypothesis of Copernicus, and the philosophy of Descartes'. [151]

In a later sequel to this pamphlet, the author signed himself 'I.G.H.' Most probably, he was the well-known mathematician and later burgomaster of Amsterdam, Johannes (Gerritsz) Hudde. His name is mentioned in a seventeenth-century manuscript index to a volume of pamphlets on Copernicanism

---

[147] Resolution of the Leiden church council, 1 Dec. 1656. Municipal archives Leiden, archives of the church council, 5. The issue gave rise to a protracted dispute on jurisdiction between the church council, the university and the Leiden burgomasters.

[148] Paulus Voet, (1656) 253-264.

[149] *Magni Cartesii manes ab ipsomet defensi* (1656). Descartes (1996) is a critical edition of the Dutch version of this text from the Utrecht archival records; see 41-44 for the history of the text. French versions have been published by Verbeek in *Querelle*, 401-437 and in AT.

[150] *Over de woorden van vader Lodesteen* (1656).

[151] [Hudde] (1656) a and b.

and Cartesianism, now at the library of Utrecht University.[152] In 1656, Hudde was quite young and still unknown (Du Bois, in his rejoinder, spoke of the 'rash and thoughtless youth' of his adversary[153] – apparently he had quickly found out), but was already active in science. In the same year he published, again anonymously, a little work on dioptrics. He certainly sympathised with Velthuysen's Cartesianism. As we shall see, in 1657 Hudde and Velthuysen made contact, probably as a result of the debate on Copernicanism.

As to the contents of the pamphlet (and this also supports Hudde's authorship), it is in no way concerned with biblical exegesis or philosophical issues, but is entirely devoted to astronomical and mathematical arguments. Du Bois prided himself on having given an invincible mathematical argument against the Copernican system, and hence against Cartesianism. The Copernicans, according to Du Bois, maintain that the planets' period of revolution around the sun depends on their distance from the sun: the further away they are, the slower they move. Mercury passes through its orbit in less than three months, Venus in about eight, and the earth in 12. As noted, the existence of such a cosmic order had been an important element in the 'Leiden interpretation'. In particular Lansbergen had adduced this as evidencing the truth of the Copernican system.

Now, according to Du Bois, Copernican theory was clearly self-contradictory here. From Lansbergen's own tables, it was evident that the period of Mercury is about 116 days and that of Venus about 19 months. Thus, Venus has a longer period than the earth, although according to the Copernican theory it is closer to the sun. Du Bois announced this refutation of Copernican theory in both his *Dialogus* and his book against a conclusive argument which would overturn all arguments of the Copernicans: 'This error in your philosophy is so immense, that I hope you, having from shame deserted that erroneous spirit of philosophy, will return at last to your mind.'[154] He proclaimed his triumph not only in print. The Leiden professor Van Schooten, who must have known Du Bois personally, complained 'that our Du Bois in nearly all assemblies and among all kinds of people, among professors, his colleagues and other learned men, even in barges and coaches, prided himself in such a stubborn and audacious way (for he had been alerted to his error before) that he had overturned the Copernican system and the Cartesian philosophy, that several prudent men believed his assertions without reserve.'[155]

As Hudde demonstrated in his pamphlet, the whole argument rests on an

---

[152] Shell number: Y qu. 79. Cf. Vermij (1995) 28-30.

[153] Du Bois (1656)b 8.

[154] Du Bois (1653) 36-38; (1655) 283-290; the quotation is from the latter work, p. 290.

[155] Van Schooten to Huygens, 30 May 1656. *OC*, I, 422 (no. 293).

elementary error. Du Bois talks about the real period of the planets with respect to the sun and the fixed stars. Lansbergen's tables, on the other hand, serve for predicting the apparent planetary positions, seen from the moving earth. Seen from the sun, our earthly point of view moves along with the other planets. Therefore, the period of Venus as calculated from Lansbergen's table is longer than the real one. One has some comprehension for Van Schooten's annoyance, and even for the uncomplimentary title Hudde gave his pamphlet. Du Bois' argument is an example of utter incompetence. On the other hand, Hudde's 40-page refutation of such a trivial argument seems somewhat overdone. Hudde demonstrates that Du Bois has not understood the argument; he shows that he does not know the literature; he gives two different proofs of the correct periods, the one algebraic, the other mechanical; and so on. The young Huygens indeed thought it a little too much. 'I agree that Mr Du Bois is rebuked and refuted as he deserves, but why at such length?' he wrote to Van Schooten – who answered that in his opinion, Du Bois had hardly been castigated enough. [156]

Probably, Hudde had expected Du Bois to withdraw in silence, filled with shame. Du Bois, however, did not feel embarrassed at all. In a vehement reaction – 'The reined-in Cartesian' – he answered Hudde's pamphlet with utter conceit. He felt compelled to withdraw his argument; however, not because of Hudde's argument, which he, as it seems, quite simply did not understand. 'That he has recourse to algebraic calculations is rather to show off that he has learned that art, than that it serves to purpose', [157] he mocked. But Hudde had referred to some Copernican authors (Gassendi and Stevin) who mentioned the periods, which Du Bois had deduced from Lansbergen's tables, in their writings. Thus, Du Bois was compelled to admit that apparently these periods were compatible with the Copernican theory. Mathematics was beyond his comprehension, but referring to authorities carried some force.

However, he went on, why ascribe so much weight to this mathematical argument? 'My own argument, by which I proceed *active*, that is, demonstrative, is the authority of God's Word. That is wherewith I actually combat that hypothesis, but he does not even touch it.' [158] Hudde had disproven one of his arguments, but all the others remained in force; and 'if this author would have known something to say upon them, he would undoubtedly have done so.' [159] And if he asserts that this one error shows Du Bois' incompetence and thus

[156] Huygens to Van Schooten, 6 May 1656; Van Schooten to Huygens, 30 May 1656. *OC*, 1, 413 (no. 288), 422 (no. 293).
[157] Du Bois (1656)b 8.
[158] Du Bois (1656)b 3.
[159] Du Bois (1656)b 8.

disqualifies his whole contribution to the debate, what should we think about the various errors Descartes has committed?

Hudde, apparently irritated by what he must have regarded as Du Bois' dullness, thereupon published a second pamphlet: 'The runaway astronomer J.D.B. hooded' (the hippic metaphor was a reaction to the title of Du Bois' earlier reply). He reproduced Du Bois' text in extenso, and added his own commentary. He reminded Du Bois of the triumphalism with which he had ventured his argument, and pointed out how Du Bois had paraded with it. And by now, Du Bois alleged that this 'conclusive argument' had never been of much importance. Hudde made his point in a parable with which he concluded his pamphlet: the tale of the stupid shepherd. This shepherd imagined himself to be an expert fencer. He boasted that he knew of a thrust with which he could vanquish any champion, and insisted on challenging the great masters of the art. In the end, when his boasting went beyond all limits, the masters ordered one of their pupils to accept the challenge. As the job was not a very honourable one, the pupil insisted on wearing a mask (an allusion to Hudde's anonymity). As expected, the shepherd was disarmed and brought to the ground at the very first attack. He then protested that this was not fair play. Apparently, one of the tactics he relied upon did not work, but he still had a multitude of others by which he would be able to gain an easy victory. He insisted that it was now the turn of the masters themselves to take up arms against him.

As is proper in case of a parable, Hudde finished with a moral. The bystanders, who had watched it all, in the end came forward to bring the shepherd back to reason. They urged him 'that you will exercise the office laid upon you after your best capacities; that you will care more properly for your sheep, that they will not be torn up by the wolf; that you will tend them not in dry and barren heaths, but in green meadows, at sweet brooklets. In that way, they would love and imitate you, and you would become a good Shepherd.' [160] A certain irritation with clerical meddlesomeness shines through. Clerics should stick to their spiritual calling and leave philosophy and other secular activities alone.

This was not the end of the discussion, however. A sympathiser of Du Bois wrote, under a pseudonym, a rejoinder in which he, in a rather peaceful way, answered the accusations levelled at Du Bois. He quoted some authorities on the point that as yet, there was no mathematical evidence for the earth's motion. This pamphlet was answered by someone who posed as an anonymous sympathiser of Hudde, but may well have been Hudde himself, who in this

---

[160] [Hudde] (1656)b 29.

way mirrored his opponent.[161] (Likewise, he had signed his second pamphlet 'I.G.H.' in response to Du Bois' signature 'J.D.B.'). This rejoinder mainly rehearsed the whole debate as it had been waged until then, in order to demonstrate that Du Bois and his friends only tried to divert attention. These two pamphlets contain no new arguments and will not be discussed here.

A belated echo

The discussion on Copernicanism and Cartesianism had gained full strength by 1656. In that year, no less than 20 pamphlets and other writings appeared on the subject. The issue was also tackled in various disputations. After 1656, however, the intensity of the debate decreased sharply. In 1657, only Velthuysen's 'Further proof' continued the debate. 1658 saw the appearance of Regius' short tract, and Wittichius' *Consensus veritatis* came out only in 1659.

The very last contribution to the debate was written in 1661 by a complete outsider: the Mennonite Dirk Rembrandtsz van Nierop, whom we met before. As a Mennonite, Dirk stood apart from developments in the Dutch Reformed Church, and initially he seems even to have missed the pamphlet war completely. Still, as a defender of Copernicanism, he took the issue to heart. When he finally learnt about the matter and read the pamphlets by Du Bois, he quickly decided to publish a reply. He justified this not only with the consideration that he might have some better arguments than the earlier contestants: 'I understand that there are still many who oppose this opinion bitterly, and try to paint it in the darkest colours, especially among those who are occupied with theology; quite as if one tried to offend, even make insecure, Holy Scripture with this feeling.'[162]

Even if one takes the Reformed religion as the only true religion, says Van Nierop, how can one be so sure that all passages in the Bible are perfectly explained? In particular those on the motion of the sun, where the theologians, after all, simply follow the received opinion, 'without investigating what is exactly the matter; as out of hundred, yea, thousand ministers, there has been before this time hardly one who knew about the difference. And even now the matter is publicly controverted, there are but a few who concern themselves with the issue.'[163]

Dirk's book is divided into four parts. The first concerns a discussion of the

161 'Irenaeus Philalethius, de tweede van die naem' (1656); 'Irenaeus Philalethius, de derde van die naem' (1656). In writing Vermij (1995), I presumed that the latter pamphlet indeed had been written by someone else. Dr W. Klever first pointed out that it probably derives from Hudde himself.
162 Van Nierop (1661) preface.
163 Van Nierop (1661) 15.

biblical sentences which seem to contradict Copernicanism. His principle is that the Holy Ghost has used a language and way of speaking which could easiest be understood by the people, or was commonly used. Still, he attaches great importance to the literal wording of the Bible. In explaining the miracle of Joshua 10:12 ('Sun stand thou still at Gibeon!'), Dirk vacillates between two alternatives. The one is that the command 'Sun, stand thou still' entails that the whole solar vortex stops moving, thus causing all planetary motions to come to rest as well (Galileo's suggestion in his letter to the Grand Duchess Christina, but now in a Cartesian jacket). The second wants the miracle to be an optical phenomenon: the earth kept on turning and the sun itself went down, but God temporarily provided a special sun to serve his people. Dirk eventually opts for the second explanation, because otherwise it would make no sense that the sun should stand still *at Gibeon*.

The second chapter deals with the credibility of the ministers in explaining Holy Scripture. It is mainly directed against Du Bois. Dirk maintains that in purely theological matters, human reason has to submit itself to revelation and that, consequently, the ministers should be believed in these matters. But in mathematical and physical things, they had better ask the experts in the field for advice. The third chapter argues that the Copernican system is in conformity with nature, and the fourth touches upon some other issues of Cartesianism, mainly in reaction to attacks by Du Bois.

14. The schism within the Dutch Reformed Church

A Voetian counter-offensive

The case of Velthuysen clearly indicates that the debate on Copernicanism and the Bible was not just a debate on the specific meaning of some particular biblical sentences, or on the merits and demerits of an abstruse scientific theory. Copernicanism, and even Cartesianism, were part of a much wider debate, and the participants were well aware of this. It was these wider dimensions which turned the issue into a major theological conflict.

By 1655, Voetius saw his enemies lining up against him. At Duisburg, and soon after at Nijmegen, Wittichius was openly propagating Cartesianism, thus challenging Voetius' authority in theology and undermining his programme of further reformation. At Utrecht, the dispute over the Utrecht canonries was quickly approaching a climax. Here, too, opposition turned into principled criticism. People like Velthuysen obligingly drew their arms from the arsenal of Cartesian philosophy. Now, Voetius was not a man to take all this lying down. But to regain the initiative, it would not suffice to emphasise the prerogatives of theology. From now on, he was to aim his attacks against his enemies in the Church itself. From a subject of academic debate, Cartesianism had to become a matter of church discipline.

The Voetians thought it wise to prepare the ground by means of a publicity campaign. Voetius used his preferred medium, the disputation. In May 1656, he presided over a disputation 'containing various miscellaneous positions'. The theses were rather heterogeneous, but the comprehensiveness of two of them is striking. One of these is the twelfth thesis on 'true Christian liberty'. Voetius is mainly concerned to show what Christian liberty does *not* consist in: not in a liberty – or even a necessity – to doubt philosophical propositions, and certainly not in liberty of philosophising. Especially the last point is very extensively documented. The other is the thirteenth thesis, on philosophy in general. Voetius defends that philosophy should be subordinated to theology. Theologians have reason to worry about the sayings which are falsely sold under the name of philosophy, as Voetius demonstrates by means of a long list of damnable propositions drawn from Cartesian philosophy. Of course, we meet familiar assessments such as that Scripture speaks according to the erroneous opinion of the people, and simply: 'The earth moves, the sun stands still.'[164] A similar list of Cartesian errors was published in a Latin pamphlet by someone using the pseudonym *Liberius Modestinus Philosophus*. It was published in Utrecht as a 'Specimen of Cartesian philosophy'. In it, 51 theses from Descartes' work were refuted, duly documented (remarkably, the author also refers to the still unpublished *Traité de l'homme*). Among these, again, is the thesis that the sun is the centre and the earth moves around it with the other planets.[165]

Then, in 1656, all of a sudden Cartesianism became a matter of concern to the various provincial synods. In July, the general synod of the United Duchies (in Germany, near the Dutch border), assembled at Duisburg, found 'that since a few years one has been hearing about a new philosophy, called after a certain Cartesius. One is notified that it is accompanied by unusual principles, injurious to sacred theology, by which the opinionated and conceited youth may easily be led astray. Learned and pious men have already warned against it in print, and experience shows that at several places the learned have attacked each other in writing; so, both parties being of our religion, it is to be feared that churches and schools may suffer still further inconveniences. (...) Now, since we hear that the synods in the Dutch Provinces will be assembled shortly (...) the general synod has thought good to ask their opinion, in order to direct themselves to it.'[166] In the Dutch Republic itself, at the same time action was taken by the synod of Gelderland against Wittichius' teaching at Nijmegen. A similar action in the provincial synod of South Holland was clearly directed against Leiden University. There is no evidence that this was a coordinated action; on the other hand, it is rather hard to believe that it was not.

One may ask whether this was perhaps an overreaction. It does not seem that by 1655 Copernicanism – or for that matter, Cartesianism – had made much headway among Dutch Reformed theologians. Lansbergen's interests were not typical of theologians of a later generation. Moreover, he was long dead and his ideas do not seem to have been regarded as much of a threat. Wittichius had been clamorously defending Cartesianism and Copernicanism, but so far he had found little support among his colleagues. As far as theolo-

---

[164] Voetius, 24 May 1656.

[165] Liberius Modestinus Philosophus (1656) 16.

[166] Synod at Duisburg, 11-14 July 1656, art. 23. Duisburg, municipal archives, archives of the evangelical community at Duisburg B6: resolutions of the general synod of the United Duchies, pp. 151-152. Also printed by du Bois (1656) preface, and 'Suetonius Tranquillus' (1656)d, 24-25.

gians were commenting on Copernicanism at all, their reactions were fairly negative.

Voetius' main theological opponent in this period was Samuel Des Marets (Maresius), the dominant theologian at Groningen University. Maresius was not a Cartesian, but initially he had sympathy for some of Descartes' ideas. He had a bitter quarrel with Voetius, which went back to a controversy about the 'Confraternity of St Mary' at 's Hertogenbosch, in States Brabant. The affair was noted enough for Descartes, in his own dealings with Voetius, to discuss it at length as an example of Voetius' immoderate and dominating behaviour.[167] After Maresius had become a professor at Groningen, some of the fiercest theological quarrels of the period were fought between Voetius and Maresius. Much of this was undoubtedly inspired by personal rivalry, but there were also more serious differences. Maresius sincerely disliked Voetius' theocratic pretensions. He took sides with the regents in various of Voetius' quarrels with secular government.[168]

As to the Copernican question, however, Maresius largely agreed with Voetius. He accepted his chair in Groningen with an inaugural address on the use and abuse of reason in matters of faith and theology. Herein, he left no doubt that reason is valuable only when it aligns itself with God's word. The Bible was to be the only source in theology. 'The heavens tell of God's glory (Psalm 19), they do not stipulate His will.'[169] Faith is to reason as reason is to the senses: reason corrects the senses, faith corrects reason.[170] Still, Maresius agreed that theology and philosophy each had its own domain, and that philosophy should be left free to a large degree.

From 1651 onward, Maresius held a number of disputations against Socinianism, later collected in his book 'The Hydra of Socinianism vanquished'. It was a refutation of a book by Johannes Volkelius 'on true religion'. Maresius reproduced the text and added his own comments. At one point, Volkelius had touched upon the Copernican issue and stated that it was of no importance whether the sun or the earth were moving. Maresius commented that this was not properly said by a theologian, who should keep to the pronouncements of Scripture, which clearly refuted Copernicanism. One could accept Copernicanism as a mathematical hypothesis, but not in an absolute

[167] Van Dijck (1973) offers a full history of the fraternity. The conflict is discussed on pp. 322-381, Descartes' part in it on pp. 359-364. See also Nauta (1935) 172-183, 241-244; Duker, II, 85-131. Some interesting remarks on the background of Descartes' intervention by Bos in Descartes (1996) 16-19.

[168] Nauta (1935) 279-281.

[169] Maresius (1643) 1.

[170] A similar metaphor in Maresius, disp. Groningen July 1667, thes. 15: '*Cedat, ut oculius corporis oculo mentis sive rationi, ita oculo fidei & revelationis, oculus sapientiae humane & rationis.*'

sense.[171] In other disputations by Maresius, the question is not touched upon at all. Only in a disputation *De mundo* is found as a corollary: 'the earth is the centre of the world.'[172]

The principal theologians at Leiden University, too, were not known as propagators of new ideas. The leading theologian, Heidanus, does not appear to have been a disciple of Descartes and never spoke out in favour of the motion of the earth. Writing on the Creation, he rejected the celestial spheres, but took care to say nothing on matters which could be controversial. He stated that in the beginning God had created a perfect world, not a chaos which had evolved only gradually into an order by the workings of the laws of nature.[173] In 1650, his colleague Johannes Cocceius called the philosophy of Descartes a new and dangerous plague.[174] The third Leiden professor, Johannes Hoornbeek, had been appointed only in 1654. He had for long been an admirer of Voetius and, since 1644, his colleague at Utrecht. Heidanus had made great efforts to get him to Leiden, which demonstrates that at this stage he was not a principled opponent of Voetius' programme.[175]

So, there seemed little reason to take large-scale measures against Cartesianism within the Dutch Reformed Church. Still, it may be that beneath the surface, there was more going on. Wittichius can be regarded as an exception, but also as representing a new generation of theologians. These people had studied at Leiden and had the chance to familiarise themselves with the doctrines of Descartes, not in the seminaries of their theological professors, but through the teachings of such philosophers as De Raei. Indeed, Wittichius and De Raei appear to have been well acquainted. It is not improbable that in this way Cartesianism was making gradual headway in the Reformed Church.

The Leiden professors, though not Cartesians themselves, apparently did not regard this development as being of much consequence and took little action to oppose it. So, the Voetians might have no reason to doubt their personal orthodoxy, but they clearly saw reason to oppose this passive attitude. In Voetius' view, a theological faculty should ensure that the students it prepared for the ministry were in no way infected with Cartesian opinions. Indeed, Hoornbeek (encouraged by Voetius) appears to have come to Leiden with the express purpose of setting things on a better footing. Heidanus very soon came to regret his support for Hoornbeek's appointment. In February

---

[171] Maresius (1651) 11. On this book by Maresius, see Nauta, 351-352.

[172] Maresius (1660) 215.

[173] Heidanus (1686) 292, 310, see also 305, 311-313.

[174] Cramer (1889) 75.

[175] Cramer (1889) 36-38. On Hoornbeek: *BLGNP*, II, 259-261. *BWPGN*, IV, 277-286.

1655, the newly appointed rector Kyper, himself an anti-Cartesian, complained to the curators that the philosophical differences of opinion had entered the faculty of theology, and were manifest in printed theses by Heidanus on the one hand and Hoornbeek on the other. The three professors of theology were thereupon interviewed by the curators, but the professors denied that they had any serious disagreements. The curators then restated the ban on teaching Cartesian philosophy, and left matters at that. [176]

Wittichius and the synod of Gelderland

Wittichius had earlier had some problems with ecclesiastical authority. The publication of his *Dissertationes duae* brought him into conflict with the Reformed Church of Cleves. The classis of Cleves (Kleve) referred the matter to the provincial synod, which in its turn referred it to the general synod of the three United Duchies (Cleves, Jülich and Berg, as well as the county of Mark). Wittichius reacted to this decision by stating: (1) that the synod should suspend judgement until it had studied the matter more closely, and had discussed the matter with him; (2) that if the synod were to show him (before the meeting of the general synod) the facts they thought were of evil consequence, he would gladly satisfy them; and (3) that the matter was beyond the synod's competence, as it had no say over the teaching at the university, especially as the matter was philosophical (but of course, he added, if the synod could convince him that he had gone too far in some respect concerning matters of faith, he would acquiesce). [177]

The churchmen appear to have been particularly offended by Wittichius' statement that 'on natural things, the Scriptures often speak according to the opinion of the people, not always according to the strict truth of the matter.' The issue was settled when Wittichius appeared ready to adapt this phrase. At the next provincial synod, in May 1655 at Wesel, the assembly read a letter by Wittichius wherein he explained that he had changed this expression into: 'The Scripture uses received expressions, even if these are founded upon false opinions' or, still more apt, 'The Scripture uses expressions which agree with the phenomena or appearances'. The synod thereupon declared itself content and decided to abstain from making further complaints to the general synod. [178] Here, the synod appears to have simply exercised its duty to supervise

[176] *BGLU*, III, 107-109.

[177] Provincial synod 2-5 June 1654, art. 14, in: *Reformierten klevischen Synoden* (1979) 49, 55-56. On Wittich's problems with his fellow theologians in Cleves, see also Trevisani (1992) 27-34.

[178] Provincial synod of 25-26 May 1655, art. 8, in: *Reformierten klevischen Synoden* (1979) 61-62, 65.

the publication of theological works, without being aware of a special danger of Cartesian philosophy in general.

By this time, Wittichius had accepted a chair at the newly founded Nijmegen University in the Dutch Republic. The province of Gelderland was unique among the Dutch provinces in that it consisted of three near-autonomous regions or quarters (a fourth quarter remained under Spanish rule). Shortly after the province had established its university at Harderwijk, in the quarter of Arnhem, the quarter of Nijmegen, not to be outdone, opened its own university in 1655. This university maintained a precarious existence for about two decades, and then disappeared into oblivion. With his move to Gelderland, Wittichius left the jurisdiction of the synod of Cleves and joined the Reformed Church of the Dutch Republic. He now came under the jurisdiction of the synod of Gelderland.

The synod assembled in July 1656. In reaction to a letter from the general synod of the United Duchies asking their advice on the issue of Cartesianism, the synod of Gelderland decided that the synodal deputies (a standing committee from the synod, which itself assembled only once a year whereafter the delegates returned to their respective churches) should urge the curators, 'that it will please them to keep, by their authority, the professors of the academies and high schools of this province therein, that they will be careful to abstain in their lessons and writings from all manners of speech which in any way might offend the truth of Holy Scriptures'. Moreover, the delegates were asked to discuss the problem in their various classes and to give their advice at the next synod.[179]

The formulation of this resolution is utterly vague and general. It concerned any manner of speech at any university. The vagueness must be deliberate. The delegates were apparently hesitant to publicly denounce a colleague. But of course, the deputies understood exactly that 'the universities and public schools of this province' referred exclusively to Nijmegen University; they never bothered to address the curators of Harderwijk University, who did issue a decree against Cartesian philosophy in August 1656, although it appears that the proposal had been made already in November 1655.[180] Moreover, whereas the synod spoke only of offending manners of speech (later resolutions discussed the problem under the heading of 'a peri-

---

[179] Synod of Harderwijk, 23-30 July 1656, art. 55. The original resolutions of the synods of Gelderland are in the State archives at Arnhem. Copies were sent to the synods of all other provinces. For the sake of convenience, I consulted the copies of the synod of Utrecht, State archives of Utrecht, archives of the Reformed (N.H.) provincial church assembly no. 142.

[180] Bouman (1844) I, 330-331, with the text of the decree. Also in du Bois (1655)a 13, cf. (1656)a preface **3^{v}.

culous way of speaking in philosophy, tending to offend the authority of the Scriptures'), the resolutions of the synodal deputies simply referred to Cartesianism.

The issue developed at a snail's pace. As the synod convened only once a year, it was a year before they could discuss any reaction to earlier measures and contemplate their following step. In September 1657, the delegates discussed at Zutphen the various recommendations of the classes (the deputies had not succeeded in addressing the curators of Nijmegen [181]). Three points were decided. First, the synodal deputies were to ask the States of Gelderland and the States General of the United Provinces to issue a decree against the dangerous way of speaking in philosophy, modelled after a similar resolution by the States of Holland (on which later more). Secondly, the synod would prepare a declaration to be signed by all candidates for the ministry, 'wherein they are to declare that they dislike such harmful principles touching theology or injurious to it, and that they will not bring forth something similar on the pulpit, or put forward elsewhere'. Thirdly, one should examine whether candidates for the ministry were contaminated by such principles. If they were, they should be refused. [182]

A year later, at Nijmegen, the delegates approved the concept of the declaration. As to examining the philosophical ideas of candidates for the ministry, the delegates asserted that this measure was being implemented. There remained the question of the decree by the provincial States. Deputies had apparently not moved so far. The point should be further clarified. [183] On 14 October 1658, the synodal deputies addressed a request to the States of Gelderland. This concerned not just Cartesianism. As a rule, provincial synods had a lot to complain about. In 1658 they asked for measures against several resurgences of popery, funerary ritual, breaks of Sunday observance, public swearing, licentious printing, pawnshops, and the dangerous new way of speaking in philosophy. The States, however, were not prepared to swallow it all without further questions. Although they did not flatly refuse the request, they preferred to temporise. They answered that if, as the synod maintained, philosophy was exercised in such a dangerous way that an express decree against it was justified, the synod should be more specific: 'that the persons whose teaching or living in the schools or universities of this province are suspect of such dangerous philosophising, should be mentioned by

---

[181] Resolutions of the synodal deputies, 1 April and 20 August 1657, art. 31. State archive Gelderland, archives of the Reformed (N.H.) synod, no. 23.
[182] Resolutions of the synod of Zutphen, 2-10 Sept. 1657, art. 38.
[183] Resolutions of the synod of Nijmegen, 18-25 August 1658, art. 35.

name, so that, after they have been heard, the States can provide in the matter as peace and unity will require.' [184]

This answer was discussed by the synod at its next meeting, a year later. Apparently, they did not like being specific, but decided that their deputies would have to persist with their request, 'So that God's precious Word, the only rule of our faith, will keep its due respect against all direct and indirect machinations of Satan'. [185] Although nothing more was stated in the resolution, the deputies were in no doubt about what had to be specified, and on 3 October they exhibited some pieces of 'dangerous speaking' to the States, i.e. six quotations, one from Descartes and five from Wittichius. Of the latter, one came from the first of the *Dissertationes duae*, and four from the work on the style of the Scriptures. From a later comment by Wittichius, one might surmise that these extracts were provided by Paulus Colonius, who had been professor of theology at Harderwijk since 1656 and was a staunch anti-Cartesian. [186]

The States thereupon decided that the deputies should first submit their objections to Wittichius himself, in order to hear his reply. [187] It seems strange that the deputies had not earlier thought of discussing the matter with Wittichius. Probably they thought a general prohibition less painful than a personal confrontation. After the State's answer, some deputies seem to have spoken to him, but these were informal discussions only. Wittichius objected strongly to the idea of being summoned and called to account at a formal ecclesiastical assembly.

Again, a year passed before the synod could decide what should be done. By now it was 1660. One of the deputies, who had spoken informally to Wittichius, explained that Wittichius seemed ready to explain his expressions in such a way that would expunge all offence. However, this did not satisfy the synod. It declared in a resolution that the expressions used by Wittichius should be deemed dangerous, for 'even if taken by him W. or some others, they are interpreted in the best and softest way, they do not leave of being offensive to others, who do not know of such an interpretation and declaration, or do not understand it; so that occasion of defamation is given to the enemies of truth.' Synodal deputies should urge Wittichius to abstain from such ways of speaking and one should ask the States for a decree prohibiting all kinds of speaking to the detriment of the Scriptures. [188]

---

[184] State's assembly at Nijmegen, Sept.-Oct. 1658, resolution of 14 Oct. State archive at Gelderland, archive of the States of the quarter of Nijmegen.

[185] Resolutions of the synod of Arnhem, 17-23 Aug. 1659, art. 44. The resolution was taken *ad notam* in the conference of the synodal deputies of 28 September 1659.

[186] See note 193. On Colonius, see Bouman (1844) I, 204-206; *BWPGN*, II, 178-180.

[187] State's assembly at Arnhem Sept.-Oct. 1659, resolution of 3 October.

[188] Resolutions of the synod at Zutphen, 15-20 Aug. 1660, art. 35.

The synod of 1661 finally saw the dénouement of the whole affair. Regrettably, little is known of the manoeuvres which contributed to the outcome. At first, Wittichius was reluctant to appear before the deputies and imposed several conditions. The deputies refused his demands that he be accompanied by two fellow professors and that the request to the States be suspended, but they were willing to agree on other points.[189] Apparently, some sort of compromise was reached. Wittichius then succeeded in giving complete satisfaction to the deputies. A formal charter was drawn up and signed by the four deputies, Wittichius and two witnesses. It read:

'As to the first point of the charge of the synodal deputies, they have, after having conferred with Mr Wittichius regarding the perilous way of speaking mentioned in the foregoing article, heard such declaration from him, wherein the said deputies have taken complete satisfaction. Which they will testify on all occasions in order to take away all suspicion and defamation. As to the latter point, regarding the request of a decree [by the States], the deputies will, in order to fulfil their charge, carry on with such discretion that they will prevent all *personalia*, from the one or the other. They will keep in communication with Mr Wittichius, in order to leave out anything that might be put in an offensive way. And as things will turn out, they may abstain from the said request.'[190]

Of course, the synod had the last word. According to a letter from Wittichius, there was much dispute over the matter, but eventually the advocates of reconciliation carried the day.[191] In the official resolutions, the synod expressed its joy about the 'means of peace'. It ratified the agreement, but did not make it public. The charter which had been drawn up was locked away in the synod's archival chest.[192] Wittichius claimed to know the reason for this secrecy: 'The urgent reason was that it was not necessary to include in the resolutions that the extracts [from my work] were not made in a honest way and did not agree with my writings. Still, this was said in public, while Colonius remained mostly silent.'[193]

---

[189] Resolutions of synodal deputies, 18 Sept. 1660.

[190] The original is not preserved, but Balthasar Bekker inserted the text in his *De philosophia Cartesiana admonitio* (1668). This edition is very rare. I have used the Dutch translation in Bekker (1693), where the piece is reproduced on p. 714. The charter is dated 20 September 1660. This may be wrong: on 18 September, deputies concluded that there was a stalemate. On 15 May 1661, it was said that a formal charter would be drawn up at '*proximum conventum*'.

[191] Wittichius to Velthuysen, 7 Sept. 1661. Leiden, university library, BPL 750. '*Acriter fuit disputatum inter amicos et inimicos, qui tamen illis tandem cedere debuerunt.*'

[192] Resolutions of the synod at Nijmegen, 14 Aug. 1661, art. 35. See also resolutions of the synodal deputies, Arnhem, 15 May 1661.

[193] Wittichius to Velthuysen, 7 Sept. 1661. Leiden, university library, BPL 750.

Preparing the South Holland synod

For the provincial synod of south Holland, which assembled in the summer of 1656 in Dordrecht, the classis of The Hague drew up the following *gravamen* for consideration:

'Whether the Christian synod would not judge it appropriate and necessary to contemplate measures serving to halt that improbable and unscriptural fanaticism that is rising so vehemently; so that the power and singularity of the Holy Scriptures will not further be injured, the teachers and professors within or between universities and higher schools no longer will be set up against each other, young students no longer will be educated in such opinions, and God's church will not be disturbed, the believers no further grieved, and their opponents will not be given occasion for joy or even calumny.' [194]

The prolix formulation is indeed specific – except on the crucial point. The participants in the synod, too, felt that the matter was proposed 'in a somewhat general and obscure way'. The classis of The Hague thereupon came with an elucidation:

'Testifies never to have thought of prejudicing the liberty of philosophising, or having aimed at somebody in particular, but has thought it necessary to watch that the principles which are put forward by many with vehemence will not be applied to theology to its prejudice and at the offence of the power of Holy Scripture.' [195]

It is all very diplomatic. The 'elucidation' appears to aim at preventing criticism rather than at achieving greater specificity. The specific aims were not put down on paper. Rumour had it, however, that some of the classis would propose that candidates for the ministry who were proponents of Cartesianism should be refused. Heidanus feared that the synod would demand an express declaration against Cartesianism, to be signed by all candidates, just as they had to sign the 'five articles against the Remonstrants' (the latter obligation was introduced by the general synod of Dordrecht in 1619, and is still in force today). As it seems, the Voetians aimed at a general ban on Cartesianism within the Reformed Church, comparable to the ban which had earlier been pronounced upon Arminian maxims.

Although the Leiden theologians strove to maintain some outward appearance of unity, it is clear that they were deeply divided over the coming synod and the gravamen, which might have serious consequences for them. In 1656, in the midst of the controversies surrounding Wittichius, Christophorus,

---

194 *Acta*, III, 517 (article 48).
195 Ibid.

Velthuysen and Du Bois, yet another pamphlet war started. Two pamphlets by an author calling himself Suetonius Tranquilus appeared in quick succession. Who was behind this pseudonym is a question which will occupy us later. His intention, as he later avowed, was to elucidate the gravamen prepared by the classis of The Hague. [196]. Moreover, this Suetonius understood the principles of propaganda. The pamphlets follow closely the rules which, in our age, Van Deursen laid down for a successful seventeenth-century publicity campaign. [197] First of all, unlike the usual prolix expositions, they were short, one running to eight and the other to 16 pages. Then, their impact was enhanced by the fact that the one succeeded the other in rapid succession. Their contents, too, were aimed to achieve maximum effect. They did not refer to earlier discussions, but tried to gain the initiative themselves. In fact, their author expressly refused to speak out on any philosophical issue.

The first pamphlet, 'State of the controversy on the Cartesian philosophy', purports to summarise the issue. It reduces the controversy to two questions. The first is whether it will do 'to introduce Cartesian philosophy in the schools and impose it upon young pupils, on the authority and fancy of some private persons, without, even against the charge and resolutions of the curators and academies, in contempt and with neglect of the common philosophy, which has been confirmed by public authority, and has been taught fruitfully until now'. [198] The way the question is posed already suggests the answer. The clever thing about the formulation used is that it does not refer even once to the *contents* of Cartesian philosophy. These turn up, however, when the author comes to the second question: whether it would do to introduce Cartesian philosophy in theology as well. (At Leiden of course, nothing of the kind had happened so far. This can refer only to Wittichius, Christophorus' case at Nijmegen. At the time, the various controversies were clearly felt to be linked.) The author then points out that the Cartesians teach doctrines which contradict the generally received Christian truth. Without further discussion, he offers a list of examples. Two points are familiar, i.e. the first – 'That the Holy Writ does not speak at all places according to truth, but in many places does not teach at all, or not according to truth, but according to the common opinion and error of the people; or that the whole of Scripture is a continuous parable' – and the sixth, 'That not the sun, but the earth turns around'. [199]

[196] Cf. 'Suetonius Tranquillus' (1656)d, 6.
[197] Van Deursen (1992) 31.
[198] 'Suetonius Tranquillus' (1656)a 3-4.
[199] 'Suetonius Tranquillus' (1656)a 7.

The second pamphlet, called 'Further opening of some pieces in Cartesian philosophy touching S. Theology', continues this way of discourse. It consists nearly exclusively of citations taken from Cartesian authors: Descartes himself, Wittichius, Christophorus, Velthuysen, the Groningen professor Tobias Andreae, and Descartes' Dutch friend, Cornelis van Hoghelande. The author purposely makes no comment. The message is clear: such blasphemy can speak for itself. Only at the very end of the pamphlet does the author make a kind of personal remark: 'I do believe that some Cartesians, when they hear or see this, will say that one might explain some of these examples so or so. But I leave it to your judgement whether one should not better abstain from bringing forth and disseminating such offensivenesses altogether, than having to do one's best to soften them with far-fetched explanations.' [200]

Suetonius was answered by an author who called himself Irenaeus Philalethius. Despite the title, his 'Reflections upon the State of the controversy... and upon the Further opening' were directed nearly exclusively against the first pamphlet. The second simply appeared just as the printing of the 'Reflections' was nearly complete. One is stupefied by the size of this reply: 90 pages against the eight of the 'State'. Irenaeus apparently followed his own tactic, which aimed at impressing by sheer size. The tone of the pamphlet is moreover very hostile and leaves an unpleasant impression. [201]

Following that, several other pamphlets were exchanged between 'Suetonius' and 'Irenaeus'. Suetonius replied to Irenaeus' 'Reflections' by issuing 'The convinced Cartesian', wherein he defended that the earlier pamphlets had summarised the matter in good faith and in accordance with truth. Irenaeus thereupon replied with 'Suetonius Tranquillus convinced of bad faith', whereupon Suetonius, in his turn, wrote 'Suetonius Tranquillus' sincerity defended'. All these pamphlets appeared in 1656. It was an extensive polemic which covered all controverted issues from Cartesian philosophy – that is to say, so far as it was concerned with philosophy at all. [202] Although Suetonius from now on put little restraint on himself concerning the size of his pamphlets, he remained the better polemicist. His attacks could be sharp indeed (although not so rude as those by Irenaeus), but on the whole he very much played the injured innocent. He consistently refused to discuss the contents of the debated points. His pamphlets were only intended to present the Cartesian theses, not to judge them, he insisted. But he never tired of repeating that the Cartesians opposed the authority of both Church and government, and that their doctrines were contrary to received theology. So, as to the motion

[200] 'Suetonius Tranquillus' (1656)b 16.

[201] 'Irenaeus Philatius' (1656)a.

[202] 'Suetonius Tranquillus' (1656) c and d; 'Irenaeus Philaletius' (1656) b.

of the earth, he stated: 'It appears a disgrace to a theologian to be moved, at the fancy of one or two wanton philosophers, who never gave any clear and firm demonstration from nature of the sun's motion and the earth's rest, to draw Holy Scripture, which affirms the contrary in plenty of places, at its hairs, distort it and bend it such as it pleases those wanton people to have it.' [203]

Who were the authors of these pamphlets? Knuttel, in his catalogue of pamphlets, identifies Irenaeus as the Leiden professor, De Raei. Thijssen-Schoute has drawn attention to a passage in the work of the seventeenth-century theologian Jacob Koelman, wherein he states that the discussion was between 'the head of the Leiden Cartesians' and 'the old anti-Cartesian theologian from Utrecht', which is generally taken as indicating the leading theologians Voetius (Suetonius) and Heidanus (Irenaeus). [204] One should add that Koelman was writing several decades after the discussion, i.e. in 1692. The annotations in Utrecht University library, which enabled us earlier to ascribe two anonymous pamphlets to Hudde, also cover the pamphlets of Suetonius and Irenaeus, but they are not completely clear. At first, the annotator wrote: 'Suetonius Tranquillus is in these treatises Mr Hoornbeeck, and mainly probably Mr Cabbeljau. Iren: Phil: for a small part Mr Heidanus, but mainly Mr J. de Raeij.' Later, someone has added under the name of Hoornbeek, 'also others'. Later yet, the names of Hoornbeek, Cabbeljau and De Raei have been crossed out, so that only Heidanus' name remains. [205]

The case for Heidanus seems pretty unequivocal. It is certain anyhow that 'Suetonius' himself regarded Heidanus as his opponent. This is clear from his many references to a work *De causa Dei*. This must be Heidanus' book *De causa Dei contra homines* from 1645, a polemical work against the Remonstrants and their leader Episcopius. Suetonius repeatedly expresses sorrow that Irenaeus – 'although he has certainly read this nice discourse' –heeds little the wise counsels it contains. Suetonius ascertains he could unmask the author of Irenaeus' pamphlets, but prefers not to do so. [206] In a later pamphlet he somewhat modifies his assertions by admitting that Irenaeus might have had some help. The philosophical material might have been handed to him by someone else, 'as we do not know the theological Irenaeus for so well versed in Cartesian philosophy, that on his own accord he would be able, without somebody prompting him, to speak so accurately about it as happens here and there in

---

203 'Suetonius Tranquillus' (1656)c, 29. The compatibility of Copernicanism with the Bible is discussed by 'Irenaeus Philalethius' (1656)a, 60-67, 86; 'Suetonius Tranquillus' (1656)c, 27-29, 33-35, 38.
204 Thijssen-Schoute (1954) 37.
205 Utrecht University Library, shell number Y. qu. 79.
206 'Suetonius Tranquillus' (1656)c 6. See also 11, 12, 19.

the 'Reflections'.' [207] He must be aiming here at De Raei, who is called elsewhere 'Irenaeus' great friend'. [208] But in the main, Suetonius asserts, the style is so recognisable, in particular at Leiden, that one cannot fail to recognise its author.

The person behind Suetonius seems less clear, but that Voetius is involved is highly improbable. The dispute has all the appearances of an internal Leiden affair. Suetonius speaks, for instance, on the troubles of 1627 concerning the moderation, 'in particular here within Leiden'. [209] Both Johannes Hoornbeek, professor of theology, and Petrus Cabbeljau, a Leiden minister who was to succeed Revius as regent of the State's college, seem pretty good candidates. Actually, both may be concerned. The debate was not a personal affair and it seems quite probable that the main authors of the pamphlets wrote them in collaboration and in discussion with their friends and sympathisers. Irenaeus asserted, when challenged over the authorship of his 'Reflections': 'This I can assert, that over the 'Reflections' have not worked eight theologians, as one of Suetonius' friends has admitted regarding the 'State of the conflict'.' [210] Clearly, two parties were concerned, and they not only opposed each other, but also stuck close together.

This said, Hoornbeek must be deemed as the central figure of the anti-Heidanus league. Not only was his position as professor of theology the most prominent, but later he emerges as the head of Voetian resistance at Leiden. It is worth mentioning that the first of Suetonius' pamphlets appears to have been published on the occasion of a disputation 'on divine attributes'. [211] So, the pamphlet war would be an extension of the strife between Heidanus and Hoornbeek as it had earlier become manifest in their disputations. Still, even if there is little doubt that Hoornbeek must be regarded as the *auctor intellectualis* of Suetonius' pamphlets, there is no evidence that he wrote them.

Although it would be going too far to discuss the contents of the pamphlets in detail, it is not without interest to remark in which direction the discussion develops. Actually, it develops in the same direction as the other dispute centred on Velthuysen: authority against Christian freedom. Irenaeus rejects Suetonius' appeal to authority. The government cannot authorise philosophy, the Church cannot put a binding interpretation of Scripture. He accuses his adversaries of wanting to have a say in everything and of trying to

207 'Suetonius Tranquillus' (1656)d 10-11.

208 'Suetonius Tranquillus' (1656)d 15.

209 'Suetonius Tranquillus' (1656)d 40. See also 'Suetonius Tranquillus' (1656)c 37.

210 'Irenaeus Philalethus' (1656)b 8.

211 'Suetonius Tranquillus' (1656)d 22, 34.

instaurate a new popery.[212] Suetonius, on the other hand, argues that Irenaeus' zeal for the freedom of philosophy comes down to the former practice of the Arminians.[213] 'Authorisation' does not mean the forcing of conscience, it just means that one kind of philosophy is made, and for very good reasons at that, obligatory in education. The new Dutch version of the Bible is authorised for use in the churches, but its marginal annotations are not, although one should honour them.[214] In a later pamphlet, Suetonius is more radical. He agrees with Streso that one should regard the marginal annotations of the Bible as authorised, 'to be the common understanding and common explanation of those places in our Dutch Reformed Churches: from which nobody should rashly diverge on his own account.'[215] Those claiming freedom of philosophy would do better to protect the freedom of churches and synods to fend off all novelties and offence.[216] To which Irenaeus replies: 'Under pretext of prohibiting 'frivolity and undetermined license', often the greatest license and frivolity is perpetrated, to wit, that people to whom such does not become or fit, lay down the law for everybody, and want to set rules and limits to someone else's freedom.'[217]

All this is in line with what we said before. It is not so much Cartesianism itself which is at stake. Heidanus had probably little interest in Cartesian philosophy and was not overly sympathetic. He might have agreed with Voetius' theological views to a large degree. What he opposed was that Voetius wanted to turn these views into points of doctrine and declare all other views heretical. The debate on Cartesianism thus laid bare a much more fundamental opposition.

The decree against Cartesianism and its aftermath

It is not necessary to describe in detail here the proceedings of the synod of South Holland. It could not be expected, of course, that the issue of Cartesianism was to be decided by rational arguments. The States' representative at the synod sensed alarm and warned Johan de Witt, the Grand Pensionary of Holland, that the issue might give rise to a serious clash. De Witt thereupon interfered and prevented a discussion of the gravamen, emphasising that the issue had the full attention of the States, who would not hesitate to take ap-

---

212 'Irenaeus Philalethuś (1656)a 15-16, 70.
213 'Suetonius Tranquillus' (1656)c 6, 16-17.
214 'Suetonius Tranquillus' (1656)c 12-14.
215 'Suetonius Tranquillus' (1656)d 20.
216 'Suetonius Tranquillus' (1656)c 8, 30.
217 'Irenaeus Philalethus' (1656)b 8.

propriate measures. The synod complied, sent a letter of thanks and left the matter at that.[218]

In order to live up to their promise, the States asked the advice of the theological faculty of Leiden on the measures which should be taken. The theological faculty was divided, as stated before, and it was only after heavy pressure that De Witt succeeded in obtaining a unanimous recommendation.[219] This led to a decree, proclaimed by the States on 30 September 'on the mixture of theology and philosophy, and the abuse of the liberty of philosophising to the detriment of Holy Scripture'. The decree forbade theologians and philosophers to interfere with each other's domains. In case of matters which could be explained from Scripture as well as from nature, Scripture should be regarded as decisive. In no case were philosophers allowed to explain Scripture according to their philosophical principles. More specifically, 'for the sake of peace and quietness' it was forbidden to propagate the tenets from Descartes' philosophy which were giving offence. All professors at Leiden were required to take an oath that they would keep the decree.[220] Eventually, all took their oath before the president-curator of the university on 8 January 1657.[221]

The States' decree against Cartesianism was not a sign of their hostility against the new philosophy. When in November 1656, shortly after the promulgation of the decree, Heereboord set out to defend some theses which clearly transgressed the limits, De Witt took no punitive action, but simply urged the theses to be withdrawn.[222] The decree was a measure of statecraft, aimed at preserving public peace. The theological discussions on Cartesianism were a disruptive factor in public life. One had to meet the orthodox in order to keep them quiet. But one should not meet them too much. After all, the States – not the churches – were leading the country. The anti-Cartesian forces should have no reason for serious complaint, but they should not have their way. For one moment, there was mention of a regulation on the books used

---

[218] States' resolutions of 14, 20 & 30 July 1656 in: *Register*, 666, 678-679, 705. Resolution of the synod in: *Acta*, III, 517-519 (art. 48). Reaction of the senate of Leiden of 12 July 1656, in: *Register*, 677-678, also printed in: *BGLU*, III, 111-112. The events of 1656 are briefly summarised by Stewart (1994) 40-41.

[219] States' resolutions of 20, 25 & 28 July 1656, in: *Register*, 679, 691, 702. Cramer (1889) 71-73. The representatives of the town of Leiden were opposed to measures against the university.

[220] Text of the decree in *Register*, 803-807; also in *Acta*, IV, 36-40. An earlier concept, dated 23 Sept. 1656, in *Register*, 787-792.

[221] *BGLU*, III, 116-118. Only Golius, at the time rector, was allowed to excuse himself, for reasons which were not put on paper. Cramer (1889) 75.

[222] De Witt (1906) 360-362. De Witt wrote on this matter to Heidanus and the rector Golius, on 22 November.

by the philosophical professors at their private colleges, but nothing more was heard of it.[223] The States preferred a condemnation in general rather than specific terms.

However, the States' measures were beside the point. They wanted to impose peace by defining the proper boundaries between the theological and the philosophical domain, as had earlier measures against Cartesianism. In 1656, however, the matter had grown into a debate between the theologians themselves, and this conflict remained unsolved. If anything, the States' intervention made it more acerbic. The Voetians, as has been explained, were rather wary of the political regime after 1650. Heidanus, on the other hand, represented a part of the Church which was much closer to the government and quite willing to cooperate. These differences emerged during the synod of 1656. Whereas the Voetians were trying to purify their Church of Cartesianism, Heidanus secretly called for the government to intervene to thwart their intentions. He was related by marriage to the Grand Pensionary Johan de Witt, and used this relation to enter into a correspondence with De Witt and urge for protection of Cartesianism. One might ask how far this had any effect on the outcome of the affair. De Witt acted from considerations of statecraft, not from personal loyalty. Probably, he used the correspondence with Heidanus to bring Heidanus into line with his policy, rather than vice versa. But by any means, it became clear that the Cartesians relied on the very regime which the Voetians regarded as ungodly.[224]

Heidanus' secret dealings with the Grand Pensionary did not remain secret for long. Suetonius Tranquillus, in the pamphlets we discussed earlier, refers to them in an unambiguous way.[225] He reproaches 'Irenaeus' (Heidanus) bitterly for it. Heidanus wants the Cartesian philosophemata, so far as they regard theology, to be judged not by the Church (as would be proper, of course, in Suetonius' view), 'but by the politicians only, because he hopes that there his advice will always be heard. Whether this will just be swallowed by the churches, classes and synods, we shall attend. It seems to us that there is as yet time for all the orthodox to wake up.'[226] Heidanus' bringing in the States of Holland will have aroused bitter feelings and contributed to making

---

[223] State's resolution of 6 Oct. 1656; *Register*, 840, also quoted in BGLU, III, 116.
[224] Heidanus' mother-in-law to De Witt, in Cramer (1889) 67. Heidanus to De Witt, 22, 24 & 27 July 1656; in Cramer (1889) 71-73. De Witt to Heidanus, 26 July 1656, in: De Witt (1906) 356-358.
[225] 'Suetonius Tranquillus' (1656)c 8, 30, and in particular 'Suetonius Tranquillus' (1656)d 12: '*Hy wilde wel, soo't schijnt, dat de Classen en Synode van Zuyd-Hollandt hem quamen bedancken, over de sorge die hy gedragen heeft voor hare vryheyd, door bedecktelijk te procureren, dat door de H.H. Staten hant-sluytinge ghedaen wiert aen de Synodus, om over de Cartesiaensche nieuwicheden geen ondersoeck te doen, noch eenige ordre tegen der selviger voortkruypinge te stellen....*'
[226] 'Suetonius Tranquillus' (1656)d 13.

the division between Cartesians and Voetians permanent. Moreover, it will have worsened the already strained relation between the Voetians and the proponents of 'true freedom', who supported the Cartesians by prohibiting the Church to take appropriate measures.

So, the decree of 1656 settled nothing. It did not end the debates on Cartesianism, nor did it restrict them to academia. The provincial synod of Utrecht took notice of the resolutions of the synods of South Holland and Gelderland. The deputy from South Holland had especially recommended the point of the Cartesian philosophy, and the Utrecht synod had conformed itself to the resolutions of the two other synods.[227] Within Holland, the provincial synod would not take any other measures, but after 1656, the Leiden church council became alert. In 1659, a certain Theodorus Callerus submitted at Leiden a doctoral disputation in theology, entitled *De veritate religionis Christianae* ('On the truth of Christian religion'). Hoornbeek, upon examining it, found that it was 'not theological, but totally Cartesian'. He protested, whereupon the disputation was retracted. The candidate submitted another one which, according to Hoornbeek, was no less Cartesian. The senate convened on the matter and decided that this disputation should be suspended, too. However, this no longer satisfied Hoornbeek who demanded stricter measures. Persons like Callerus should not be allowed to graduate at all. The university should publicly disavow the theses, which had been made public already. As it seems, Hoornbeek was afraid that he was no longer being taken seriously at the university and that Cartesianism would be allowed to slip in behind his back. He turned to the States with a request in order to obtain full satisfaction. This, of course, caused a lot of annoyance.[228]

This affair raised the alertness of the church council of the Reformed Church at Leiden. In December 1659 it decided that, as among theologists the troubles of the Cartesian philosophy had reared their heads again, one had to think of measures to prevent further offences. On 2 January, the church council came to the following resolution. The theological students at Leiden were, as a matter of fact, members of the Leiden church. If they presented themselves to a classis to be examined for the ministry, they needed an attestation by the Leiden church council regarding their membership and their conduct as such. Now, the council decided that henceforth, it would mention in these attestations 'that the church council trusts that the student, to whom it grants the testimonium, is willing, in conformity to the resolution

[227] Resolution of the provincial synod of Utrecht, 2-6 Sept. 1656, art. 55. State archive of Utrecht, archive of the classis of Utrecht no. 347.
[228] *BGLU*, III, 102*-107*. Cramer (1889) 80-82. Callerus' disputation is at Leiden University Library, 236 A 36. It is quite comprehensive and indeed appears rather Cartesian.

of the Christian synod, to abstain from philosophemata which, taken from the philosophy of Descartes, may offend some people; or to teach any other harmful principles, taken from any other philosophy, in the church and mix them up with S. Theology.'

The point is that the church council was entitled, or rather obliged, to refuse an attestation if it had doubts about someone's sincerity or orthodoxy. Without such an attestation, one could not be accepted as a member of another congregation, let alone be admitted as a minister. Heidanus, who as professor of theology was also a member of the church council, was 'not pleased' with the resolution, as was noted with some understatement in the consistorial book. He protested and appealed to the classis, which, however, took no action. Action was taken, however, by Leiden University. A serious and rather protracted conflict ensued. Rector and senate accused the church council of infracting upon the jurisdiction of the university. The church council was not entitled to act with greater authority in the case of students than in the case of other members of the congregation. According to rector and senate, the church council actually suggested that professors and curators of the university did not do their work well, that even the States of Holland had been negligent and had not settled the matter, so that the church council should take some ecclesiastical measures in addition. The church council, for its part, declared that the attestations were a purely ecclesiastical affair, with which the university had nothing to do. Just as the church should not meddle with the affairs of the university, so the university should leave the church on its own. We need not follow this dispute in detail. After protracted negotiations, the States of Holland decided that the attestations should remain on their former footing.[229]

Building a Cartesian faction

The South Holland synod of 1656 loomed large over Dutch intellectual life. It does not seem coincidental that the dispute over Copernicanism reached its climax in this year, nor that it quickly died out thereafter. The various pamphlets do not seem to have been written for the education of the general public. (As we saw earlier, Dirk Rembrandtsz van Nierop got wind of the dispute only in 1661.) They rather served the end of convincing the various deputies

---

[229] Cramer, *Heidanus*, 82-89. The pieces are printed in *BGLU*, III, 161-162, 102*-107*, 121*-137*. The latter section comprises extracts from the resolutions of the Leiden church council. It should be noted that not all the relevant resolutions are reproduced. See in particular the resolutions of 12 Dec. 1659 and 2 Jan. and 5, 12 & 19 March 1660, in municipal records Leiden. Not pleased: resolutions of 2 Jan. 1660 ('*D. Professor Heidanus heeft geen gevallen in dese resolutie konnen nemen*').

and influential ministers who were still wavering over the subject. Once the issue was settled, there was no longer much need for an open dispute. The pamphlet wars served not so much to reach a conclusion as to draw the lines clearly.

The debate on Copernicanism and Cartesianism was part of one large, connected debate and the participants were well aware of this. It was generally felt that the decision of the synod would not just be a local affair. Heidanus reminded De Witt that the synod of Cleves – that is to say, the general synod of the United Duchies at Duisburg – had resolved, in a similar matter, to attend the resolution of southern Holland, 'from which you by the way may remark, how much will depend on it, what position our synod will take.' [230] Against the Voetian offensive, Cartesians, defenders of the new science, and people who simply rejected clerical interference in their philosophising, became aware of their common cause. In 1656, the young Christiaan Huygens wrote to Van Schooten that the latter might show his broadsheet proclaiming his recent discoveries concerning Saturn to Heidanus, 'who shall be pleased to learn of a new corroboration of the Copernican system.' [231]

Now, the Voetian camp was a self-conscious unity, formed for many years around Voetius himself, bound by loyalty to their programme of further reformation. Voetius' adversaries up to this moment did not present such a unity. They did not behave according to a common programme, but appear to have been a loose collection of individuals. However, once they found themselves in an all-out ideological war, they quickly made up arrears. At the beginning of 1656, when the pamphlet war between Velthuysen and Du Bois had just started, Wittichius, Christophorus sent Velthuysen a letter from Nijmegen. 'The common cause we are defending for the propagation of truth, against which its opponents are resisting powerfully, seems to demand that we and all cultivators of the true philosophy keep mutual friendship, and consult each other regularly on our business. With the present, I wanted to make a start between you and me, and lay a foundation for our friendship.' [232]

This was not just a declaration of intention. Wittichius, Christophorus had a pertinent question as well. 'My occasion is, that I have seen that our common opponent Du Bois, in the work he has written against you, occasionally mentions my name as well and slanders my reputation among the illiterate. It is not fitting, nor allowed, for me to answer him in Dutch. So I would ask you whether you, in the answer you are preparing, will occasionally take on my

---

230 Heidanus to De Witt, 27 July 1656, in Cramer (1889) 72.

231 Huygens to Van Schooten, 10 March 1656, *OC*, I, 389 (no. 269).

232 Wittichius, Christophorus to Velthuysen, 28 Feb. 1656. Leiden, university library, BPL 750. ('*Hujus colloquii hic praesentibus inter Te et me volui facere initium, ac nostrae amicitiae jacere fundamentam*.')

defence.' He listed the places were Du Bois had distorted his words, and sent Velthuysen a copy of his work on the style of the Scriptures, 'as I am not sure whether you have received it from Mr De Raey.' For the rest, he left it all to Velthuysen who, as he wrote courteously, without any doubt could deduce and explain these things much better than he himself could possibly do in a letter.[233]

Wittichius, Christophorus also warned that they should not allow their opponents to play them off against each other. In some matters, notably the infinity of the world, he and Velthuysen held different opinions. 'So, it seems that we must care that we settle this disagreement in a convenient way, so that we rob our opponents of that chance to glorify.' Velthuysen heeded these councils. In his second pamphlet, he referred approvingly to Wittichius, Christophorus' views.[234] Later, Velthuysen followed Wittichius, Christophorus' example by approaching an ally. Shortly after the appearance of the *Mathematical proof*, Velthuysen entered into correspondence with its probable author, Hudde. Velthuysen sent him some booklets. Hudde sent a reply, wherein he explained his scientific projects. In a postscript, he sent his regards to Professor Johannes de Bruyn.[235]

The most curious attestation of the mutual solidarity which rose inside the upcoming Cartesian 'party' is a four-page appendix to Velthuysen's pamphlet of 1656. It presents what it claims to be extracts from two letters, one from Velthuysen himself, the other from the Cartesian professor of philosophy, Johannes de Bruyn. The addressees are not known. Both letters describe the same event, a conversation which had occurred shortly before between Velthuysen, De Bruyn, a certain Nieuwstadt (probably Joachim Nieuwstadt, secretary of the town of Utrecht), and the Leiden professor of philosophy, Henricus Bornius. Some details are different, but on the whole the two versions agree.

The four people had assembled at the house of a trumpet maker, a certain Scot or Scotten, a cousin of Bornius and Nieuwstad, to view a wind gun. Wine was ordered and the conversation turned to the philosophy of Descartes. Velthuysen and De Bruyn were, of course, convinced Cartesians. Joachim Nieuwstad, too, is known as an adherent of the new philosophy.[236] Bornius was a singular case. In 1646, he had been awarded his doctorate at Leiden under Heereboord on a clearly Cartesian dissertation. When ap-

---

233 Ibid.

234 Ibid. Cf. van Bunge (1995) 52, who notes it as striking that Velthuysen's later approval of Wittich's view on certain points contradicts his earlier statements.

235 Hudde to Velthuysen, 13 Oct. 1657. Amsterdam, University Library, ms. D. 29.

236 On him: van Maanen (1983); Meinsma (1983).

pointed professor at Leiden in 1652, he seems to have been known as a Cartesian. Probably, he was so only in a general sense. Based on a study of his works from this period, Sassen describes him as sympathetic to Gassendi rather than Descartes. He had a sincere interest in the investigation of nature, but was on the whole a traditional philosopher. Soon, he was found to oppose Cartesianism as much as he possibly could. He attacked it in several disputations. During the disputes at Leiden on the occasion of the gravamen of the classis of The Hague, Bornius was one of three professors (the others were Hoornbeek and the historian Thysius) who disavowed the protest by the Leiden senate against clerical interference with teaching. [237]

It was not strange that Velthuysen, De Bruyn and Nieuwstadt wanted to hear an explanation of Bornius' change of mind. Bornius then confessed, according to De Bruyn, 'that it was against his opinion, yeah, that he was still the same he had always been.' According to Velthuysen, he expressed the hope that within a year the philosophy of Descartes could be taught freely. De Bruyn reports that Bornius said that, if circumstances were to turn to his favour, people would see who Bornius really was. These 'circumstances' are not to be identified with the official prohibition of Descartes' philosophy. He did adduce this argument ('that it was the order of the curators, and that he was ready to teach the fables of Aesopus, if he would be ordered to'), but this clearly was a subterfuge. More sincere seem his statements 'that this declaiming against the Cartesians served to cure them of that use of authority', and 'that the wrong he had suffered demanded that revenge'. That is to say, Bornius had personal motives.

Velthuysen's letter gives chapter and verse: Bornius was angry with Heidanus. He opposed Heidanus personally, not so much Cartesian philosophy. This seems quite plausible. Indeed, not everybody liked Heidanus. He was 'the Pope of Leiden', [238] just like Voetius was the Pope of Utrecht. The exact reasons for Bornius' grudge are not mentioned, but they may have to do with his difficulties in obtaining an ordinary professorship. Heidanus from his side felt little sympathy for Bornius. He described him to De Witt as: 'A renegade Cartesian, who now pours forth his venom against the very philosophy which he formerly praised so highly, and who behaves as if he were full of furies.' [239]

Velthuysen, De Bruyn and Nieuwstad disapproved of Bornius' attitude. A

---

[237] On Bornius: Sassen (1962) 16-32, 69-91; see in particular 78-81, on his campaign against Cartesianism and its background. Disputations by Bornius have not been preserved.

[238] The term is used in a letter by A. Rehoorn to Velthuysen, Amsterdam 3 Dec. 1665, Leiden, university library, BPL 750, who is scathing about Heidanus' intolerance of people who disagree with him.

[239] Cramer (1889) 69.

personal grudge should not be allowed to harm a good cause. Their disapproval, however, had tactical as well as moral grounds. Bornius' campaign suppressed truth and was harmful to the liberty of philosophising. When at Leiden philosophy was antagonising the authorities, it was easily predictable that the affairs at Utrecht would suffer, too. That was the reason to make the two letters public: 'in order that his disputations, sent hither, will do less harm to that liberty which we do still have.' Bornius' case illustrates that a philosophical stance sometimes could serve other, more worldly ends. As the philosophical antithesis developed into factional strife, it obtained all the properties of factionalism. On the other hand, the reaction of Velthuysen & co. shows that by this time the philosophical issue was no longer the field of abstract speculation: it was an element in a struggle for supremacy, whereby one could not afford to break rank.

Wittichius, Christophorus was probably one of the main originators of this fabric of mutual support. He established the first contact with Velthuysen and thereafter remained in touch with him. His letters clearly attest to an attempt to build up a coherent Cartesian block in Church and university. They extensively discuss candidates for several ministries or professorships. 'I should like to learn from you what kind of a person this is, what his conduct and learning, under whose teaching and at what university he has laid the foundations of his theological studies, and whether he is an opponent of the saner philosophy or not,' Wittichius, Christophorus writes with respect to a certain Potheunck. [240] He also conducted an elaborate correspondence with other prominent Cartesians, such as the Groningen professor of theology, Johannes Braun. When Braun got the staunch Voetian Van der Marck as his colleague, the letters speak about this appointment as a general calamity. All friends are deeply sympathetic to Braun's misery, Wittichius, Christophorus assures him. [241] It does not seem just paranoia when the young Tennulus attributed his problems in obtaining an appointment to 'the league... of the Cartesians, who try to put their friends in all places. [242] One should add, of course, that the Cartesians were just following the example set by Voetius himself, who for years had been nominating only those who would support his Further Reformation programme.

---

[240] Wittichius, Christophorus to Velthuysen, 2 Feb. 1661. Leiden, university library, PBL 750.

[241] Wittich to Braun, 26 March, 12 July and 7 Aug. 1682. Copies in Leiden, univesity library, BPL 1961.

[242] Tennulus to Gronovius, 19 Sept. 1666, published in 'Témoignages', 240: '*Vous scavez la ligue et les maximes des cartesiens, qui cherchent pour toutes les villes leurs camarades; ceux là me font tort, de croire que je ne suis pas de leur famille.*'

The dispute in a wider setting

Although it may seem that we have departed somewhat from the Copernican debate in the last paragraphs, they are quite pertinent to it, as they show us why people at the time got so agitated over the issue and what function these debates served. The polemics of Wittichius, Christophorus, Velthuysen and Hudde – and of Du Bois, Streso and Voetius on the other side – were statements in a far wider debate on the question how the Dutch state should be run. It is only as part of this wider debate that Copernicanism became an issue in the Reformed Church.

This wider debate is a long and complex story and to rehearse it here, even in outline, would be going much too far. Suffice it to say that the deepening of the quarrel had much to do with the political situation, especially in the province of Holland, which saw mounting polarisation between the Voetian wing in the Church, who were hoping for a restoration of the stadholderate, and the republicanism of the States of Holland. Tensions became especially acute when in the late 1660s William of Orange, the son of the late stadholder William II, reached his majority and the States still refused to give him the position of his father. They even decided to abolish the stadholderate 'in eternity'. Proponents of the States' party, such as Lambert van Velthuysen, formulated radical republican ideologies, which denied the churches all influence in the state. These conflicts did not remain confined to the cabinets of politicians and pamphleteers. For instance, by 1660 the conflict at Utrecht had become so bitter that the town government banned two ministers from the city for their seditious behaviour. It is telling that the regents only dared to impose these measures after they had quartered seven companies of troops in the town. The situation changed when in 1672, the shock of a French invasion brought about the downfall of the regime of 'True Freedom' and the restoration of the stadholderate. However, the new regime proved no more inclined to give the Voetians their way than the old, and faction struggles remained the order of the day.

A person like Voetius would probably have opposed Copernicanism in any circumstances. But such opposition might well have remained an incident, like many controversies within the Church, had it not been for the political circumstances in the United Provinces after 1650. Until then, nobody had bothered very much about the matter. And, as other theologians were to prove, Cartesianism was very compatible with Reformed orthodoxy. Although the political tensions after 1650 did not automatically lead to a dispute about Copernicanism, they did make some kind of clash almost inevitable. And as Voetius' anti-Cartesianism had been well established by this time, it is not surprising that Cartesianism was at the heart of the conflict.

The Voetians – who supported the House of Orange – and their opponents – who favoured the State's regime – had many reasons to resent each other. Very succinctly stated, they did not have a quarrel because they disagreed on such issues as Copernicanism, they had a disagreement because they had a quarrel. The anti-Copernicanism of the Voetians was not an unlucky incident, a clumsy manoeuvre they might have evaded had they better thought it over. Instead, they consciously and deliberately opted for confrontation. As they set their hopes on a change of government and the instauration of a truly god-fearing regime, they had no interest in accommodating their views to the factual state of affairs, to secular learning or to theories supported by their opponents.

It started with the debate on Copernicanism and Cartesianism, but over the course of the years the opponents found many more issues to disagree about. Just after the pamphlet war on Cartesianism was over, there arose a controversy on Sunday observance. The reason was Voetius' stress on *precisitas* (strictness of morals) which was partly inspired by English Puritanism. Voetius' opponents thought he carried such strictness too far. A disputation on Sunday observance, in 1658 defended under Heidanus, caused a real pamphlet war, so that by 1659 the States of Holland intervened and forbade further writings on the subject. The subject itself was not new. In fact, it had been discussed for decades in the Dutch Reformed Church, and sometimes rather vehemently. But now, this old discussion got a new meaning. It is not that Heidanus and his followers were opponents of Sunday observance as such, quite the contrary. Here again, it was not so much Voetius' strictness which aroused opposition, but the fact that he decried anybody who held different opinions. However, this debate was put on a higher theological plane by Johannes Cocceius, who in the course of this controversy emerged as Voetius' main theological opponent. [243]

After 1659, the dispute moved increasingly to theological issues. In the 1660s, Cocceius developed a new idea on the Covenant and other topics. Although his ideas were perfectly in line with Reformed orthodoxy, the Voetians took grave offence. In 1665, Voetius started a series of disputations on the problem of justification; these disputations implicitly but quite clearly attacked Cocceius' theological ideas. [244] This caused a new round in the theo-

[243] See for details of the dispute, H.B. Visser (1939) 115-149. See also Schotel, *Openbare eeredienst*, 163-199; van Veen (1889) 118-129; J. Th. de Visser (1926) 452, 470-471; Cramer (1889) 38-41; Duker, II, 207-219; Evenhuis, III, 118-127. The last two mentioned authors tend to regard this dispute as the origin of the Voetian-Cocceian quarrels. Probably, they regard the earlier debate on Cartesianism as not theologically relevant. On Voetian *precisitas* in general, see Duker, II, 230-270.

[244] Duker, II, 221-229.

logical disputes. From now on, the opponents disagreed not only on Cartesian philosophy, but also on fundamental theological dogmas. To Cartesian theologians, who were looking for a theological legitimation to defend their Cartesian hermeneutics, Cocceius' ideas proved quite useful. Cocceius' theology enabled a sharp distinction to be made between theology and philosophy, between the domains of reason and of revelation. Whereas Voetian theology aimed at subjugating all to Holy Scripture and regarded philosophy as the handmaid of theology, Cocceius left natural knowledge alone and occupied himself only with the theological meaning of the Bible. [245] Cocceius' ideas therefore formed an alternative theological programme, which allowed the Cartesians to disavow Voetius' ideas without falling in any way into the pitfall of 'Arminianism'. Leading Cartesian theologians, like Johannes Braun or Fransiscus Burman, based their theological handbooks on Cocceius' theology. In the end, two parties had emerged and the Dutch Reformed Church was virtually split into two. From now on, against the 'Voetians' stood the 'Cocceians'. This whole development demonstrates how much in the religious disputes of the time, differences of opinion were sometimes only secondary. [246]

It has been a much disputed question in Dutch church history why so many Cocceian theologians enthusiastically accepted Cartesian philosophy. [247] As it seems to me, this question has been badly posed. The question should be why these Cartesians came to adopt Cocceian theology, not vice versa. A discussion of this point would be going too far here. What needs some emphasising is that Cartesianism did keep a crucial position in the debate. The Groningen professor Maresius, Voetius' main opponent before 1656, had remained aloof during the struggles between Leiden and Utrecht (one might add that at Groningen, political tensions were less acute; this province still had a stadholder, from a sideline of the House of Orange). He finally took sides a decade or so after the rift between the two parties had become manifest. However, he did not opt for the Leiden group with whom he shared his aversion to Voetius' over-preciseness and theocratic pretensions. On the contrary, he made peace with Voetius. This happened in 1669. In the past, friends had many times tried to reconcile the two uncompromising men, but only now did they succeed. Maresius started a campaign against Cartesian philoso-

---

[245] Van der Wall (1996).

[246] On seventeenth-century Cocceian theology after Cocceius, see van Asselt (1988) 138-144. On purely theological points, there is sometimes considerable overlap and borrowings between the parties. Cf. Broeyer (1994); cf. also Bisschop (1993) 33-34. van Sluis (1994) also argues that we should regard Voetians and Cocceians rather as factions bound by contact and interest than as adherents to a theological system. Duker, II, 229 agrees that *precisitas* was more important than doctrinal differences in clearly separating the parties.

[247] Cf. Cramer (1889) 75-77; Trimp (1987) 139; van Asselt (1988) 33-37 (see also note 24 on p. 175).

phy and Cocceian theology. He engaged in controversy with Wittichius, Christophorus and began what was to be a protracted quarrel with his colleague at Groningen, Jacob Alting, who was a Cocceian. This led to a new round in the dispute over Sunday observance.[248]

As Nauta has argued, it is probable that the peace with Voetius was signed because Maresius in the end had spied a more dangerous enemy. In 1667, he had read Lodewijk Meijer's book on philosophy as interpreter of Holy Scriptures. Meijer was a follower of Spinoza rather than of Descartes, but he could be seen as drawing the consequences of Cartesian hermeneutics to its extremes. His work caused great commotion in the Dutch Reformed Church. As it seems, it alerted Maresius to the dangers inherent in Cartesianism. According to Nauta, his peace with Voetius should be regarded as an offensive alliance against a common enemy.[249] One could argue that Meijer's book was just one expression of the growing tensions in Church and society in the late 1660s which enhanced polarisation, but in essence the explanation seems correct.

As it seems, it is only when he discovered the work of Meijer that Maresius really became worried about Copernicanism. Until then he had not paid much attention to the subject. Although rejecting the motion of the earth, he maintained that one could defend it as an astronomical hypothesis. Once he came to regard Cartesianism as an enemy, he became more pertinent. From 1667 onwards, he refuted Meijer's book in a series of disputations. In the very first of these, he asserts that the Cartesian philosophy has its merits, but one should not regard it as certain. The motion of the earth (about which Meijer did not say a word, by the way) according to Maresius has not been established beyond doubt. The Cartesians themselves assert that God could have created the world in a different way, had He so desired. So, one might follow the Tychonic model as well as the Copernican, since both save the appearances. The simplicity of the Copernican system is not a valid argument. God is not obliged to create the world in such a way that it is easiest for us to understand.[250]

Finally, in his work *De abusu philosophiae Cartesianae*, Maresius extensively refuted the Copernican world system. Descartes himself, he maintains, had been cautious and had not proclaimed his ideas as absolute truths. It is intolerable if theologians try to correct the Scriptures on the basis of mere philosophical hypotheses. True, the Holy Ghost wanted to speak our human language, and not deal with physics and philosophy. But He did not want to

[248] Wittichius, Christophorus: Nauta (1935), 361-369; Bizer (1958) 347-357. Alting: Nauta (1935), 369-380; H.B. Visser (1939), 189-200. Eventually, Maresius came into conflict with Cocceius, too, Nauta (1935) 377-378.

[249] Nauta (1935), 359.

[250] Maresius, disp. Groningen 19 Feb. 1667, thes. 17-18.

accommodate to errors which result from our fallen state, as if He would encourage such views or could not express Himself in another way. Maresius reviews the common biblical sentences and concludes that Copernicus is definitely wrong. It is a pity to say that the Roman Catholics, in their condemnation of Galileo, have acted more prudently than the Reformed. [251] The case of Maresius is interesting, because it makes particularly clear what motives in the end were decisive. Apparently, he could swallow Voetius' over-preciseness and his theocratic pretensions, even if he did not like them and had combated them the better part of his career. Cartesianism was the issue which united him with Voetius, once he had become aware of the dangerous consequences some were drawing from it.

On the other hand, more liberal or humanist-minded theologians could come over to the Cartesian side. Johannes de Meij, a learned minister from Zealand, was very interested in the works of nature. In 1655 he published his *Sacra physiologia*, a work on the biblical sentences dealing with natural things. The work has the character of a baroque encyclopaedia, more concerned with textual exegesis and scholarly references than with empirical research. De Meij appears sceptical about astronomical theories. He thinks it improbable that the stars are as big as the astronomers claim. Among the topics dealt with is the question of the motion of the earth. De Meij states that the sun moves and the earth stands still. Decisive here are the biblical sentences. Saying that Scripture speaks according to the appearance or popular opinion would undermine the Bible. [252]

By 1667, however, De Meij appears to have changed his mind. In a book on the works of God in nature, he gives an overview of the solar system according to the Copernican hypothesis. [253] This change appears linked with a more general change in outlook. De Meij in the meantime appears to have studied Cartesian physics, referring to such authors as Regius and Descartes himself. Although these recent insights are combined in various ways with more traditional ideas, he appears convinced of the basic tenets of Cartesianism. [254] When this book appeared, De Meij appears to have been heavily attacked for his Copernicanism. In the second volume, which appeared later that year, he partially retracted. Natural reason teaches that the motion of the earth is probable. However, De Meij admits that the motion of the sun and the rest of the earth accords better with some biblical sentences, and hence that it is easier to defend the authority of the Scriptures from this opinion. So, he had

[251] Maresius (1670) 26-29.

[252] De Meij (1655) 202-204.

[253] De Meij (1667)a 81-82, 252-253.

[254] Cf. de Meij (1667)a 31-45, 188.

defended this view in his earlier *Physiologia*.[255] It is doubtful whether this convinced his opponents. In later works, De Meij continues upholding heliocentric astronomy and Cartesian philosophy.[256]

In an earlier chapter we noted that at the universities, in the second half of the seventeenth century the Copernican cause became closely linked with Cartesian philosophy. As it seems, among theologians, too, the question of the system of the world became closely linked with Cartesianism – although this 'theological Cartesianism' may have had a somewhat different character than the Cartesianism of the academic philosophers. In reaction to the motion of the earth being decried an exponent of this pernicious philosophy, a form of Cartesianism adapted to theology arose, which gave Copernicus' theories a bridgehead amongst Reformed theologians. Voetius' attacks made the bond between Copernicanism and Cartesianism much stronger than it otherwise would have been.

Theologians on the system of the world after 1656

The outcome of the debate of about 1656 was a clear division of Dutch theology into two factions: the Voetians and the Cocceians. The factions diverged on several points, some of them political, some philosophical, some theological. The exact positions upheld were formulated in the debates of the 1650s and did not change thereafter. For Copernicanism, this means that the motion of the earth was consistently rejected by the Voetians as being contrary to the Bible, whereas the Cocceians maintained that the Bible did not speak against it. A full discussion of the theological literature of this time would be going too far here, but a number of examples from both sides may illustrate this point.

On the Cocceian side, we might first turn to Franciscus Burman, professor of theology at Utrecht. He had been a close friend of Wittichius, Christophorus since their student days at Leiden. His *Synopsis theologiae* is regarded as the main summary of Cocceian dogmatic theology. The book was reprinted, and also translated into Dutch. In the part devoted to 'the Creation in particular', Burman explains that common prejudice has it that the earth is the centre of the universe. However, we do not know the limits of the universe, as it consists of an immense number of vortices. So, we cannot speak about its centre. As to the planetary system, 'the astronomers are all but unanimous' that the sun, and not the earth, is the centre. The earth is

[255] De Meij (1667)b 138-139.

[256] De Meij (1681). See pp. 650, 661, 735-740. De Meij doubted, however, Descartes' vortex theory: see p. 659.

transported and turned around by the liquid celestial matter, 'although it is resting in its own heaven'. As to the various places in Scripture which seem to speak against the motion of the earth, one should bear in mind 'that when speaking about natural things, Scripture often uses phrases taken from popular use, and adhibits common language, which is not apt for expressing the accurate truth of things'.[257]

Another prominent theological professor on the Cocceian side was Johannes Braun, a man of German descent. After studying at Leiden, where he matriculated in 1654, he made a grand tour through France, Germany and Switzerland. He served as a minister in several Walloon congregations in the Dutch Republic, before becoming professor at Groningen in 1680. In 1688, he published his *Doctrina foederum* ('Doctrine of the covenants, or summary of doctrinal and controversial theology').[258] Its chapters had originally been defended separately as disputations. The original Latin edition saw several re-editions in the following years, in both the Republic and Germany. A Dutch translation, which appeared in 1694, was quite popular, too. Part of the work deals with the works of Creation. In his discussion of the works of the fourth day, Braun stated: 'it is more credible that the earth is moved around the sun with the other planets, than that the sun, with the planets and the fixed stars, moves around the earth.' This agrees better with the laws of motion which God has impressed into the corporeal bodies. A second argument is that the earth is only a small body in the vortex of the sun. Braun refuted the arguments drawn from some, by now well-known Scriptural passages.[259]

However, the matter was touched upon not just in Latin and in academic works. The many local ministers writing edifying or pious works in the vernacular also made clear their opinion on the Copernican question. A prominent Cocceian spokesman was Henricus Groenewegen.[260] One of the later currents within Cocceianism was called 'Green Cocceians', a pun on his name (*groene wegen* = green roads). He had studied with Cocceius and Franciscus Burman and later became a minister in Enkhuizen. He was a prolific writer, and many of his books were translated into German. One of his more popular works was a Dutch 'Practice of the catechism of Heidelberg, or foundations of Christian theology'. In 1684 the book saw its fourth and in 1706 its seventh edition. In his discussion of question and answer nr. 26 of the catechism, he extensively discussed the system of Copernicus:

---

[257] Burman (1681) 296-297; (1697) 334-335. (book 1 Ch. 42, par. 49.) See on this work, Broeyer (1994) 116-125; see in particular 118 (on Copernicanism).
[258] The work is discussed in Graafland (1994).
[259] Braun (1688) 170-171. (1694) 190-191.
[260] Graafland in *BLGNP*, III, 151-154.

'*Question.* Is it an error to state that the earth is moved and that the sun is the centre?
Answer. It has been proved for long that this is not an error which contradicts the nature of the matter. But in vain some try to prove that it would be contrary to Holy Scripture.
Question. With what arguments?'

The arguments and their refutations are familiar by now and do not need an elaborate exposition. It is striking that, apart from exegetical arguments, we also find a reformulation of Descartes' argument from 'true motion', i.e. the earth is at rest because it allows itself to be swept along by the celestial matter, and in order to resist being swept along, there should be motion in it.

Frederik van Leenhof, a minister in Zwolle, became a rather controversial figure when in 1703 he published his book 'The heaven on earth'. It earned him the reputation of being a Spinozist, and provoked considerable scandal. But until then he had been a very popular author of pious and edifying works. His 'Chain of biblical theology', a Dutch work, was reprinted several times. In his introduction, he stated explicitly that his work was based primarily on the writings of Cocceius. He moreover made an eloquent plea for the separation of reason and revelation, and of theology from philosophy. 'Philosophy, in particular scholastic philosophy transposed to theology, has wrought much harm, and transformed the rational creature into an animal, and taken away under pretext of mysteries all virtue of reasoning, as nearly everybody by now becomes aware. (...) If a philosopher desires to be master over the faith, he desires a pernicious evil, and if a theologian wants to teach a philosopher, he shows his reproachable delusion and ignorance (...).'

After this, one does not expect van Leenhof to discuss a philosophical issue such as the system of the world. Still, in discussing first of all the way God is known from nature, he cannot resist making clear his physical stance, be it only in passing: 'We shall not investigate so much where the earth is set, and how it is moved with a triple motion, as is commonly assumed nowadays, without the Holy Scriptures contradicting it, but how 'he formed it to be inhabited', Isaiah 45:18...' And again: 'from the fixed order and certain course (...) (which we explain from Copernicus or rather Descartes' world system), all set times, feasts, years and days are derived'.[261]

We conclude with the Cocceian minister Johannes d'Outrein, who was renowned for his learning but had no scholarly ambitions and just wrote a large number of edifying works for the laity. In 1700-1701 he published 'Essays of holy symbols', wherein he gave a typological, symbolical contemplation of the

[261] Leenhof (1684) 31, 32. See also 3, where Leenhof speaks of the vortices from which the heavens are made.

world. For his work he had recourse to classical and Christian wisdom, but also to contemporary physics. In his introduction, he kept aloof as to the system of the world. God's greatness should be evident from both Copernicus' and Tycho's systems. However, in his discussion of Maleachi 4: 2, he is more explicit. He states that physical theory is in favour of Copernicus and that the Bible does not contradict it. For the former, he refers to Varenius, and for the latter to Salomon van Til. Thus, the sun is at the centre and the earth and the planets revolve around it. His exposition seems to imply a vortex theory.[262]

Turning then to the Voetian side, it is evident that those who were active in the discussions in the 1650s stuck to their opinion and played their part in disseminating an anti-Copernican stance. We may turn for an example to the Utrecht professor, Essenius. In his *Synopsis controversiarum theologicarum* (1661), the thesis that the sun rests in the centre and the earth is moved is one of the many doctrinal errors he discusses.[263] He also lists all the Biblical sentences explained wrongly by those he considered unorthodox. Among the most frequently denounced abuses is the thesis that, when speaking of natural things, the Bible often speaks according to the appearances and the erroneous opinion of the people.[264] A variant is Essenius' denunciation of the thesis that one cannot obtain knowledge of natural things from Holy Writ.[265]

Voetius died in 1676 and Essenius died six months later, in 1677. This was not the end of Utrecht as a centre of Voetian theology, however, for they were succeeded by two convinced Voetians who faithfully continued the work of their predecessors. One of the two was Petrus van Mastricht, whom we met earlier as a participant in the theological debate in the 1650s. Before he came to Utrecht in 1677, he had been a professor at Frankfurt an der Oder and at Duisburg. He remained active in combating Cartesianism throughout his career. In 1677 he published *Novitatum Cartesianarum gangraena* ('Cancer of the novelties of the Cartesians'), a title that is eloquent enough about the intention of the work.[266] In the chapter on the system of the world, he refutes four Cartesian errors: 1) that there are probably mountains, woods and humans on the moon, and generally, that there are more worlds;[267] 2) that the moon is

---

[262] On D'Outrein's cosmological theories, Schenkeveld-van der Dussen (1993). On D'Outrein, see also Bisschop (1993) 160-179 and Evenhuis in BLGNP, I, 237-239.
[263] Essenius (1661) 21.
[264] Essenius (1661) 106, 132, 133, 134, 138, 139, 144, 145, 147, 148, 150, 153, 154, 157, 159, 163, 178, 185, 191, 195, 200, 204, 207, 249, 251, 254, 255, 262, 272, 276, 287, 291. See also 4 and 308-309.
[265] Essenius (1661) 130, 139, 217, 222, 240, 241.
[266] On this book: Bizer (1958) 357-362.
[267] This question had been discussed by Braun (1688) 163, (1694) 182, and Burman (1681) 281, (1697) 316-317.

not a luminous body (whereas the Bible calls the moon a great light, Gen. 1:16); 3) that the earth is a planet; and 4) that – the main error discussed in this chapter – the sun stands still and the earth is moved with a double motion.[268]

The same views turn up in works not just devoted to polemics. Van Mastricht published a 'Theoretico-practica theologia', which saw a much expanded edition in 1698. In his discussion of the works of Creation, he spoke of the Copernican system several times, both under the heading of the world (universe) and the earth. Van Mastricht explains the system of the three parts of the world according to both the Cartesians – who use their hypothesis – and the Reformed, who derive their knowledge from Scripture. As it seems, mentioning (correctly) that the Cartesians do not refer to Scripture, is enough of a refutation for Van Mastricht.[269] Under the heading of the earth, Van Mastricht mentions the controversy regarding the earth's alleged diurnal and annual motion. The Copernicans and the Cartesians stand against the Reformed and the Tychonians. The former want to explain the phenomena, while the latter refer to God's Word. It is an obnoxious practice to oppose philosophical conjectures to the testimony of the Scriptures.[270] As to the difficulties which adhered to the traditional system according to the Cartesians, he answered: 'God will solve them, and those who uphold the Tychonian system have solved them for long.'[271] Van Mastricht also refuted, as being against Scripture, the opinion that the earth is a planet just like the moon. In that case, the earth should have been created only on the fourth day of Creation, when the stars were created, while the Bible clearly states that it was created on the first day. Moreover, in that case Hell, which is within the earth, would be in heaven.[272] Finally, Van Mastricht refutes the view that the sun is at rest in the heavens.[273]

Voetius' direct successor at Utrecht was Melchior Leydekker, who had earlier been a minister at Renesse (Zealand). His work *Fax veritatis* ('The torch of truth'), which had been written at Renesse, appeared in 1677.[274] It is a collection of controversies. In the chapter on the Scriptures, Leydekker discusses ten controversies. Among them, he discusses and refutes the opinion that the Scriptures speak not according to the truth of the matter, but according to the

---

[268] Van Mastricht (1677) 384-396.
[269] Van Mastricht (1698) 324-325.
[270] Van Mastricht (1698) 333-334.
[271] Van Mastricht (1698) 325.
[272] Van Mastricht (1698) 334.
[273] Van Mastricht (1698) 337.
[274] The book is discussed by Bizer (1958) 363-371. On Leydekker, see BWPGN, V, 775-785.

erroneous opinion of the people.[275] In the chapter on the Creation, he discusses 'whether a theologian, with due regard to the authority of Holy Scripture, can believe that the system of the world is Copernico-Cartesian, according to which the sun rests in the centre, whereas the earth is moved just as the other planets'. Leydekker thought this impermissible and quoted a number of Scriptural passages to substantiate his claim.[276] Also in his *Idea theologiae reformatae* he flatly dismissed the Copernican system as being contradicted by Holy Scripture.[277]

In the case of the Voetians, too, we do not have to limit ourselves to professors at the theological faculties. Many local ministers expressed similar feelings in the vernacular. Franciscus Ridderus, a Reformed minister at Rotterdam, where in the 1660s he had some difficulties because of his outspoken Orangism, wrote 'Scriptural light' (1675-1680), a five-volume work which aimed at elucidating obscure passages in the Bible. In his preface, he explicitly rejected the Cocceian exegesis. As he saw it, Cartesianism and Cocceianism covered each other neatly by that time. He refuted the Copernican world system on the basis of Joshua 10: 12-13, 1 Samuel 28: 12-20, 2 Kings 20: 11, Psalm 19: 5-7, Ecclesiastes 1: 4-5, Joshua 38: 8, Matthew 5: 45 and Acts 7:22. The work consists of several thousand pages and I am not sure that I have not overlooked anything, but these places may suffice.[278] On several other occasions, Ridderus does not mention the Copernican issue itself, but argues in general that the Bible is not speaking according to popular prejudice, so on the occasion of Genesis 1: 16, 1 Samuel 28: 12-20 (explicitly against the Copernicans) and 1 Kings 7:23 (the text on King Salomon's molten sea).[279] It would be going too far to summarise all his arguments; also, they are not very original. Let me just mention one of his proofs that Joshua's words ('Sun, stand thou still!') should be taken literally. Joshua could not speak any untruth, still less write it down, 'As he should be understood not just by God, but also by the Church, which cannot understand him in another way but by his clear historical words, which attest that the sun interrupted its course, and stood still.'[280]

The many polemical works against the Cocceians yield a rich harvest of anti-Copernican statements. The minister Leonardus van Rijssen (Ryssenius)

---

275 Leydekker (1677) 24-29.
276 Leydekker (1677) 291-295 (quotation on p. 291).
277 Leydekker (1696) 60.
278 Ridderus (1675-1680) II, 24-30, 145, 296-302, 599-606, III, 75-78, 434, IV, 118, 590.
279 Ridderus (1675-1680) I, 12, II, 145, 213-214. See also I, 8 (Genesis 1: 2), II 221-227 (1 Kings 8: 27, against the infinity of the world).
280 Ridderus (1675-1680), II, 27.

published in 1676 his 'Death struggle of the Cartesians and Cocceians', which clearly tried to capitalise upon the new political situation after the restoration of the Prince of Orange. In the preface, Ryssenius asks for a national synod. The book enumerates the usual objections against Cartesianism. The Cartesians, he states, turn their philosophy into an idol. Ryssenius also refutes the theory of the earth's motion, remarkably with natural rather than scriptural arguments. [281]

A polemicist from a somewhat later period is Henricus Brink. In 1685 he published his 'Touchstone of truth and errors'. Somewhat altered and expanded, the book was reissued in two volumes in 1690-1691. He too regarded Cartesianism and Cocceianism as essentially similar. He rejected vehemently the view that the Bible might sometimes speak according to the opinion of the erring people. [282] 'It has been an uncontroverted truth for all the centuries of Christianity that natural knowledge is fallible, but the supernatural, which is taken from God's Word, is infallible.' [283] As to the Copernican system, this was introduced by Descartes as a convenient hypothesis only, but by now it is unanimously accepted by the Cocceian theologians. However, it can be disproved both from Scripture and from nature. Something which everybody has seen happening before his eyes for centuries should not be dismissed without utter necessity. Among the astronomers, the system of Copernicus is still a matter of dispute. Tycho's system might save the appearances as well. 'Which should induce any Christian believing the authority of Scripture to maintain the old opinion.' [284]

Jacobus Koelman was a minister who was dismissed from his ministry because of his radical views. Afterwards, he acted as an independent preacher and was much admired in some quarters. He was a zealous controversialist who waged an all-out war against Balthasar Bekker. In his work 'The poison of Cartesian philosophy discovered' (1692) he also dealt with Cartesian physics. Descartes' followers among the Reformed differed in many respects, but 'the principal part, the pith of this false philosophy, is accepted by all the Cartesians, even the theologians; to wit, that our earth is turning around the unmoved sun, and that the earth is as much a planet as the moon, or Jupiter, Mars, Venus, Mercury and Saturn.' Koelman then demonstrated that this opinion was contrary to the Bible. [285]

This stance continued well into the eighteenth century. A work of long-

---

[281] Ryssenius (1676) 51-53.
[282] Brink (1685) 4, 43-49.
[283] Brink (1684) 2.
[284] Brink (1685) 187-195; quote on p. 191.
[285] Koelman (1692) 224 (quotation), 228-230, see also 230-233.

standing popularity was 'Reasonable religion' by the Rotterdam minister, Willem à Brakel. Originally published in 1700, the work has remained popular among certain groups of Reformed Christians to this very day. [286] The title should not be mistaken for the proclamation of some Enlightenment. 'Reasonable religion' (the term is taken from Romans 12:1 in the Dutch version) simply means that one should worship God *in geest en waarheid* ('with understanding and reason'). A Brakel was pertinent on the issue of Copernicanism: 'God says that the earth stands still and immovable and that the sun makes its revolutions. So this is a fixed and uncontroverted truth.' That sun and stars were fixed and the earth moving, is 'the fancy of people whose head turns around.' [287]

In 1713 another Rotterdam minister, Jacob Fruytier, published 'Zion's struggles'. This work, too, was reprinted many times, and for the last time in 1868. To Fruytier, Cartesian philosophy had lost nothing of its actuality. 'Who ignores how that philosophy [sc. of Descartes] is praised and abused? Has not this philosophy taught its pupils that Holy Writ in many instances only speaks according to the foolish opinion of the people? Dares one not to maintain on that foundation that the sun stands still and the earth turns round, even if their Maker attests that it is untrue? (...) Scripture is in their view just capable of teaching moral lessons, but one cannot learn higher wisdom about natural things from it.' [288]

Copernicanism, then, remained a controverted issue among Dutch Reformed theologians well into the eighteenth century. However, after the bitter disputes of the 1650s, it was no longer the main subject of a separate discussion. It was just an element in a discussion which comprehended a large number of subjects: Cartesian philosophy, Cocceian theology, Sunday observance, and the relation between Church and government. Still, it was an important element which was seldom overlooked.

One should not carry this too far, however. The controversy we have seen so far is one among theologians. How far the common church-goers were divided over the matter is another question. Undoubtedly, both the Voetians and the Cocceians had a number of devoted followers. On the other hand, it is quite probable that in many cases such division as there unmistakably was will have been caused mainly by loyalty to particular ministers, not to a set of ideas. There is a clear separation between Voetian and Cocceian books, even books written for the laity. But it is far from certain that the common layman

---

286 A new edition (*Redelijke godsdienst...*), was published in Leiden in three volumes in 1881-1882. A fifth impression of this edition was published in Utrecht in 1979, certainly not for scholarly use.
287 À Brakel (1707) 52, 231.
288 Fruytier (1715) 591-592.

or laywoman, looking for spiritual guidance in some edifying work, had a marked preference for books from one or the other direction. There is still some research to be done in this field.

Part v. God back in nature: Copernicanism in the 18th century

15. Newton's theories and Copernicanism at Dutch universities

The last stage in the Copernican discussion was opened by the publication of Isaac Newton's *Philosophiae naturalis principia mathematica* in 1687. Unlike Descartes, Newton deduced the motion of the planets by mathematical calculations from the basic laws of nature. Newton's work is justly famous and needs no rehearsal. What matters here is the question how and when his theories came to bear upon the cosmological debates in the Dutch Republic. It should be pointed out that it would still be some time before his theories were accepted. In particular, one should not think that the downfall of Cartesianism coincided with the rise of Newtonianism or of British experimental philosophy. Although many people were finding fault with Descartes in the period around 1700, they were not prepared to regard Newton's theories as an alternative. The case of Huygens – who fell out with Cartesianism and still considered Newton's ideas as 'absurd' from a physical point of view – is well known. One could also point to Nicolaas Hartsoeker, who combined criticism on Descartes' natural philosophy with a violent polemic against Newton's ideas.[1] In some cases, the discontent with Descartes' ideas initially led people to a stricter rationalism rather than to a more experimental approach.

At Dutch universities, Newton appears to have been hardly known initially. In his inaugural address as professor of philosophy at Harderwijk in 1700, Adriaan Reland praised Ramus, Bacon, Kepler, Galileo, Torricelli, Harvey, Gilbert, Digby, Boyle, Guericke, Gassendi, Descartes and several mathematicians, but did not even mention Newton.[2] At Leiden, however, Newton was known at an early date. Burchard de Volder even received a free copy of Newton's *Principia* in 1687. But, despite all his misgivings about Descartes, there is not a single indication that in natural philosophy he was moved even an inch

[1] On Hartsoeker see above. On his anti-Newtonianism see Berkvens-Stevelinck (1975); Aiton (1972) 109-110; see also Hartsoeker (1710) 24-38 (especially 29); see ibid. 102-116 for a general statement regarding his philosophical paradigms.
[2] Reland (1700).

by Newton's ideas.[3] His student G.H. Casembroot graduated in 1696 as doctor of philosophy with a disputation on high and low tide. Casembroot defended the Cartesian theory that it was caused by pressure of the moon's orbit, and in the corollaries he emphasised the truth of Descartes' system of the world. Still, he was aware of the elliptical form of the moon's orbit, referring for that to both Kepler and Newton. According to him, however, the elliptical form could be explained by a corpuscular model. Actually, he did not speak of the moon's orbit, but of its vortex.[4]

Another student of De Volder was Wyer Willem Muys. He was very interested in the new science and subsequently became professor of medicine at Franeker. Muys, too, was familiar with the *Principia*. One section of his textbook of physics, which appeared in 1711, was devoted to a refutation of Newton's book.[5] The title of a book from 1712 by a certain Yvo Gaukes seems to promise some kind of Cartesio-Newtonianism, as it is said to contain 'various opinions on philosophical matters, among others especially from Descartes and Newton'. On closer inspection, however, the author sides with Descartes and refutes Newton on all important points.[6]

Leiden: the transformation of natural philosophy

It was only after the publication of the second edition of Newton's *Principia* in 1713 that the importance of his work dawned upon Dutch universities. The first to take a public stance was the famous Leiden professor of medicine, Herman Boerhaave, in a 1715 oration. He reserved the highest praise for Newton's work, calling him 'the miracle of our time', even though he did not profess any 'Newtonian' philosophy.[7] Moreover, there are no indications that he introduced Newton's ideas in his courses.

The change in the university curriculum was brought about by two men: Willem Jacob 's Gravesande and Petrus van Musschenbroek.[8] Both had familiarised themselves with Newton's theories while they were staying in Eng-

---

[3] On De Volder, see above. Two letters from De Volder to Newton, from 1684 and 1687, have been published in English translation by A.R. Hall: Newton (1982) 11-12.

[4] Casembroot, disp. Leiden 1696. Thesis 12: '*Hunc quidem vorticem ellipticam habere figuram Keplerus supposuit, sed Newton eleganter demonstrat*'.

[5] Muys (1711) praefatio; scol.gen. at section I.

[6] Gaukes (1712).

[7] Boerhaave (1983) 158, 160-163. See the perceptive remarks by van Berkel in his review of this edition in *Tijdschrift voor de geschiedenis der geneeskunde, natuurwetenschappen, wiskunde en techniek* 8 (1985) 74-77.

[8] On 's Gravesande, see the introduction by de Pater in 's Gravesande (1988). On Musschenbroek, see de Pater (1979).

land. Both later obtained a professorship in the Netherlands, 's Gravesande at Leiden in 1717 and Musschenbroek at Utrecht in 1723 (in 1740, Musschenbroek became 's Gravesande's colleague at Leiden). Musschenbroek taught Newtonian astronomy in his courses, but he was foremost an experimentalist. In his writings, he hardly touches astronomy. As, moreover, 's Gravesande had by far the greater influence, we will concentrate on him.

's Gravesande (born 1688) had originally been trained as a lawyer, but his main interest was mathematics. In The Hague, where he set up a juridical practice after his graduation, he joined a group of friends with whom he launched a learned periodical, the *Journal littéraire*. His moving in Huguenot circles may have helped him keep in touch with foreign, in particular English science, but not much is known about it. The *Journal littéraire* appeared from 1713 onwards and quickly became a leading periodical in its field. 's Gravesande was for many years one of its main editors. Thanks to him, the journal allotted ample space to physical and scientific subjects.

In 1715-1716, 's Gravesande was a secretary to a Dutch embassy which, headed by Arent van Wassenaer van Duyvenvoorde, went to England to negotiate a treaty with the English. This proved to be a turning point in his career. His duties in England left him ample time to familiarise himself with English science and English scientists. He was elected a fellow of the Royal Society and introduced to Newton, Desaguliers and other leading scientists. In this way, he obtained first-hand knowledge of Newton's new theories. Indirectly, moreover, the trip settled his scientific career in a more material way. Van Wassenaer van Duyvenvoorde knew how to reward his collaborators and, being a man of much influence, proposed 's Gravesande for the chair of mathematics and astronomy at Leiden. He was nominated for the position, and in 1717 occupied the chair.

's Gravesande, whose responsibility was later widened to comprehend all philosophical disciplines, turned Leiden into Europe's leading university in the field of natural philosophy. For one thing, he continued and elaborated upon De Volder's and Senguerd's teaching of experimental physics. In his teaching, theory was consistently based upon demonstration and experiment. Moreover, he proved to be an ingenuous inventor of scientific instruments. The physical cabinet of Leiden university was enriched with a large number of new devices designed to illustrate all kinds of physical phenomena. These instruments became widely known by means of 's Gravesande's textbooks and were imitated all over Europe.

His largest significance, however, lies in his restructuring of physics teaching in general. Descartes' *Principia philosophiae* were no longer the model to be followed or emulated. Instead, 's Gravesande used Newton's theories as a point of departure. Although several English authors and lecturers, notably

Desaguliers, had already started popularising Newton's ideas, it may be claimed that 's Gravesande was the first to frame a new systematic approach to physics as a whole. This approach was popularised in his textbook 'Mathematical elements of natural philosophy, confirmed by experiments, or an introduction to Sir Isaac Newton's philosophy'. The original Latin edition appeared in 1720-1721. Subsequently, several reprints and translations appeared, in Dutch, English and French. The English translation by Desaguliers was particularly popular and was reprinted seven times.

In this book, 's Gravesande devoted ample space to the system of the world. The entire last part (of four) was devoted to it. This part is divided into two sections, the first of which is titled 'On the system of the world', and the second 'The physical causes of the celestial motions'. The first section is a rather straightforward description of the Copernican universe. Only in the very last sentence does 's Gravesande devote some attention to contrary views: 'Those who maintain that the earth is at rest are not supported by any astronomical or physical argument. That is, they are not arguing from the phenomena. Ignoring the simplicity of the system and the analogy of the motions, they defend that their opinion is not incompatible with the observations; wherein they are wrong, as we shall show in the next part.'[9] In the second section, the Copernican system is shown to derive logically from Newtonian physics. The motion of the earth around the sun is 'a necessary consequence of the laws of nature, which are deduced from the phenomena.'[10] 's Gravesande did not bother to discuss the current objections against the motion of the earth. Only the objection that bodies should be flung from the earth because of the centrifugal force is mentioned, and answered in a very succinct way.[11]

Within natural philosophy as transformed by 's Gravesande, the genre of disputing on 'the system of the world' – whereby the systems of Copernicus, Tycho and Ptolemy were rehearsed and discussed – ceased to exist. As mentioned above, 's Gravesande's textbooks are completely silent on the systems of Ptolemy and Tycho. They simply explain the 'true' Newtonian universe. The disputations by his students closely follow suit. They presuppose the Copernican system rather than discuss it.[12] When they do argue for it, they do so in a straightforward way, rather than by discussing its advantages over competing world systems.[13] Such arguments should clearly be seen as academic

---

[9] 's Gravesande (1721) 114. In the original, this is indeed only one sentence.

[10] 's Gravesande (1721) 158.

[11] 's Gravesande (1721), 160.

[12] Nieuwenhuis, disp. Franeker 1 July 1734. Dinckgreve, disp. Groningen 10 Dec. 1777.

[13] Mangard, disp. Leiden 3 July 1744.

exercises, rehearsing matters of fact, rather than as contributions to an actual debate. Debates were waged not on Copernicanism as such, but on such questions as the reality of gravitational attraction. Oosterdijk Schacht devoted a section to refuting the attacks on Newton by Hartsoeker and the French Jesuit Castel.[14] To 's Gravesande and his pupils, the discussion on the system of the world was really a thing of the past. They did not entertain any doubt that the Copernican system, as transformed and explained by Newton, represented the true state of affairs in the universe.

Leiden professors in defence of Copernicus

Even if 's Gravesande did not use his textbooks to wage a polemic against the opponents of Copernicanism, it is clear that he was deeply committed to its cause. On one occasion, he wrote an elaborate treatise to defend Copernicanism for the benefit of a non-academic public, on behalf of Jacques Saurin, minister of the Walloon congregation of The Hague. Saurin was a famous preacher whose sermons were, out of curiosity, eagerly attended by foreign visitors. Orthodox in his beliefs, his theology was rather intellectual. When in the early years of the eighteenth century the Amsterdam publisher Van der Marck undertook the project of a large and expensive, multi-volume work (in French) on the most memorable episodes of the Bible, lavishly illustrated with engravings by such prominent artists as Hoet, Houbraken and Picart, it was an obvious choice to ask Saurin to take care of the texts.

The text was, from a publisher's point of view, probably less important than the illustrations (indeed, one could buy a set of engravings without text), but it was an asset of course to have another famous name on the title page. However, the whole undertaking proved too much for Van der Marck. The project lingered on for years and eventually the work was published not by him but by Pierre de Hondt, in The Hague. The first two volumes, with text by Saurin, appeared in 1728. (There was also an edition of Saurin's text without illustrations, in small octavo volumes.) Four more volumes were to follow between 1735 and 1739, but for these De Hondt had to turn to other authors.[15]

One of his essays in these volumes would lead to Saurin being accused of

---

[14] Oosterdijk Schacht, disp. Leiden 26 Sept. 1726, par. 31-33 (pp.37-46).

[15] On the whole project, see van Eeghen, I 104-105; IV, 43; Kossmann (1937) 194; Kleerkooper and van Stockum (1914-1916) 249-250. Van Eeghen, basing herself on advertisements in newspapers, maintains that the first edition was published in 1720 by Picart, whereas the octavo version of the text was produced, also in 1720, by the publisher Du Sauzet, also from The Hague. This can hardly have been the same text, as the letter by 's Gravesande included in it and dealt with below is dated 22 Dec. 1721. The preface by Saurin (1728) admits that the edition was delayed and that the text had already been in print for four years (but had not been published).

unorthodoxy and brought about an elaborate conflict within the Walloon churches. However, this had nothing to do with Saurin's view on Copernicanism or on nature in general, but on the 'officious lie'.[16] Saurin's views on astronomy went unnoticed. Still, from the start Saurin shows himself as open to modern views in this respect. In the first discourse (on the Creation), he explains that the universe is immeasurably large and that it is probable that the stars serve as so many suns in their own worlds. About Genesis 1:16 (the famous text on the two great luminaries), he stipulates that 'in the system which he [Moses] has given us of the works of nature, he has followed not so much physical exactness, as the impression they make upon our senses.'[17]

In Volume II, Saurin devoted a chapter to the text of Joshua 10, and here he had to be explicit on the Copernican issue. First of all, he emphatically rejected the view that the standstill of the sun and the moon at the command of Joshua could be ascribed to some natural cause, as Grotius, Spinoza and La Peyrère had suggested. This was clearly due to some supernatural intervention. The question, however, was whether God had intervened in the motion of the sun and the moon, or had suspended the rotation of the earth. In general, if biblical passages, taken literally, 'imply contradiction and shock the demonstrations, one should take them in a figurative sense.'[18] In that case, the prophets apparently have accommodated themselves to the understanding of the people to whom they spoke. The problem is whether such is the case in Joshua 10.

As Saurin could not answer this question, he asked the view of an expert in the field. The expert Saurin had recourse to was, quite naturally, his fellow citizen 's Gravesande. In reply to Saurin's demand, 's Gravesande wrote an extensive letter, which Saurin inserted *in extenso* in his book.[19] Apparently, 's Gravesande regarded this as more than just an occasional trifle. When the *Journal literaire* – of which 's Gravesande himself, as will be remembered, was an editor – reviewed Saurin's work, it drew special attention to this letter from 's Gravesande, and quoted the greater part of it verbatim.[20] Still later, Allamand was to republish the letter in 's Gravesande's collected works,[21] so that the letter was printed in three different contexts during the eighteenth century.

's Gravesande asserted that the question about the system of the world was not a difficult one. As far as he knew, there was no dispute on the matter

[16] Laursen and McDonald (1994).

[17] Saurin (1728) I, 4-5.

[18] Saurin (1728) II, 79. The treatise on Joshua 10 is the fifth treatise, pp. 61-94. The accompanying plate, by G. Hoet, shows Joshua ordering the sun and the moon to stand still. Most prominent in the plate, however, is the big hailstorm.

[19] Saurin (1728) II 79-91. The letter is dated 22 December 1721.

[20] *Journal littéraire* 14 (1729), première partie, 107-144 (review), with 's Gravesande's text on pp. 113-128.

[21] 's Gravesande (1774) II, 298-310.

among experts in astronomy. The main argument in favour of the motion of the earth was that it was a necessary effect of the constitution of the universe. If one took account of the laws by which God governed the sun and the planets, the matter could not be doubted. 's Gravesande demonstrated his point by giving a brief outline of Newtonian physics, which showed that gravity necessarily caused the planets to move around the sun. This is the same argument he put forward in his textbook, of course. In the letter, however, he did not stop at that but continued with a discussion of the objections from the Bible. Saurin printed this part of the letter as well, although at least on this part he must have deemed himself sufficiently competent. On the whole, 's Gravesande's view agreed fully with the one expressed by Saurin himself: 'the aim of Holy Writ in the cited passages is not, as a thousand authors have demonstrated, to teach us what is; it is concerned with appearances, and the authors express these.' [22] In the rest of the letter, 's Gravesande mixed astronomy with exegesis. He explained in detail that the text of Joshua itself gave sufficient proof of its figurative character: the sun could not really stand 'at Gibeon', as the Holy Land was not between the tropics; and that there has never been another such day (as the Bible says) is true only with respect to the Holy Land, as some other regions on earth (i.e. the poles) have natural days which last for months.

's Gravesande was succeeded at Leiden by Johan Lulofs. In 1742, he was appointed to the chair of mathematics and astronomy, and two years later, the chair of philosophy was added to his charge. Lulofs can be regarded the most influential astronomer of his generation. [23] He appears to have been a pupil not of 's Gravesande but of Petrus van Musschenbroek. Still, his dealing with astronomy is quite in line with the way which had become common under 's Gravesande's, as is clear from his textbook on geography: 'Introduction to a physical and mathematical description of the earth' (1750). The work was written in Dutch, as Lulofs took special pride in being useful to his fellow countrymen. It contained a chapter on the annual and daily rotation of the earth, wherein these motions were argued in the by now well-known, straightforward manner, referring to the laws of Kepler. Religious arguments were completely absent. [24] Lulofs' 'Introduction' was quite a successful work, and was translated into Latin and German. Several editions of the Latin version were published.

However, as in the case of 's Gravesande, the fact that Lulofs took the Copernican system for granted in his textbooks did not mean that he did

[22] Quoted in 's Gravesande (1774) II, 307.

[23] See on him Vermij (1999).

[24] Lulofs (1750). Lulofs' preface states that he started the work in 1740. The chapter on the earth's motion (Chapter III): 87-123. See also par. 60 in Chapter IV.

not feel urged to defend it at other moments. Because of his ideal of usefulness, he directed himself not just at an academic audience. Even before his professorship, in 1741, while he was still a lawyer in his native Zutphen, Lulofs published two Dutch translations of astronomical works. The first is less relevant in this respect. It was a translation of the physical and astronomical textbook by John Keill (which Musschenbroek seems to have used in his courses). The Leiden publishers dedicated the book to 's Gravesande, by the way. Lulofs not only translated the book, but also added his own annotations and elucidations, almost all of them of a technical-astronomical nature. Most of them concern further references to the literature. [25]

The other work Lulofs translated into Dutch has an obvious Copernican tenor: *Copernicus triumphans* ('Copernicus triumphant') by the Danish astronomer Peder Horrebow. [26] In this work, which was first published in 1727, Horrebow claimed to have discovered, on the basis of his and his teacher Rømer's observations, the annual parallax of the fixed stars, and thus to have provided proof of the earth's annual motion. The work elicited much interest at the time, though it was challenged. Later, Horrebow's claims were generally dismissed. It was not until the nineteenth century that astronomers demonstrated annual parallax in a way that was deemed beyond reproach. Lulofs, too, was aware of the criticisms brought forward against Horrebow's work. In 1740, he wrote Horrebow a letter wherein he announced his intention to translate Horrebow's work into Dutch, with annotations and a historical essay on the attempts to establish parallax. In this letter, he asked Horrebow for comment on the criticisms levelled at his work, in particular those from the Italian astronomer Manfredi. Horrebow failed to answer the letter. When Lulofs wrote a second time, Horrebow sent a reply (or so he claimed), which went astray, so that Lulofs had to ask a third time before he finally obtained an answer. It was duly printed with his translation, as part of the historical essay which preceded the main work. [27]

---

[25] Keill (1741). It is a translation of the *Introductio ad veram physicam et veram astronomiam*, published by the Leiden publisher Verbeek in 1625. Verbeek combined two series of Oxford lectures by Keill, which had earlier been published separately. Lulofs' translation was also published by Verbeek.

[26] Lulofs' translation: Horrebow (1741)b. The original work was reprinted by Horrebow in his mathematical-physical works in the same year: Horrebow (1741)a 241-304. See Moesgaard (1972) 141-144 for a discussion of Horrebow's work.

[27] The three letters from Lulofs (21 Nov. 1740 and 4 Feb. & 9 may 1741), which have largely similar contents, are printed with Horrebow's reply (7 Feb. 1741) in Horrebow (1741)a 291-304; Horrebow (1741)b contains a Dutch translation of Horrebow's reply (no page numbers). The Dutch version contains a postscript to the text of *Copernicus triumphans,* containing some corrections. These apparently arrived too late to be used by Lulofs in the translation itself. In Horrebow's own edition, these corrections have been digested into the text; consequently, the postscript is omitted in his version of the letter.

Lulofs' many notes to the work are again mainly of a purely technical nature. In his very extensive preceding essay, however, he tackles the problem in a more general way. He underlines the importance of parallax in the Copernican debate. With the current precision of instruments, he maintains, it is no longer believable that one would not be able to discern the effect of the earth's annual motion upon the position of the stars. He gives a long enumeration of earlier attempts to measure parallax. He declares himself fully convinced by Horrebow's observations, 'so that I would count the motion of the earth fully demonstrated by them even if there were no other proofs at hand; which, however, are not lacking.' [28] In the textbook on geography, he still referred to Horrebow's observation of parallax as proof of the annual motion of the earth.

Although, according to Lulofs, the observations are convincing, parallax measurements should be continued in order to achieve greater accuracy. As to Manfredi's objections, he asserts that Manfredi's observations support those by Horrebow rather than refute them. Manfredi's refusal to accept the latter's observations must be due to fear of the Inquisition, which prevents him from disclosing his true feelings. Were Manfredi living in the free Dutch Republic, he undoubtedly 'would have valued one single cogent astronomical ratiocination, supported by observations, more than thousand decisive places from the most holy fathers of the Church.' [29] It is clear that to a Protestant like Lulofs, sentences from the Church fathers are not decisive. As to the arguments from the Bible itself, Lulofs does not appear to be impressed by them either. In one exceptionally non-astronomical note to Horrebow's main text, Lulofs comments on the exegesis of Joshua 10:12. 'If one wants to stick in a superstitious way to the literal sense of Scripture, this [Joshua's] saying is false.' But the Bible should not be read that way. Everywhere in the story 'occur figurative and concealed expressions, which are however well suited to the external appearances.' [30] There is no contradiction between the Bible and Copernicanism. Lulofs' harmonising of Scripture and astronomy deserves to be noted, for in religious matters he does not appear to be especially liberal -minded. At the end of his essay, Lulofs refers those who still maintain their literal exegesis to Kepler's advice in *Astronomia nova*, i.e. that they may glorify the Creator without concerning themselves with astronomy. One may serve God as well with one's eyes without using reason.

---

[28] Lulofs in Horrebow (1741)b, introduction, ******^{v}.

[29] Lulofs in Horrebow (1741)b **4.

[30] Lulofs in Horrebow (1741)b 76-77.

Copernicanism and anti-Copernianism at Utrecht

In the following years, students of 's Gravesande held a number of chairs of philosophy, thus spreading their master's programme throughout the Dutch Republic: Johannes Oosterdijk Schacht and Godefridus du Bois at Franeker, Johan Hendrik van Lom at Harderwijk, and Elie de Joncourt at the illustrious school of 's Hertogenbosch. At Leiden itself, 's Gravesande's work was continued not only by Lulofs and Musschenbroek, but also by Jean-Nicolas-Sebastien Allamand, who had been a close friend of 's Gravesande. [31] Most of these people left few writings on astronomy, apart from disputations and similar set exercises. Oosterdijk Schacht, for instance, took his degree on a disputation on the motion of the planets in elliptical orbits, wherein he rehearsed the proof that Kepler's laws derive from Newton's law of gravity. As it seems, they were quite content to use their master's textbooks and felt no need to write on the subject themselves.

Still, it would be going too far to state that 's Gravesande's hold on Dutch astronomical teaching was total. Not all Dutch academic life suddenly turned 'Newtonian'. At Groningen, a tradition of Leibnizianism and Wolffianism in natural philosophy was established and then maintained throughout the eighteenth century. Amsterdam, too, was rather late to accept the new physics. Utrecht – the long-time main rival of Leiden in the Dutch Republic – is a case in point. It not only allowed Musschenbroek to establish his alternative programme of natural philosophy, but also harboured a professor of astronomy who maintained an independent position and could boast of a certain following of his own. This was Jacob Odé. Born in 1698, Jacob Odé was of the same generation as 's Gravesande and Musschenbroek, but he never reached the stature of his famous colleagues. The textbook he wrote was never translated, and even in its original Latin form it was not reissued. Odé's teaching career was beset with difficulties and he never received the prestigious chair at Leiden. Still, his case is instructive as it demonstrates the impact of Newton's physics on a man who had not been initiated into it as a student, and even initially rejected it. [32]

Odé was not the product of a prestigious teaching institution and never made a voyage to England. Instead, as a native of Gelderland he received his education mainly at the provincial university of Harderwijk, where he matriculated in 1716 in the faculty of theology. Soon, however, interest in natural philosophy, in particular meteorology, seems to have prevailed. He went to Utrecht, where in 1721 he took his doctorate in philosophy. Two years

---

[31] De Pater in 's Gravesande (1988) 24-25; de Pater (1990) 143.

[32] See on him also Vermij (1999) 140-141.

later, the same university appointed him extraordinary professor of philosophy. As a matter of fact, Odé was a Copernican, but at this stage his view of the universe appears to have been largely inspired by Descartes. He defended the view that the tides were caused by the pressure of the moon,[33] and seems to have been a proponent of the vortex theory of planetary motions. In 1721, he stated that 'the planets do not move in empty space'; according to him, they were moved all round by some matter.[34] These statements might be read as an implicit criticism of Newton, whose views Odé must have heard of by this time. Also in 1721, he stated that Newton's theory of comets was defective.[35] Still, Odé was not a die-hard Cartesian. In his inaugural address of 1723, he confronts Descartes and Aristotle, and admits that Descartes too has committed many errors.[36] Descartes' followers have turned his philosophy into a dogmatic system by swearing on their master's words, exactly what they rebuked the followers of Aristotle for. Odé clearly belongs to the large group of philosophers who were dissatisfied with Cartesianism, but saw no real alternative to it. Newton's theories were not regarded as such; in his inaugural address, Odé does not so much as mention his name. But in a sense, he was ripe for them. As he got Musschenbroek for a colleague in Utrecht, he will soon have become more familiar with Newton's ideas. In any case, only a few years later his ideas appear markedly changed.

Odé published a textbook of natural philosophy in 1727, remarkably enough a year after Musschenbroek, who was his colleague at Utrecht, had his first textbook printed. There are no indications of intellectual disagreement. Odé rather seems to have tried in these years to make a name for himself. He was still only extraordinary professor and had plenty of reason to try to further his career. The following year he published a textbook on natural theology, put in geometrical order. As for the book on natural philosophy, in its general part Odé is critical of Descartes, but still retains various Cartesian elements, for example in his definition of motion. Decidedly un-Cartesian, however, is his definition of gravity as a special quality of matter, which is not effected by natural means but by the will of God.[37] (A similar view emerges in the corollaries of a disputation, defended the previous year under his presidency.[38]) The second, special part of the book deals with the earth

---

[33] Odé, disp. Utrecht 17 June 1719, coroll. 9, 11.

[34] Odé, disp. Utrecht 21 Nov. 1721, coroll. 97, 98: '*Planetae in spatio vacuo non moventur*; *Nos illorum sententiam defendemus, qui haec corpora materia quadam circumduci arbitrantur.*'

[35] 21 Nov. 1721, coroll. 92: '*Illust. Newtoni de Cometis sententia vitio suo non caret*'.

[36] Odé (1723) 34-44.

[37] Odé (1727), 1, 83 (motion), 61 (gravity).

[38] Odé, disp. Utrecht 10 April 1726.

and the order of the heavenly bodies. Like 's Gravesande's textbook, it culminates in a discussion of the solar system, but Odé's treatment is much more traditional in that it reviews the three great world systems of Tycho, Copernicus and Ptolemy. However, his discussion contains many Newtonian elements. By now, he has come to accept Newton's theory of comets. He extensively discusses the arguments against the vortex theory of Descartes, those of Newton taking pride of place. He also explains Newton's theory of gravity.[39] The theory of Ptolemy is, and the theory of Tycho appears to be, 'contrary to reason and observations'.[40] Copernicus' theory can save the phenomena, and that without such auxiliary constructions as a *primum mobile*, solid orbs and epicycles.[41]

In short, all arguments support the Copernican system. Copernicanism appears in Odé's definition of the earth, where he states that 'the earth is a planet, inhabited by humans, which, as the modern astronomers state, is moved around the sun in an orbit larger than Venus' orbit...'[42] Remarkably, however, Odé stops short before the final conclusion. Whether the earth itself is moving around the sun or not cannot be ascertained on the basis of present observations. The daily and annual motion of the earth are accepted by all important astronomers and indeed do appear very probable. Still, all arguments are liable to certain exceptions. Therefore, 'I as yet have not dared to affirm for sure that the case is so in reality.' According to Odé, only when at least two arguments – i.e. one demonstrating the daily rotation, the other the annual motion – are proved in such a way that there remains no room for any reasonable subterfuge, would this, long longed after in disputations, be warranted.[43]

This seems rather an unnatural move by Odé, especially considering his earlier views on the Copernican issue. Rather than expressing his real opinion, all this was much more plausibly inspired by tactical motives. In 1727, Odé was appointed extraordinary professor of theology at Utrecht. Now, what befitted a philosopher might still be found objectionable in a theologian, especially at Utrecht, where the spirit of Voetius was still very much alive. Keeping an eye on his career prospects, Odé probably thought it wise not to give offence to conservative theologians. And then he may have felt that the concession after all was minimal. Still, it is interesting that in the case

[39] Odé (1727) II, 225-235 (comets), 256-257 (vortices) 259-261 (gravity). Newton's theory of comets is also called probable by Arntzenius in Odé, disp. Utrecht 10 April 1726, coroll. 19.

[40] Odé (1727) II, 248, 265.

[41] Odé (1727) II, 251.

[42] Odé (1727) II, 31.

[43] Odé (1727) II, 264, cf. 256.

of Odé, the conversion from a Cartesian-inspired to a Newtonian physics coincides with a greater awareness of religious sensibilities. Odé retreated to a more biblical stance on still another point in his theory of the earth, that is, the earth's constitution and history. Earlier, he had explicitly refused to condemn Descartes' theory of the earth as being unbiblical, as Descartes had presented it simply as a hypothesis.[44] In his textbook, his criticism is still rather mitigated. But he definitely condemned Burnet's theory of the Flood and commented favourably on some recent attempts to reconcile philosophy with the orthodox exegesis of Genesis.[45] A corollary, defended a year earlier under his presidency, expressly condemned Descartes for mixing his philosophy with so many hypotheses opposed to both Scripture and reason.[46]

Even though Odé showed some caution in his textbook, his teaching of Copernicanism appears to have been unqualified. Though somewhat in the shade of his more famous Leiden colleagues, he imbued a number of young men with the principles of Newtonian physics. When Musschenbroek left Utrecht for Leiden, Odé at last became ordinary professor, of mathematics, astronomy and physics. In his inaugural oration in 1742, he rejected Descartes' theory of vortices and commended the theories of Newton.[47] In 1744, a certain Casparus van Oort took his degree in philosophy at Utrecht. The laudatory poems accompanying his thesis make it clear that he was a student of Odé. Among the corollaries, there are several defending the theories of Copernicus, Kepler and Newton.[48] In 1750, Petrus van Meerwyk held a disputation under Odé wherein he elaborately defended the daily and annual motion of the earth. His main argument was Newton's proof that Kepler's laws presuppose a central force.[49] Theological exceptions are absent from these works. Van Oort explicitly states that the Copernican issue should not be decided from Scripture, but by philosophers and mathematicians.

That there remained some real opposition to Copernicanism at Utrecht, even within the philosophical faculty, is demonstrated by a disputation held as late as 1750. The president was Professor Johannes Horthemels, an Aristotelian who had succeeded Musschenbroek to the chair of natural philosophy; the respondent as well as author was a certain Willem Hendrik s'Jacob. This 'Physical astronomical disputation containing an investigation of the presently most famous system of the world, the Copernican one' set out to de-

[44] Odé, disp. Utrecht 21 Nov. 1727, coroll. 46.
[45] Odé (1727) I, 13-17, II, 39-42.
[46] Odé, disp. Utrecht 10 April 1726, coroll. 9, 10.
[47] Odé (1743) 42, 44-46.
[48] Van Oort, disp. Utrecht 22 June 1744, coroll. 5, 6, 10, 13-14.
[49] Odé, disp. Utrecht 17 June 1750.

molish the arguments which argued for Copernicanism. In fact, s'Jacob's enumeration of these arguments rather serves to give an idea of the strength the Copernican argument had gained by this time. s'Jacob rehearses an impressive list: the laws of motion as established by Newton', the annual parallax claimed to have been observed by Horrebow, the variation with geographical latitude of the period of a pendulum, and so on; it seems odd to all of a sudden encounter here Galileo's argument from dominant Western winds, which by then had become rather obsolete. Anyhow, s'Jacob appears to be unimpressed.[50]

As to the argument from the laws of motion and gravity, he argues that what has been found true for some planets, need not be true for all of them. And then, people may find at a later date some as yet unknown laws of nature to account for these phenomena. Horrebow's discovery of annual parallax is dismissed with an appeal to older authorities, who claimed that parallax cannot be observed because of our great distance from the fixed stars; as well as with an appeal to the Bible, which states that the heavens are immeasurable (Proverbs 25: 3, Jerem. 31: 37). In short, all arguments from natural philosophy are uncertain. The biblical passages, on the contrary, are perfectly clear and leave no room for doubt. There is no indication that they should be taken in a tropical, non-literal way. One should add that s'Jacob's disputation was really a rearguard action. By this time, Copernicanism and Newtonianism had become generally accepted at all Dutch universities, even at that one-time fortress of Voetianism, Utrecht.

---

[50] Horthemels / s'Jacob, disp. Utrecht 10 June 1750.

16. New tendencies in theological and apologetic thinking

The impact of Newton's theories went beyond the teaching of physics or astronomy at universities. In England, as is well known, they were used as a foundation for Christian apologetics. The same happened in the Dutch Republic. People here were well aware of developments in England, but apologetic arguments were largely determined by local debates and circumstances. It will be remembered that by 1700 the Copernican issue had been largely mixed up with Cartesianism, and Cartesianism had become an issue in Dutch ecclesiastical strife. A climate had been created wherein one had to choose between Cartesianism – which by many was deemed irreligious – and faithfulness to the Bible, which implied among other things the rejection of Copernicanism. The advent of a new physics, one deemed superior to the Cartesian theories, made it possible to depolarise the situation and make Copernicanism acceptable.

It seems unlikely that Newton's theories were inherently more in accordance with religious orthodoxy than Descartes'. They were based on mathematical demonstrations in a way Descartes had only dreamed of. If they could be used to attenuate tensions, that was probably just because people were tired of the continuous struggles. Orthodox theologians had come to recognise that they could not anathematise scientific theories unpunished. On the other hand, the propagators of scientific theories were by now willing to put these theories in such guise as not to provoke religious sensibilities. The main merits of Newton's theories from a religious point of view were that they were untainted by previous denunciations and provocations, and that there were no ecclesiastical reputations at stake in their acceptance or rejection. Any other new theory could have done; in fact, Boyle's experimental philosophy in some instances served just as well (albeit that Boyle offered no arguments for Copernicanism). In the following, we shall investigate how the new physical theories were put forward as an alternative to Cartesianism, with the express intent of bringing science in accordance with religious feelings.

Jean Le Clerc's reception of the Principia

Jean Le Clerc (or Johannes Clericus) – a Swiss theologian, philosopher and scholar – settled in the Dutch Republic and became professor at the Remonstrant seminary in Amsterdam. He was a man of some importance in the republic of letters and had many connections with England. He admired the philosophy of Ralph Cudworth and was a friend of John Locke. Le Clerc published a large number of scholarly works and edited some of the earliest learned periodicals to appear in the Netherlands. These were written in French and addressed a fairly large, international circle of educated laymen.

Being committed as he was to modern developments in philosophy, Le Clerc inevitably was a Copernican. Remonstrants usually had little difficulty with Copernicanism, and Le Clerc did not have to be afraid of giving much offence. He professed Copernicanism in his Bible commentaries. Commenting on Job 9:6, he explained that the earth is constantly moved around the sun in a fixed orbit.[51] In his commentary on the famous text in Joshua on the standstill of sun and moon, he argued for a non-literal interpretation. For one thing, 'it is quite evident by now, among all expert astronomers, that not the sun is moved, but the earth'. However, with respect to the miracle as told in Joshua, this fact was of little relevance. 'It is not probable that the daily motion of the earth, or if you prefer the sun, has been interrupted for such a small cause as the defeat of a number of Canaanites on that very day rather than the following. God used not to be so spendthrift with miracles, if that expression is allowed, that He perturbs the natural order in such a large part of the universe for an insignificant cause.' According to Le Clerc, following Grotius in this respect, the history should be taken as a poetic hyperbole. Le Clerc agreed that, by divine providence, the sun had illuminated the place longer than usual, but this could have happened by some kind of refraction or some preternatural luminary.[52]

Le Clerc was not only a Copernican, but was also quite open to Newtonian physics. In 1688, he published in the journal he was editing a review by John Locke of the first edition of Newton's *Principia*. Locke and Le Clerc were good friends and had common philosophical interests, so we may be quite sure that from that time on Le Clerc at least had a basic idea of Newton's work. That this indeed is the case is clear from the textbook on physics he published in 1700. The work is rather eclectic, Le Clerc offering simply the views of various authors on various subjects. Newton's *Principia* is quoted at some length where it refutes Descartes' theory of vortices. Le Clerc also gives

---

[51] Le Clerc (1731) 678.
[52] Le Clerc (1708) 23-24.

a very rough sketch of gravity and of the planets as moved by central forces; curiously, he describes these forces as inversely proportional to the distance (rather than to its square). Moreover, the nature of these forces is unclear – Newton does not define them. Le Clerc appears to prefer some kind of plenist universe, whereby gravity is caused by some kind of corpuscular mechanism.[53] It would be going too far to call Le Clerc a 'Newtonian' at this time. He was interested in Newton's results, but he used them on the same footing as other theories and did not regard them as a viable alternative to Cartesianism. As is clear from his work, he was much more interested in the experimental philosophy of Robert Boyle or in the ideas of the Cambridge Platonist, Ralph Cudworth.

This changed with the second edition (1713) of the *Principia*. Upon reading it, Le Clerc apparently came to realise the full impact of Newton's ideas. In the journal he was editing, the *Bibliothèque ancienne et moderne*, he wrote a review of this second edition. He focused mainly on Cotes' preface and Newton's newly added Scholium, with the philosophical and religious implications he found there. He stressed particularly the anti-Cartesian and anti-materialistic tenor of the work. Descartes had assumed that only matter and motion governed the world. Newton had now proved that the world was governed by a universal principle – gravitation – which could not be explained mechanically.[54]

As Le Clerc explained: 'Mr Newton ... has not only discovered most sublime and very important truths, in his work on the *Mathematical principles of natural philosophy*; he has also given openings to go further, by applying his principles to three sorts of sciences, which one can perfect in that way, and on which diverse ingenious men have already worked in a felicitous way.' These three sciences were astronomy, physics and natural theology. As to the last-mentioned, Newton's principles 'show that it is impossible that the world has been made, and remains in its present state, by purely mechanical forces and movements. This leads us to recognise that there is a fully immaterial God, who is the Creator of the world. (...) This is quite different from the principles of Descartes, who believed that it sufficed for God to have given motion to matter just once to see everything in the world, or at least everything material, come forth from it.'[55]

This natural theology was what interested Le Clerc in Newton's work. The quoted passage serves as an introduction to a series of reviews of British works on physico-theology, by John Ray, William Derham and George

---

[53] Le Clerc (1700) 48-51.

[54] *Bibliothèque ancienne et moderne*, 1 (1714 part 1), pp. 69-96.

[55] *Bibliothèque ancienne et moderne* 3 (1715 part 1), pp. 42-44.

Cheyne.[56] The attention Le Clerc paid here to English 'Newtonian' literature was not just an occasional diversion, but rather the launching of a philosophical crusade. English philosophy was propagated, expressly and purposely, as an alternative to Cartesianism. Le Clerc really had discovered a philosophical alternative, an alternative moreover which would, or so one might hope, be less repulsive to orthodox Calvinists. The Newtonian philosophy which he helped to shape would be a major influence in the eighteenth century.

Le Clerc's work did not go unnoticed. His long summary of Cheyne's physico-theological book was consequently translated into Dutch by the Mennonite merchant Lambert ten Kate and published under the title 'The Creator and His government known from His creatures; according to the light of reason and mathematics. Aiming at a devotional religion, destruction of the foundations of atheism, and an orthodox use of philosophy'.[57] Although based on the extract by Le Clerc, Ten Kate added a large number of remarks and footnotes of his own. Most of the annotations refer to Newton's *Principia* as well as to Gregory's textbook on astronomy. Ten Kate also wrote a preface, wherein he stressed the anti-Cartesian tenor of the book. Descartes, he stated, appears to have been mainly interested in constructing a physical system which would explain the world as a continuously ongoing machine. This, however, seems not possible according to the physical laws as they really are. Ten Kate commended Newton's theories, which demonstrated God's hand in nature in a mathematical way. To this end, he referred to Cotes' preface to the *Principia* and quoted the *Scholium generale*. Finally, he added some appendices, among them a translation of Kepler's introduction to *Astronomia nova*, with its argument that heliocentrism is not contradicted by Scripture.

Bernard Nieuwentijt and Dutch physico-theology

For all the merits of Le Clerc's work, one should keep in mind that he was an outsider in the Dutch Republic. His stance may have carried much weight in the republic of letters, but the people who still had to be convinced of Copernicanism were not to be found there. To Voetian theologians, Le Clerc, as a Remonstrant, was not a trustworthy authority. They could only be convinced by authors of unsuspected orthodoxy. The one person who probably contributed most to alleviating the suspicion within the Dutch Reformed Church about the new philosophy was Bernard Nieuwentijt, a medical doctor and a regent of one of the smaller towns in Holland, Purmerend. Nieuwentijt was

[56] *Bibliothèque ancienne et moderne* 3 (1715 part I), pp. 41-158. Ibid. 4 (1715 part II), pp. 352-450. Cf. Evers (1988).

[57] [Cheyne] (1716).

himself a member of the Reformed Church and appears to have taken his belief very seriously. By his pietist inclinations and moral scruples, he appears a spiritual heir to Voetius and the Further Reformation.[58]

However, religion was not the only force in Nieuwentijt's life. Not much is known about his education; he entered Leiden university only to be dismissed within a year for serious misbehaviour. But it is clear that in his youth, under the influence of Cartesianism, he had fallen in love with the new science. That love would remain for all his life, but Cartesianism soon lost its appeal to him. The reasons for this must have been partly – and probably mainly – religious. Later in life, he appears to have looked upon his early Cartesian years as a period of irreligiousness. Though strict proof is lacking, it is generally assumed that he went through some kind of religious crisis. This left him with the problem of justifying his scientific activity without having recourse to Cartesian philosophy. He found such a justification in the experimental philosophy as it was propagated in England by Robert Boyle and others.

Nieuwentijt's programme is in some respects similar to Ten Kate's, as explained in his preface to the Dutch version of Cheyne's book. Ten Kate aimed at a mathematical foundation of religious truth. Cartesianism was bad, not so much because it was irreligious, but because it was bad science. Moreover, it gave rise to the fatal errors of Spinozism. Nieuwentijt maintains the same line of argument. Despite clear misgivings, he did *not* denounce Cartesianism as a dangerous heresy. He did think Cartesianism was bad science, and apt to lead people astray from a right insight into God's Creation, but he was ready to accept the Cartesians' declarations of orthodoxy at face value. Thus, he did not allow himself to become involved in the Voetian-Cocceian controversy. On the other hand, people who *did* use science or philosophy in such a way to combat orthodoxy, were inexcusable. These were not to be found among the reformed Cartesians, however, but among sectarian groups, mostly antinomians. Most dangerous of all were the Spinozists. Against this danger, Nieuwentijt felt urged to take on the defence of Christianity.

In 1715, Nieuwentijt published his work 'The right way of contemplating the works of the Creator'.[59] The express purpose of this work, and of a companion volume which appeared posthumously five years later, was to demolish the system of Spinoza. As Nieuwentijt explained, Spinoza pretended to

---

[58] Nieuwentijt's career is more fully described in Vermij (1991) and Bots (1972) 23-48.

[59] Nieuwentijt (1715). The work was translated three years later by John Chamberlayne as *The religious philosopher: or, the right way of contemplating the works of the creator* (London 1718). One should note, however, that Chamberlayne's text, which was at the base of a French translation as well, skips about half of the original, distorting Nieuwentijt's thoughts in a systematic way.

build his 'atheistic' philosophy on mathematical and scientific principles. In this way, Spinoza succeeded in deluding some weak spirits, who were led to believe that there was a contradiction between the results of science and the Christian revelation. In reality, Spinoza is simply applying false principles. A contemplation of nature according to sound scientific principles not only does not contradict revelation, but also demonstrates at all places God's wisdom, goodness and power, and is even able – according to Nieuwentijt – to show that the wisdom revealed in the Bible is more than human. The 'sound scientific principles' Nieuwentijt appealed to were those of British experimental philosophy as propagated by Boyle and others. Nieuwentijt argued the same on a more theoretical level in the second book (1720).

Descartes and his followers had rigorously cut the links between philosophy and theology, thus creating the possibility that the two would arrive at contradictory assessments. Nieuwentijt aimed at demonstrating the impossibility of any such conflict. To him, the attractiveness of British experimental philosophy largely lay in the fact that it made a sharp distinction between facts and hypotheses. By delineating the boundary between the two in an appropriate way, one could dismiss as a mere 'hypothesis' any scientific result which seemed unacceptable from a theological point of view. On the other hand, the biblical passages were regarded by Nieuwentijt as experimental 'facts'.

Somewhat surprisingly, Nieuwentijt also reckons Newton's theories under the heading of unverified hypotheses. He acknowledges that Newton had a philosophical alternative to offer, but this was exactly the reason for him to be cautious. He knew Newton's work and quoted both the first and the second edition of the *Principia*. Nieuwentijt admits the apologetic potential of the theory of gravitation,[60] but is still unwilling to commit himself. Recent observations, he explains, make it very probable 'That this Natural Law of Gravity extends itself thro' the whole visible Universe, and seems to prevail over all, even over the great Heavenly Bodies... upon which Foundation the whole Physical System of Sir Isaac Newton, who seems to be chiefly followed by the Great Men of this Age in many Things, is built. But I don't here undertake to found myself upon the bare Opinions of any Philosopher, forasmuch as they are often contradicted by others, so long as the Experiments are not only not incontestable, but likewise not sufficiently known.'[61].

The full implications of this stance become clear in the way Nieuwentijt deals with the Copernican question. As one of the main points where science and revelation appeared to be in conflict, Nieuwentijt of course could not

[60] Nieuwentijt (1715) 479; (1724) 299.
[61] Nieuwentijt (1715) 762, quoted in Chamberlayne's translation (Nieuwentijt (1724) 477).

overlook it. In his chapter on the visible heavens, he explains that there are two systems which try to explain the motions of the planets, the Copernican and the Tychonic system. He mentions briefly the semi-Tychonic system, but feels that this combines the disadvantages rather than the advantages of both. Tycho's system is rather an adaptation of Copernicus' view, undertaken because Tycho and his followers had problems diverging from Scripture, which seems to deny the motion of the earth. At this point, however, Nieuwentijt is not concerned with investigating their accordance with reality, but with demonstrating how they display God's greatness. On the basis of Copernicus' system, one can demonstrate God's greatness as it confirms the immeasurableness of the heavens.[62] In the same chapter, Nieuwentijt refers favourably to Newton's criticism of Descartes' vortex theory of planetary motions, but then draws the quite un-Newtonian conclusion that it cannot be disproved (nor proved, as a matter of fact) that the heavens are made of solid matter, as the old Aristotelian theory had it.[63]

The main discussion on the system of the world is in Chapter 30, 'Of the unknown things', the greater part of which is devoted to it.[64] The heading is not there for nothing. Nieuwentijt does not openly commit himself to one or the other astronomical system. According to him, one cannot decide the question on astronomical grounds. The point is argued by referring to the failure to observe the parallax of the fixed stars, and above all by referring to a large array of acknowledged astronomical authorities who testify to this effect. The list became a *locus classicus* for the anti-Copernicans of the eighteenth century. For instance it is referred to by Odé, in his 1727 textbook, and in the disputation by s'Jacob. One should admit that Nieuwentijt's reading is rather selective and at some places comes close to bending the evidence. Even such staunch Copernicans as Isaac Newton, David Gregory and Christiaan Huygens are made to support his view – the latter in a personal interview, for which we have to take his word.[65]

Nor does Nieuwentijt – and herein he is rather unique – go so far as to have the Bible decide the issue. Referring to the bitter disputes on the matter which raged some years before, he explains that some exegetes have maintained that the passages on the motion of the sun should be taken literally; others, 'who adhere to a philosophy with which this does not agree', maintain that the Bible speaks figuratively in those places. Nieuwentijt leaves the issue in the

[62] Nieuwentijt (1715) 686-687; (1724) 427-428.

[63] Nieuwentijt (1715) 690-691; cf. (1724) 431.

[64] In Chamberlayne's translation, this is Chapter 29.

[65] See the tenth paragraph of this chapter, 904-906, or the eighth section in Chamberlayne's translation, 578-580.

balance, 'as we do not intend to meddle ourselves in the arguments of differing exegetes.' He does object strongly, however, to the opinion, defended by some, that the Bible adapts itself to the erroneous opinion of the common people, which opinion 'has given occasion to many unfortunate people, to have irreverent thoughts on the Divinity of this Word.'[66]

Although he does not disavow them openly, Nieuwentijt's sympathies are clearly not with those who defend figurative interpretation. Throughout his book, he seeks to demonstrate that the literal sense of the Bible is strictly true and in accordance with modern science. For instance, the text on King Salomon's basin – which had been adduced as an example that the Bible does *not* speak according to strict mathematical truth from Lansbergen onwards – is explained by him in such a way that the Hebrew words denote not a round but a hexagonal form. In that case, the said ratio of circumference to diameter (30:10) indeed would be geometrically exact.[67] Many other examples could be given. Nieuwentijt is clearly keeping the Cartesian and Cocceian Bible interpretation at bay. His literal interpretation could offend nobody.

Nieuwentijt, then, takes a rather comprehensive position. In his stance on the system of the world, he deliberately avoids taking sides in any factional or theological strife within the Dutch Reformed Church. The enemy was clearly localised in a group which was nearly universally despised: the Spinozists. He did not go so far as to decry any philosophical opinion which might seem to have dangerous tendencies, but denounced any openly 'atheistic' use of the new science. In this way, he emphasised the compatibility of science with the whole of the Reformed tradition.

The end of the theological strife

Nieuwentijt's work was avidly read. Eight editions of 'The right way' appeared between 1715 and 1759, not to mention translations. It must have done much to bridge the gap between the new science and established Christianity. Probably, it was not in the first place the theoretical discourses which made Nieuwentijt's work so persuasive. In 'The right way', he tried to prove his point mainly by example; that is, he reviewed a large amount of experimental knowledge to demonstrate how in each case the experiments demonstrated God's greatness as well as His providence. In this way, the book became a kind of overview, in Dutch and on a popular level, of experimental science in general. As such, it appeared at the right moment. Many people

[66] Nieuwentijt (1715) 912. These passages are not included in Chamberlayne's translation.
[67] Nieuwentijt (1715) 597-602.

found the new science a fascinating topic, and performing experiments was rapidly becoming fashionable among the populace at large.

But Nieuwentijt's apologetic and theoretical excursions, which make up a very substantial part of the book, were important. They allowed people to feel that they could read this scientific work, or toy with science itself, in good conscience. Nieuwentijt made it perfectly clear that any possibly suspect philosophy, such as Cartesianism, had nothing to do with the new science itself. So far as the Spinozists and other atheistic philosophers claimed a basis for their ideas in modern science, they were simply deluding themselves (if not purposely deluding others). Experimental science itself was a pious, God-pleasing enterprise, fully compatible with reformed orthodoxy.

Thus, Nieuwentijt contributed in an important way to making the new science socially acceptable. But clearly he would not have achieved much had the Copernican issue still been important as an element in factional strife. However, in the course of the eighteenth century, the bitter hostilities between Voetians and Cocceians gradually ceased. Both parties continued to exist, but their overall significance changed. They became more purely ecclesiastical currents, without much bearing on the political situation. This was largely due to the calming down of political agitation. The regents agreed that their conflicts of interest should not get out of hand, and that the common people and the Church should not have an opportunity to meddle with them. Instead of playing on the ecclesiastical differences to outwit their opponents, they actively tried to pacify them. In 1694, the States of Holland accepted, at the instigation of William III, new regulations aimed at pacifying the two parties – Cocceians and Voetians – by allotting each an equal share in the material affairs of the Church. Cocceian and Voetian ministers should be appointed in fair proportion, without favouring either party. On the local level, many similar regulations came into force. They did not dispel all suspicion at once, but in the long run they did help to neutralise the bitterness of the former contests.[68]

Although Voetians and Cocceians were still being used as party names, the real divisions within the eighteenth-century Dutch Reformed Church appeared to follow quite different lines. Huisman has divided eighteenth-century Dutch ministers into three groups: enlightened (or intellectual) ministers, pietists and traditionalists. 'Enlightened' ministers were those who concentrated on the scholarly side of theology. The pietists continued and developed some points from the programme of the Further Reformation, but they had largely withdrawn from the world. They did not meddle with

[68] Van der Bijl (1994) *passim*. Evenhuis, III, 133.

the realms of politics or science. As there was probably only a small number of enlightened and pietist ministers, the group of 'traditional' ministers comprehended the majority of both Voetians and Cocceians. The point is, however, that the traditional ministers were very much aware of the enlightened and pietist tendencies and as a group felt one in resisting them.[69] Voetians and Cocceians should by this time probably be viewed as factions rather than as groups divided by real differences of opinion. The earlier covenants which allotted both groups an equal share in the ecclesiastical appointments may well in the end have served to perpetuate them artificially.

As stated, Nieuwentijt was not a Copernican. In order to be convinced of heliocentricity, people needed other authorities, most notably Newtonian physics. Still, Nieuwentijt was important in making these authorities acceptable. First of all, he rid the Copernican issue of its polemical overtones. Although unwilling to accept the motion of the earth, he did not denounce the idea as heretical. Secondly, he offered a framework which could be used to accommodate science and religion in a more general way. Unlike Cartesian Bible interpretation, Nieuwentijt's 'physico-theology' was acceptable to all parties.

How Copernicanism came to be fitted into this physico-theological framework can be demonstrated by the case of Hermanus Johannes Krom, a minister living in the province of Zealand. He wrote his works at the end of the century, so that he demonstrates at the same time the viability of the physico-theological solution. Like Nieuwentijt, Krom was intent on demonstrating the harmony between science and the Bible. He therefore wrote a treatise on the Creation, wherein he emphasised that the Creation story is fully compatible with natural reason. He also discusses King Salomon's basin and arrives at the same conclusion as Nieuwentijt: the basin was not round but hexagonal. But, contrary to Nieuwentijt, Krom was a Copernican. He even maintained that, probably, there were many other inhabited worlds in the universe. As for the miracle of Joshua 10:12, this was a divine intervention in the ordinary course of nature. One can take the text quite literally, according to Krom, as long as one keeps in mind that the text speaks in a relative and optical way. So, the literal interpretation of the Bible is saved mainly by stretching the sense of the word 'literal'. Clearly, good and pious intentions counted more than rigorous ratiocination.[70]

---

69 Huisman (1983) 37-41.
70 De Pater (1987) 188-194.

17. The cosmological debate in society at large

A fashion for astronomy

As in the eighteenth century science lost its association with Cartesianism and became religiously safe, the general interest in science, which had started in the seventeenth century, increased considerably. New insights reached an ever-larger segment of the population. Science now became part of the general culture of the age. Astronomy in particular became something like a fashion, exercised not only by acknowledged scholars, but also by all kinds of amateurs.

Some of these amateurs were in touch with international developments and performed scientific research on a high level. Actually, most astronomical research at the time was exercised outside the universities, by independent amateurs. Some of these amateurs built their own observatories, such as the Middelburg town architect Jan de Munck, the Amsterdam merchant Jacobus van der Wall, and Johann Maurits Mohr, Reformed minister in Batavia, in the Dutch East Indies. Some, such as the surveyor Jan Klinkenberg, observed comets (Klinkenberg discovered five comets and co-discovered a further ten). Others calculated eclipses. Some, such as the Amsterdam amateur Nicolaes Struyck, achieved European renown. In 1749, Struyck became fellow of the Royal Society and in 1755 correspondent of the Paris Académie des Sciences. One of his main fields of interest was the calculation of comets. He reconstructed the orbits of historical comets, presenting them in a comprehensive overview and dividing them into classes. He also constructed a '*cometarium*', which offered a three-dimensional model of cometary courses.[71] For these people, whose world-view was based on Newtonian physics, the Copernican debate was all over. They occupied themselves with the actual scientific questions of their days. Their very existence attests that modern cosmological insights apparently were no longer offensive. But here, a detailed description of their ideas or activities would be superfluous.

---

[71] On these amateurs see the comprehensive work by Zuidervaart (1999).

However, the fascination with science and astronomy extended beyond the educated segment of the population, which could afford to follow the activities of foreign scientists and academies. Planetaria became favourite showpieces. The most famous example is the planetarium the textile manufacturer and astronomical autodidact, Eise Eisinga, built with his own hands on the ceiling of his living room in Franeker in the period 1774-1781. Eisinga invested much time and effort in representing the celestial motions in the most accurate way. His planetarium is an impressive monument to the astronomical interest during the eighteenth century.[72]

Lay philosophers who defended their somewhat fantastic systems were a familiar phenomenon by now. One of the most notorious of eighteenth-century amateurs (he waged an extensive polemic with Lulofs) was Meindert Semeyns, regent in the town of Enkhuizen. He developed a 'magnetic system' to explain the whole of nature. He was definitely a Copernican: the motion of the planets was caused by the rotation of the sun on its axis. One could even say that Semeyns regarded himself as a kind of second Copernicus. The problems Copernicanism had encountered were prefigurations of the problems Semeyns encountered regarding his own ideas – and their final triumph, of course.[73] A certain Andries van der Poest wondered how it was possible, the earth's axis being tilted, for the earth to move in a circular rather than a conical orbit. He found the solution by putting the earth's axis perpendicular and tilting its orbit instead. He proposed this solution in 1775 to the *Zeeuwsch Genootschap der Wetenschappen* (Zealand Society of Arts and Sciences) and had it printed a few years later.[74]

Semeyns and van der Poest belonged to the social elite. Even if poorly educated, they based themselves on common learning and did not doubt the Copernican world order. Other people of more humble descent had less opportunity to familiarise themselves with modern insights, but still were enraptured by modern science. Their efforts often appear not just over-ambitious but downright fantastic. Such people concocted their systems from a lot of disparate elements: daily experiences, some ill-understood scientific data and – not in the least place – biblical passages.

Very few of these cases are documented, of course. A revealing instance is offered by the Friesian farmhand, Tjerk Jansz de Boer. De Boer had very little schooling. His parents taught him to read, and when he was about 19, he

---

[72] Terpstra (1981) 209-283.

[73] Semeyns (1760) 1-2.

[74] A. van der Poest to Zeeuwsch genootschap, 31 Jan. 1775 (this letter was brought to my attention by H. Zuidervaart at Middelburg). [Van der Poest] (1778). The name is also written as A. van der Poest Clement.

taught himself to write. At about the same time he learned how to calculate from a boy with whom he ploughed. A new world appears to have opened up to Tjerk, and he continued his studies by devouring books on arithmetic, navigation, astronomy, surveying and geometry; 'all studied while I had to earn my living by heavy peasant's work, without the tuition by any master, but by the enlightenment of the mind and by exercising during my free hours', as he explained later. One might ask where de Boer would have ended up had he been taken care of by a competent teacher. As things went, he felt himself in the end capable, by Divine inspiration, of explaining the whole of nature and lecturing the great Newton, about whose fame he had heard, for having badly understood the law of nature in the motion of the planets.[75]

Writser Roelefs de Smith published his ideas on cosmology in Groningen in 1722. De Smith defended that the earth was resting in the centre of the universe, turning all round on its axis once every year; that the moon was half mirror-like, half sky-blue; that the sun's diameter was about a fifth of the diameter of the earth; and so on. De Smith claimed that he could demonstrate his propositions by means of a 'sphaera', apparently some kind of planetary or other model of the universe he had made for the purpose. He had recourse to the Bible, too.[76]

It thus appears that the way science and ideas on science influenced general culture among the lower strata of society is rather complicated. Scientific theories were not rejected out of hand; rather, people seem to have been fascinated by them. Still, their full significance was not grasped and they became combined with all kinds of other elements. Here, too, ideas about science were harmonised with what one took to be biblical truths. As the significance of the Bible in all questions of learning had been stressed for centuries, biblical thinking had got a strong grip on general culture. As a consequence, even in the enlightened and science-minded eighteenth century, Copernicanism still met with disbelief and resistance.

Some final skirmishes

After the demise of Cartesianism, there no longer seemed to be any ground for disputes on matters of natural philosophy. However, this is only partly true. Paradoxically, the first half of the eighteenth century saw the polemics on Copernicanism flare up again, be it on a much smaller scale than before. It will be remembered that, after the big controversy of 1656, there had been no

[75] Zuidervaart (1995) 167-169.

[76] De Smith (1722). His sphaera: 19-20. On the Bible: 14-15, 28-29.

major theological conflicts over the issue. This did not mean that the two parties lived on in peace, but as they had organised themselves along some mutually agreed borderline, the need for large-scale combat was over. The Copernican issue remained part of the much larger controversy between Cocceians and Voetians, but as such it no longer gave rise to separate discussion. However, when the old dichotomy between Cocceians and Voetians began to fade away, people felt the need to define their position anew. Not everybody immediately felt at home in the new circumstances, and old controversies were stirred up again.

The first controversy to be mentioned here occurred in 1714. Someone hiding behind the initials 'I.S.' published a pamphlet entitled 'Amusing countrymen's talk, on the moving or not moving of the sun, and the turning and not turning of the earth'.[77] The author is generally taken to be Jacob Schuts, a Reformed lay preacher from The Hague. Schuts spent all his energy on propagating and confirming Christian piety among the population of Holland, and in this way became a rather well-known figure among his contemporaries. His earnest and strictness remind one of Voetius. He wrote a large number of tracts, often of a polemical nature. For instance, he was a bitter opponent of Balthasar Bekker, who had denied the power of the devil and the existence of witches.

The pamphlet is in the form of a dialogue between three Dutch peasants – Jaap, Kees and Piet, each a very common Dutch name. It is preceded by an introduction. A preface, signed 'N.N.', states that the tract is published without the knowledge and consent of its author. It is not clear how serious this assessment should be taken. The dialogue part is written in a language which is supposed to imitate in some way the dialect of the country-folk, probably to achieve a comic effect. Kees upholds the Copernican theory, while Jaap and Piet combat it. Against Kees' appeal to Copernicus, Galileo and others, the other two refer to the Jesuit author Athanasius Kircher: 'Against all those you mentioned there, I put the one man Athanasius Kircherus, ordinary professor of mathematics at Rome, who has so flatly refuted those others that you and anybody of your mind should be ashamed.'[78] Kircher's *Mundus subterraneus* had been translated into Dutch. It had been a rather popular work in the seventeenth century, but by 1714 it was somewhat out of date. (There is also a short reference to a certain 'Lieve Symons'; this might be Lieuwe Willemsz Graaf, but more probably some unknown local celebrity.) Kircher's arguments are not discussed in much detail and the author's knowledge of physics

---

[77] I.S., *Vermakelyke boere-praat* (1714).
[78] Ibid., 7.

Figure 11: Frontispiece of a Dutch translation of Bernard Le Bovier de Fontenelle, *Entretiens sur la pluralité des mondes*, edition 1765. The philosopher shows the lady the plurality of worlds in the universe. (Amsterdam university library.)

and astronomy is clearly very scant. Still, as can be expected, Jaap and Piet succeed in utterly ridiculing Kees' arguments.

In the dialogue itself, there is no explicit discussion of or appeal to biblical passages. Still, religion looms prominently in the background. This is evident in the passage where Kees proposes that the other two should put aside all prejudices and investigate the matter as though nothing had been said about it in the Bible. Of course, Jaap and Piet are not taken in and point out that this would be perfect godlessness.[79] The introduction, on the other hand, is much more explicitly religious – just as one would expect from a person like Schuts. The motion of the sun and the immobility of the earth are proved with a long list of biblical passages. The author moreover complains bitterly about ministers who neglect to follow God's Word in every detail and even try to impose on their flock that thus they are not only not contradicting God's word, but even rendering its true meaning. He emphasises that the motion of the sun may not be a religiously important issue, but that contradicting Scripture is.

The pamphlet was answered by a certain C. Mennink. His pamphlet has not been preserved. All we know about it comes from a second pamphlet by Schuts, wherein he defended 'Countrymen's talk' against Mennink's attack.[80] The second pamphlet, too, is anonymous, but its author admits that he is the author of 'Countrymen's talk'. Moreover, the publisher's preface points out that he had received the defence from a friend who had got it from a good friend in The Hague.

Mennink is an utterly obscure person, about whom no further information is available. Schuts describes him as a follower of the sectarian leader Willem Deurhof. In another pamphlet, published the same year, he attacked Mennink on other points as well. Schuts' polemic against Mennink's Copernicanism cannot be separated from the crusade against Deurhof's influence. The Copernican issue, once again, was rather a stick with which to beat a dog. The replacement of the Voetian-Cocceian polemic by the construction of an opposition between orthodoxy and 'atheism' is a general tendency of this period, also to be discerned in Nieuwentijt. Schuts, however, had too little feeling for science to have much influence upon it.

Limiting ourselves to the parts of the discussion which concern Copernicanism in a strict sense, we will not be surprised to find that a large part of the discussion concerns biblical exegesis. The various relevant texts are interpreted in various ways. But Mennink also introduced a physical element. As

---

[79] Ibid., 7-8.

[80] J.S.K.B., *De vermakelyke boere-praat... verdedigt* (1714).

far as one can deduce from Schuts' second pamphlet, which quotes Mennink rather extensively, Mennink defended a Cartesian-inspired world-view, with vortices and so on. But he does not seem to follow Descartes very closely and may well have enriched his general world-view with some theories of his own. Schuts may have been justified in not taking Mennink's utterances very seriously, but on the other hand his own views were not really up to date either. A reference by Mennink to the pressure of the air, which occurs in a passage on the lack of resistance to a sphere moving in air, is answered by Schuts by a flat denial that a phenomenon like air pressure, which people try to impose upon us by air pumps, exists. Mennink's reference to the telescope is answered by a reference to a discussion he had been present at on the barge from Utrecht to Leiden. Someone ventured the idea that the phenomena observed in the sky by means of the telescope are really in the earth, and are merely mirrored by the sky. Schuts apparently thinks this probable, as the Bible itself compares the heavens to 'a molten looking glass' (Job 37:18).

'Countrymen's talk' was discussed again in the cadre of another small controversy, which took place in 1723. It was provoked by a certain Petrus de Laat a person about whom little is known.[81] Judging from his works, however, he seems to have been of a sensitive nature, tending to mysticism and to going to extremes. He studied theology for some time, but does not seem to have finished his studies and never became a minister. It is reported that he interrupted his studies for some time when he came under the influence of the mystic Antoinette Bourignon, but quickly returned to the Reformed fold. He published several works, most of them during his years as a student (he added 'S.S.Th.St.' to his name). In 1722, he published a book on Sunday observance, vehemently denouncing any moderation on the subject, and in 1723 a refutation of 'the Cartesians' ideas attacking Scripture, concerning the turning or turning all round of the earth, and the standstill of the sun'. Cocceianism and Cartesianism were still two sides of the same coin to him. His 1723 pamphlet was directed at Cartesianism in general, but, as he explained, he limited himself to discussing the Cartesians' assessment that Scripture often adapts itself to the erring opinion of the common man, with which they tried to save their darling, the Copernican system. The argument mainly concerns the discussion of biblical texts. Its wording is extremely sharp. 'Truly, unless God possesses treasures of an infinite and for us unexplainable patience, it would not be possible that His formidable Majesty could endure for even an hour this Cartesian brood of Satan, which dares to attack the very heart of Godness'. Cartesians, according to De Laat, are worse than Socinians. The

[81] See the article on him in BWPGN, V, 448-450.

latter err on the mysteries of faith, the Cartesians on things which one can very well understand by natural reason.[82]

De Laat was answered in an equally sharp manner by an anonymous author. The work, which also appeared in 1723, is in the form of a letter to a friend. The authors signs it 'J.J.' and indicates that he lives in Rotterdam. He does not seem to have been a person of note. His answer was not inspired by a desire to defend Descartes, as he expressly disavows Cartesianism. In particular, he disapproves of Descartes' mixing philosophy with Holy Scripture.[83] Instead, he was inspired by his disgust of clerical intolerance. He accuses such authors as De Laat of trying to impose a new papal authority on the believers. They would like to reintroduce the Inquisition into the Netherlands, 'not as in the time of the Duke of Alba [the notorious Spanish governor of the Netherlands at the time of the revolt], but in a more hypocritical way'.[84]

The pamphlet also revisited the earlier pamphlet by Schuts. As the anonymous 'J.J.' explained, 'Countrymen's talk' had been handed to him some time before in the village of Alblasserdam (which is 12 kilometres from Rotterdam). This led to a discussion wherein he pointed out that Lansbergen had refuted such views already in 1629. For the time being, he could only summarise Lansbergen's arguments in as far as he could recall them, but later he wrote (in response to his interlocutor's demand) a comment on 'Countrymen's talk'. I will skip the contents, as they are not really interesting. His friend was satisfied, but thought it too aggressive with respect to certain ministers. J.J. thereupon in a letter, the text of which is also printed in the pamphlet, defended his sharp tone and for the time being refused, as useless, to have it printed. Apparently, it was the appearance of de Laat's pamphlet that spurred him to make these old notes public, together with a refutation of the pamphlet by De Laat. J.J.'s implication that tracts like those by Schuts circulated and were discussed in the countryside has the ring of truth. Otherwise, it would be difficult to explain that Schuts' tract was not utterly forgotten by 1723. Schuts himself, by the way, had included a doggerel against the Copernican system, which, he asserted, had been made some time before.[85]

The last controversy is the one about which we have the most fragmentary evidence. It took place in Groningen in the period 1737-1738. In fact, it is not even clear whether the two fragments we possess really belong together. We will discuss them one by one. The first is a pamphlet by Nicolaus Engelhard, professor of philosophy at Groningen, concerning the compatibility of Co-

---

[82] De Laat (1723) 4, 14.

[83] Ibid. 26-27.

[84] J.J., *Brief aan een vrint*, 28; see also 9-10, 27.

[85] J.S.K.B., *De vermakelyke boere-praat... verdedigt*, 36.

pernican theory with the Bible. The argument does not interest us here, but the occasion is of some interest. As Engelhard explains, a Groningen minister had proclaimed from the pulpit 'that those who teach that the earth moves around the sun, call God a liar to His face.' Herewith, he was aiming at Engelhard, who would have preferred to let it pass in silence, 'but judging it intolerable that ignorance would domineer and gain the upper hand in the congregation of believers, I felt constrained to serve this warmed-up dish once again.'[86]

The expression 'warmed-up dish' seems to indicate that Engelhard considered the whole discussion as hopelessly lagging behind its time. On several occasions, he reinforces this impression. He feels it necessary to stress that the words of the said minister have really been spoken this way, 'although not without much offence of many pious people'.[87] At the end of his pamphlet, he remarks that the reverend would have done better to keep silent on matters beyond his competence, 'the more, as one would be laughed at by the whole of the learned world, were it known that here in Groningen one decries from the pulpit such views, which are everywhere considered as sufficiently established nowadays.'[88]

There is nothing known about a continuation of the discussion between Engelhard and his opponent. Still, the discussion on the system of Copernicus lingered on in the province of Groningen. This is clearly shown by the second pamphlet, from 1738. The title is very long, but too telling not to quote the first part: 'Course of the sun and rest of the earth, according to the certainty of the doctrine of Salomon: Eccl. 1 vs. 4, *The earth abideth for ever*, and vs. 6, *(The sun) goeth toward the south, and turneth about unto the north etc.*[89] Against the system of Copernicus as the one by Ptolemy...'. It is written by a certain Alaricus (which clearly is a pseudonym[90]) and presents itself as a link in a larger discussion. The title page summarises the earlier contributions. As it appears, 'Alaricus' had earlier written a 'Peasant's philosophy, or unlearned remarks against Renatus Dubitation or doctrine of doubt'. A pamphlet by a certain 'Stuid:' (another pseudonym) had thereupon passed a 'satirical sen-

---

[86] Engelhard (1737) 8. It is not clear at which minister Engelhard is aiming. Examination of the archives of the Groningen church council and classis yielded no further details. On Engelhard, especially on his ideas on metaphysics, see Wielema (1997).

[87] Engelhard (1737) 7.

[88] Engelhard (1737) 22.

[89] 'The earth abideth': in the Dutch version this reads more or less as 'the earth stands'. As for the quote from Eccl. 1:6, most versions agree that the subject of this passage is the wind, not the sun. The then standard Dutch version (or *Statenvertaling*), however, took it to be the sun.

[90] At the end, the work is signed 'P.H.P.'.

tence' (*spot-vonnis*) on this work. Both these works appear to be lost. The pamphlet we have is Alaricus' rejoinder to 'Stuid:'.

The work is a poem, hardly an ordered argument and difficult to follow. Very little is said on Copernicanism itself. The pamphlet widely digresses, going into all kind of philosophical and religious subjects. It appears that 'Alaricus' was one of the many lay philosophers of the time and that he had presented in his earlier book a world system of his own.[91] As can be seen from the title, he refuted both Copernicus and Ptolemy. Moreover, he prided himself on not following any philosopher – whether a pagan like Aristotle or a modern Epicurean like Descartes or Spinoza, for whom he has some very sharp words. The only authority he acknowledges is biblical revelation.

The 1738 pamphlet is the last known instance of printed polemics on the system of the world in the Dutch Republic. Whether the matter was settled or not, there were no further public quarrels after that date. It was only more than 100 years later, in the wake of the anti-Copernican publications of the German Schöpffer, which also attracted some attention in the Netherlands, that the subject again obtained some relevance. In 1853, some new pamphlets on the Copernican issue appeared.[92] However, one can hardly speak of the old polemic flaring up. There is some serious disagreement on cosmological matters between the authors of the 1853 pamphlets, but all of them defend Copernicanism. Interesting though they are, however, I shall not further elaborate upon these nineteenth-century developments, as this would carry me beyond the proper boundaries.

The cases mentioned concern rather minor or local affairs. In no case did the issue rally a significant number of combatants. Most of the participants in the discussions are rather obscure people – lay theologians rather than established ministers. Only in the case of Engelhard and his opponent are two more prominent people involved. Nor do the disputes focus on the important cultural centres. The debate on Copernicanism seems to have moved to the countryside. It does seem, indeed, that Engelhard was right in stating that the

---

91 Cf. Alaricus (1738) 7-8.

92 C. Schöpffer, *De aarde staat stil. Bewijzen dat de aarde zich noch om hare as noch om de zon beweegt*, translated from the German by M.J. van Oven (Utrecht 1853). There were two reactions: 'Joshua', *De aarde staat niet stil. Duitsche onzin in het Hollandsch wederlegd* (Utrecht 1853), and L.H. Verweij, *De aarde staat niet stil. Betoog dat de aarde zich beweegt* (Amsterdam 1853). Verweij's pamphlet was also translated into German (Amsterdam and 1854). Verweij was a physician from The Hague, who, on other occasions, also came out in defence of freemasonry, homeopathics and spiritism. His astronomical knowledge was not up to date, to put it mildly, and from a scientific point of view, his ideas were untenable. This earned him a public reproach from his fellow Copernican 'Joshua': *De betogen, voorkomende in de brochure van Dr. L.H. Verweij, De aarde staat niet stil, getoetst aan rede en ervaring, en onhoudbaar bevonden* (Utrecht 1853).

anti-Copernicans were fighting a rearguard action by this time, and that the motion of the earth was no longer an issue among the learned. This might even be one of the reasons behind this relative upsurge in anti-Copernican polemics in the first half of the eighteenth century. Convinced anti-Copernicans felt the need to make some noise as they realised that their party was becoming weak.

Still, before we can conclude that anti-Copernicanism was dead and buried but for some last convulsions among a few obscure people, mostly in backward places, we should consider some further matters. Local disputes like those described are difficult to trace. As noticed, several of them have left only very fragmentary evidence. Others will have been lost altogether or buried in places hard to find. (Engelhard initially considered publishing his tract in a larger volume of miscellanea, [93] in which case it would have been much more difficult to trace.) Many local disputes of course never reached the stage of being put into print. So, the few cases we have reviewed probably represent a much larger number of quarrels. Although anti-Copernicanism may have retired from the centres of learning, it was not quite dead.

On the contrary, in the country at large it appears to have gained considerable strength since the seventeenth century. Semeyns wrote in 1760: 'Though this has been demonstrated simply and without contradicting Scripture, and though the theory of the annual (around the sun) and daily (on its axis) motions of the earth have been accepted by all astronomers without any hesitation, one still finds people who, either from dullness of wit or sometimes from other insights, cannot acquiesce in this.' [94] The anti-Cartesian and anti-Copernican offensive of the Voetians had succeeded in gaining the support of a large part of the population. The 'popular prejudice', so much decried by the enlightened, was largely the result of a systematic campaign by a part of the educated elite. Many people had become aware of the problem and now supported the strict biblicism of the Voetians, even after it had been abandoned by the leading theologians. One should add that, as has been demonstrated before, the general climate of Newtonianism in itself was not unfavourable to a biblicist outlook in natural philosophy. On a popular level, nobody laid down the Copernican system of the universe without uttering the ritual formula that, of course, it was not contrary to Scripture. The lawyer Joan Jacob Mauricius – even if he clearly expressed his view that God wants us to learn physics by our own efforts and not by revelation – admitted that the question

---

93 Engelhard (1737), preface.

94 Semeyns (1760) 2.

of the biblical passages concerning the sun's motion was a sensitive (*teder*) one and for that reason he did not want to elaborate on it. [95]

In 1772, an anonymous 'amateur of mathematics' published a small work wherein he defended the Copernican system. In the preface, he justified his project as follows: 'Although we live in a very enlightened century, wherein all arts and sciences have been elevated nearly to their summit, one still finds many, even wise and prudent people, who cannot believe the motion of the earth. Some of them agree that such a motion would be a more convenient way than the motion of the sun, but they feel that it is contrary to Scripture. However, those who read this booklet somewhat attentively will find not only that such is not true, but quite the contrary, that Scripture asserts the same...' [96] The author seems reasonably at home with astronomy and physics, which makes his arguments rather predictable. Most interesting is his listing of the arguments which were apparently put forward against Copernicus at this time. The work is divided into two sections. The first ('about which commonly falls most dispute' [97]) deals with the daily rotation of the earth. The author discusses and refutes both physical and biblical objections. The biblical objections refer to the stock texts of Joshua 10:12-13, Ecclesiastes 1:5-6, Psalm 19:6-7, Genesis 1:16, Job 37:17, and Song of Songs 4:16. Among the physical objections is the argument that according to Copernicus, the sun should make a screw-like motion, which is against the rules of mechanics. 'Others state that they cannot imagine that the earth is turning on its axis every day, so that we would turn upside down within twelve hours...' [98] The second part deals with the earth's annual motion. Herein, the author shows how the seemingly irregular motions of the planets are, in the Copernican system, perfectly regular. A final, concluding part demonstrates, with reference to Varenius and Christian Wolff, the dimensions of the earth, the solar system and the universe, with the aim of showing God's greatness and human paltriness.

In the end, the attitude of the proponents of Copernicanism may appear more telling than the number of its adversaries. Even if there were no longer any anti-Copernican voices to be heard in the second half of the eighteenth century, people appear to have felt guilty about defending Copernicanism. In 1779, a work was published wherein a certain Rutgerus Ouwens, rector of the Latin school in The Hague, defended the view that one cannot prove the motion of the sun from Joshua 10:12. Originally, Ouwens had pronounced

---

95 Mauricius, (1765-1766) 69-70 noot 28. See also, a separate part in the same book: *Vaarwel aan den zangberg*, 4-5 note 8.

96 *Beweeging der aarde* (1772), 3.

97 Ibid.

98 Ibid., 20-21.

it as a Latin oration on the occasion of the prize-giving festivities at his school on 6 September 1778. The printer's preface asserted that it had been translated and printed at the insistence of some persons of judgement, knowledge and taste. In his text, Ouwens stated that the earth's motion around the sun is so certain, 'that presently nobody, who is gifted with some small part of sound reason, will not allow this opinion; although on the other hand, people are not wanting who are reluctant [*beschroomd*] to accept a doctrine which seems to run counter to Holy Writ.'[99] However, as Ouwens continued, the Copernican theory is not contradicted by Holy Writ at all. If anything, Joshua 10: 12-13 teaches the standstill of the sun. Ouwens' exact argument here appears similar to Galileo's in his Letter to the Grand Duchess, although he does not refer to it. The contents of his oration do not deserve further discussion. The work is of interest, however, as an attestation of the continuing worry people felt over the Copernican system. They asserted that it was true and did not contradict the Bible; however, they needed to be continually reassured that this was indeed the case.

[99] Ouwens (1779) 9.

General conclusion

A central element in our story has been the importance of the universities in the Dutch setting. Even though none of the important intellectual developments originated at a university, the universities picked from current ideas what suited them best and systematised them into academic programmes of learning. In this way, intellectual developments became institutionalised, visible and much more influential. Without the universities, the development undoubtedly would have appeared much more blurred and chaotic. However, this role could be played only in constant interplay with the outside world. For instance, the humanist ideal was accepted at Leiden at an early date. Starting from these foundations, Leiden scholars developed a certain approach to cosmographic questions, which stressed among other things the harmonious constitution of the universe. However, it was only when this approach was no longer confined to academia and reached astronomers whose loyalty to the Leiden programme was not very deep, that its full potential was discovered and it was used to support heliocentric astronomy.

On the other hand, the telescopic discoveries by Galileo failed to elicit a noteworthy reaction at the universities. Because the meaning of these phenomena was not altogether clear at the time, they remained the field of speculation and intellectual experiments. No university programme could be built upon such unstable foundations. These speculations therefore remained the field of independent thinkers, and their speculation was for the time being very chaotic (although some elements would reappear in the philosophy of the second half of the century). Only very slowly, and very cautiously, did university professors begin to pay attention to these things.

It was Descartes who succeeded in formulating the various new insights into a coherent and convincing intellectual programme. Once this had been accomplished, the universities enthusiastically adopted the new learning. The humanist intellectual programme was largely abandoned and the initiative was seized by the professors of philosophy, notably physics. Even such philosophers who decidedly rejected Descartes' ideas could not resist taking on a new role and asking new questions. It does not seem that this transition was prepared or facilitated by a social shift. The humanists did not fade into oblivion with the rise of Cartesianism. Classical philology was to remain a much respected discipline for centuries to come. But after 1650, people no longer believed that philologists had anything to say on the constitution of the universe, or the workings of nature in general.

It was foremost as an academic programme that Cartesianism could be accepted by society at large. Still, here again, independent thinkers and philosophers acted more freely. University professors who supported the new philosophy either chose to stick to the letter of Descartes or tried to accommodate his ideas with Aristotelian notions. Outside academia, the variety of ideas, either caused by misunderstanding or by daring, was much greater.

Newton's ideas were seized upon at the universities and in society at large at about the same time. One should add that in the two cases, people sought to achieve different aims. The academic Newtonians in the first place wanted to build a new mathematical physics and astronomy; outside academia, people were looking for an alternative to Cartesian philosophy, mainly for religious reasons. It may appear a lucky coincidence that the two programmes tallied. Anyhow, it was the authority Newton's theories acquired as a result of being turned into a university programme which helped to carry the day outside academia as well.

Important in all this is that in the Dutch setting, the universities were the only institution where a programme of learning could be shaped. Moreover, the faculties were to a large degree free to regulate their own business. The secular character of the Dutch state minimised theological influence. This may largely account for the fact that only a very few people claimed that one should regard Copernicus' theories as a mathematical hypothesis for saving the phenomena. People who did were mostly theologians: Lubbertus, Rivet and Maresius (the last-mentioned before he turned into an enemy of Cartesianism). A philosopher like Schoock appears also guided by religious motives. The argument that heliocentrism could be accepted as a mere hypotheses seems first and foremost a theological manoeuvre, aimed at neutralising or suppressing the dangerous consequences of the theory. The principal propagators of this interpretation in an international setting were also theologians (e.g. Osiander and Bellarmine) and theologically minded people (e.g. Melanchthon) who for some reason or other had to come to terms with Copernicus' theory. In the early Dutch Republic, it simply was not the theologian's province. The system of the world was left to astronomers and humanists. (Lubbertus' and Rivet's stance is known only from private letters.) The difference between the 'Leiden' and the 'Wittenberg' interpretation largely stems from their different institutional settings.

This also explains why theological positions after 1650 suddenly became so extreme. Theologians had little chance to influence physics. If they wanted to have a say on the issue at all, they first had to impose their authority. The debate on Copernicanism was really on the subordinate position the Church had in the Dutch state. As such, it was largely motivated by political considerations. Theologians entered the fray because they felt they had powerful allies and sensed that they could destroy their opponents. Small wonder, then, that they were not very interested in a compromise.

Perhaps the most striking aspect of the development sketched in the foregoing is its discontinuity. There were two rather sudden transitions. The first, which occurred in the middle of the seventeenth century, was a major one: the introduction of Cartesian philosophy. The second, which occurred at the beginning of the eighteenth century, was a minor one: the introduction of Newtonianism. These were mainly changes in intellectual outlook, and as such had a much wider bearing than suggested by the terms 'Cartesianism' and 'Newtonianism'. Even those who had remained aloof from Descartes' ideas were deeply influenced by the 'new philosophy'.

To anybody familiar with Thomas Kuhn's concept of 'paradigm shift', this will not come as a surprise. According to Kuhn, a paradigm shift is brought about by a crisis in the old science, as scientists become aware that they are unable to solve the problems they encounter in their day-to-day practice. Kuhn himself regarded the transition from the geocentric to the heliocentric system as an example of such a shift. Probably, few of today's historians of science would agree with that. Copernicus can hardly be said to have solved some new problems which astronomers had met in the course of time. It would rather seem that some age-old problems had got a new meaning, so that by the sixteenth century a solution seemed more pressing than in Ptolemy's time. The fundamental transition around the middle of the seventeenth century could be described as a paradigm shift, although one may still doubt whether Kuhn's more detailed description is applicable.

What the story seems to tell is that the big changes were brought about by transitions in more fundamental layers rather than in the sciences themselves. The curious thing about the change that occurred around 1650 is, of course, that it did not solve any of the problems which had made the Copernican system unacceptable up to that time, save for one: from then on, the heliocentric system could be regarded as being in line with more general philosophical ideas, on the world, nature, God and man. Nor did Cartesian physics offer any solution to the original problem, viz. the many anomalies in the motions of the planets. On the other hand, most earlier arguments which had favoured the Copernican system (such as those regarding cosmic harmony) also proved incompatible with the new physical ideas and were soon forgotten. The Copernican system became just the exponent of this new view of nature. It was no longer debated for its own merits or demerits.

It seems more apt to regard the sharp discontinuities as effected by institutional constraints. Judging from the Dutch evidence, the position of the Copernican system during the scientific revolution was that of an icon, a rallying-cry, a point of reference, or even a shibboleth. The various arguments for or against proved less important than its simple presence as an ideal, a challenge to traditional learning or a symbol of the new. It does not seem that the many discussions centred on the subject had in themselves an important impact on the development of the new learning.

Abbreviations

ARA – Algemeen Rijksarchief, The Hague.
AT – *Oeuvres de Descartes*, C. Adam and P. Tannery eds. (2nd edition, 11 vol., Paris 1973-1978).
BMGN – *Bijdragen en mededelingen betreffende de geschiedenis der Nederlanden.*
BGLU – *Bronnen tot de geschiedenis der Leidsche universiteit.*
BLGNP – *Biografisch lexicon voor de geschiedenis van het Nederlandse protestantisme* (Kampen 1978-).
BWN – A. J. van der Aa, *Biografisch woordenboek der Nederlanden* (1852-1878; repr. 7 vol., Amsterdam 1969).
BWPGN -*Biographisch woordenboek van protestantsche godgeleerden in Nederland* (5 vol., The Hague 1903-1949; not finished).
DMS – *Duisburger Mercator-Studien.*
IBJ – *Journal tenu par Isaac Beeckman de 1604 à 1634*, C. de Waard ed. (4 vol., The Hague 1939-1953).
JHA – *Journal of the history of astronomy.*
Knuttel – W.P.C. Knuttel, *Catalogus van de pamflettenverzameling berustende in de Koninklijke Biblioheek* (9 vol., The Hague 1889-1920, repr. 1978).
MML – *Maandelyke mathematische liefhebberij.*
NNBW – *Nieuw Nederlandsch biografisch woordenboek*, P.C. Molhuysen and P.J. Blok eds. (10 vol., Leiden 1911-1937).
OC – Christiaan Huygens, *Oeuvres complètes* (22 vol., The Hague 1888–1950).
Opsomer – C. Opsomer ed., *Copernic, Galilée et la Belgique. Leur réception et leurs historiens / Copernicus en Galilei in de wetenschapsgeschiedenis van België* (Brussels 1995).
PW – Simon Stevin, *Principal works* (5 vol., Amsterdam 1955-1966).
Reception – J. Dobrzycki ed., *The reception of Copernicus' heliocentric theory* (Dordrecht 1972).
TGGNWT – *Tijdschrift voor de geschiedenis van de geneeskunde, natuurwetenschappen, wiskunde en techniek.*

Bibliography

As to the alphabetical order, Dutch readers will notice that the vowel 'ij' is regarded here as a combination of 'i' and 'j'. Non-Dutch readers should note that prefixes are disregarded; for instance, 'J. van den Berg' is not listed under V for 'van' but under B for 'Berg' (i.e. as 'Berg, J. van den').

Archives and manuscripts

Amsterdam, University Library:

B b 63: Dutch overview of the planetary theories of Ptolemy, Copernicus, Tycho and Lansbergen.

D 29: letter from Hudde to Velthuysen, 13 Oct. 1657.

Amsterdam, Maritime Museum:

ms. C. 55: Jacob van Veen, *De loopende werlt.*

Arnhem, State Archives:

Archives of the reformed (N.H.) synod, no. 23: Resolutions of the synodal deputees.

Archives of the States of the quarter of Nijmegen.

Groningen, University Library:

ms. 106: [Nicolaus Mulerius], *Tabularum Frisicarum pars altera. Singulorum planetarum abacum complectens, e fontibus Cl Ptolemaei Regis Alfonsi, anno 1612 constructa.*

ms. 445: Dutch treatise on the use of the globe, 1656.

Leiden, University Library:

BPL 902: Jacob van Veen, *De loopende werelt*, 1670.

BPL 907: Johannes de Raeij, comments on Descartes' *Principia philosophiae* and other works.

BPL 908: N.N., comments on Descartes' *Principia philosophiae.*

BPL 1558: Dutch treatise on astronomy and navigation.

Pap. 2: Martinus Hortensius, *Responsio.*

Leiden, Boerhaave Museum:

ms. 12,731: Jacob van Veen, *Astronomische raadtselen*, 1673.

Leiden, municipal archives:
Archives of the church council, 5: resolutions.
The Hague, Algemeen Rijksarchief [Dutch State Archives]:
1.01.46 (admiraliteitscolleges) nr. 1369: resolutions of the Amsterdam admirality board.
Archief Ned. Herv. Kerk, oud-synodaal archief 1566-1816, nr. 5, 321-326: '*Notabel verhael van een hoochgeleert ende vermaert persoon [Mulerius], seer dienstich tot waerschouwinge van alle Christenen, die enige affectie draegen tot de Christelijcke religie*'.
Haarlem, State Archives:
Coll. R. Hooykaas.
Leeuwarden, Provincial Library:
Coll. Gabbema, letters from N. Mulerius.
Utrecht, University Library:
Hs. 8+ C.13: Cornelius Valerius, *Physicae seu de naturae philosophia institutio* (Utrecht 1613) with annotations by Cornelis Booth.
Hs. VI G 14: *Astronomia* (Latin).
Utrecht, State Archives:
Archives of the reformed (N.H.) provincial church assembly, no. 142: Copies of resolutions of the synod of Gelderland.
Middelburg, Provincial library
Koninklijk Zeeuwsch Genootschap der Wetenschappen, manuscript collection no. 4092: letter from Andries van der Poest, 31 Jan. 1775.

Printed sources

As some of the works cited, especially disputations and pamphlets, have only been preserved in rare copies, I deemed it convenient in some cases to indicate where the copies I consulted are kept. This is done by means of an abbreviation between brackets behind the title. The following abbreviations are used:
A – Amsterdam, University Library
BL – London, British Library
FS – Halle, Frankesche Stiftungen
G – Groningen, University Library
H – Herborn, Evangelisch-Theologisches Seminar
L – Leiden, University Library
PBL – Leeuwarden, Provincial Library
U – Utrecht, University Library
UX – Utrecht, University Library, collection of xerocopies
W – Wolfenbüttel, Herzog August Bibliothek

Maps are referred to only in the notes. In particular the Bodel Nijenhuis collection at Leiden University Library has been consulted.

a. *Disputations*

(The names of the respondents are written as they occur on the title pages. The same applies to the indications 'resp[ondens]' and 'auctor' (where no indication is given, which happens in a few very early disputations, I put 'resp.'). Only original disputations are listed; later editions, in collections, textbooks, etc. are not mentioned here.

Leiden

Johannes Murdison

1600 *Theses ... de mundo* (resp.: Leoninus Leo).

1600 July 5: *Themata physica de elementis* (resp.: Abraham Antonides Merius).

1600 Nov. 18: *Disputationum logicarum decima-sexta, de analysi, complectens introductionem ad libros analysicorum Aristotelis* (resp.: Ioannes Vitellius).

1601 June 2: *Theses philosophicae de anatomia universae naturae* (resp.: Richardus Bland).

Anthonius Trutius

1598 July 25: *Theses miscellaneae de terra* (resp.: Johannes à Lodensteyn).

1599 Jan. 13: *Theses ethicae de liberalitate* (resp.: Joannes Martini F. Lydius).

1599 July 24: *Φῦσιολογία de mundo* (resp.: Joannes Georgius Holtzapfel, dictus Milander).

1603 Jan. 22: *Theses physicae de coelo* (resp.: Carolus C.F. Ryckewardus).

Petrus Bertius

1598: *Theses physicae de elementis* (resp.: Theodorus Oortius).

1600: *Theses philosophicae* (resp.: Ioannes Narsius).

1603: *Theses peripateticae de natura coeli* (resp.: Cornelius a Nieren).

1604: *Theses miscellaneae de terra* (resp.: Joannes Wyringius Marci F.)

1605: *Προβληματα logica desumpta ex parte propriâ* (resp.: Dionysius Spranckhusius).

Gilbertus Jacchaeus

1604 July 4: *Theses physicae de natura* (resp.: Isaacus Iunius).

1604 Dec. 4: *Theses physicae de principiis, et natura* (resp.: H. van der Myle).

1607 Dec. 22: *Disputationum physicarum septima de coelo* (resp.: Gosardus Dremmius).

Franco Burgersdijk
1627 June 16: *Disputationum physicarum septima, de coelo* (resp.: Marcus Mamuchet). (L)

Adriaan Heereboord
1661 June 22: *Disputatio philosophica, continens postiones miscellaneas, nobiliores, ex universa philosophia depromptas* (resp.: Michaël Hendrix).

Johannes de Raei
1651 March 6: *Disputatio physica de motu locali* (resp.: Petrus à Couwenbergh). (L)
1661: *Disputatio philosophica de mundi systemate and elementis. Prima* (resp.: Carolus a Loten).
1668: *Disputatio philosophica, de terrae rotunditate* (resp.: Johannes Gruterus).

Jacob Revius
1647 Sept. 14: *De Terrae stabilitate* (resp.: Cornelius Keetmannus).
1647 Sept. 18: *De stabilitate Terrae, altera* (resp.: Joannes ab Appeldorn).
1647, Oct: *Quaestiones aliquot de Terra & Coelo continens* (resp.: Johannes Kischius).

Burchard de Volder
[1682: *Disputatio philosophica de systemate mundi, prima* (resp. Casparus Langenhert).] This disputation I have not seen.
1694: *Disputatio philosophica quae est de mundi systemate* (resp.: Gysbertus Henricus Casembroot).

Pro gradu
[The Leiden doctoral disputations up to 1610 are in the university archives, inv. nos. 347, 352-354. They are available at the University Library.]

Gerardus Joannes Vossius, *Universalis philosophiae ακροτηριασμος*, 1598 Feb. 23. (See also Vossius (1955).)
Petrus Pilius, *Disputatio physica de cometis*, 1598 Feb. 26.
Thomas Johannides Bergensis, *Positiones ex praecipuis philosophiae partibus*, 1601 March 25.
Richardus Bland, *Φιλόσοφος seu positiones philosophicae*, 1601 Dec. 13.
Lamouralius de Stembor, *Disputatio inauguralis*, 1606 Dec. 31.
Thomas Erpenius, *Aphorismi aliquot philosophici*, 1608 May 5.
Willebrord Snellius, *Theses philosophicae*, 1608 July 12.
Thomas Carbasius Hornanus, *Theses philosophicae*, 1609 March 18.
Cornelius Uythage, *Disputatio philosophica inauguralis, continens conclusiones de divinis & humanis rebus in omni philosophiae genere.*

Johannes Uythage, *Disputatio philosophica inauguralis, conclusiones nobiliores, per universam philosophiam continens*, 1662 April 25.
Philippus de la Fontaine, *Disputatio philosophica inauguralis de aestu marinis*, 1662 Nov. 8.
Cornelius Uythage, *Disputatio theologica inauguralis de veritate religionis Catholicae contra Arminianos*, 1663 Nov. 26.
Cornelius Sevenhuysen, *Disputatio philosophica inauguralis de libero arbitrio, virtute, & summo bono*, 1664 May 13.
Antonius à Coppenoll, *Disputatio philosophica inauguralis continens varias ex philosophia desumptas quaestiones*, 1664 June 30.
Wolferdus Senguerdius, *Disputatio philosophica inauguralis de tarantula*, 1667 Dec. 7.
Gerardus de Vries, *Disputatio philosophica de mundo*, 1671 May 28.
Johannes Annesley, *Disputatio philosophica inauguralis, continens varias considerationes de astronomia vulgari*, 1671 July 8.
Samuel Koleseri, *Disputatio philosophica inauguralis de systemate mundi*, 1681.
Gilbertus Henricus Casembroot, *Disputatio philosophica inauguralis de aestu marino*, 1696.
Jacob vander Ghiessen, *Disputatio philosophica inauguralis de mundo*, 1703 Sept. 20.
Joannes Oosterdijk Schacht, *Dissertatio astronomico-physica inauguralis de motu planetarum in orbitis ellipticis*, 1726 Sept. 26.
Petrus Mangard, *Disputatio astronomici inauguralis, de systemate Copernicano*, 1744 July 3.
Johannes Samuel Creutz, *Specimen philosophicum inaugurale exhibens varias theses philosophicas*, 1773 Oct. 27.

Utrecht

Jacob Ravensberg
1640 Nov. 25: *Disputatio astronomica de mundi systemate, affinibusque materiis.* (H)
1642: *Encyclopedia mathematica* (resp.: Al. de Bie, Barth. a Wesel and Andr. Lansman).

Henricus Renerius
1635 June 10: *Disputationum physicarum tertia, de mundi et coelo* (resp.: Ludovicus à Vosbergen). (H)

Johannes de Bruyn
1653 May 14: *Disputatio physica de mundo* (resp.: Gerhardus Bornius). (BL)
1655 June 20: *Disputatio physica de natura et origine fontium* (resp.: Gedeon Deutz). (UX)

1655 Oct. 13: *Disputatio physica quarta, de corporis divisibilitate* (resp.: Johannes a Swanevelt). (UX)
1655 Nov. 14: *Disputationum physicarum quinta, de atomorum affectionibus, & corporum continuitate* (resp.: Johannes Kalander). (UX)
1656 Feb. 27: *Disputationum physicarum octavo, de vacuo* (resp.: Abrahamus Marius). (UX)
1656 April 19: *Disputatio physica de coelo* (resp.: Samuel Gelei). (UX)
1656 May 21: *Disputationum physicarum nona, de metu vacui & studio conservandae plenitudinis* (resp.: Petrus Hulsman). (UX)
1657 April 25: *Disputatio philosophica de libero hominis arbitrio* (resp.: Johannes a Swanevelt). (UX)
1660 (s.d.): *Disputationis physicae de alitura, pars quinta* (resp.: Nicolaus Shepheard). (UX)
1660 April 14: *Disputationum physicarum decima-octava, de mundo visibili pars prima* (resp.: Petrus Clerquius). (UX)
1661 Dec. 11: *Disputatio philosophica aliquot nobilissimas materias ex physica & mathesi selectas continens* ('Responsurus Author': Henricus a Rhee). (UX)
1662 May 7: *Disputatio physico-optica de presbytia & myopia* (resp.: Martinus Martens). (UX)

Gisbertus Voetius
1638 Sept. 21, Oct. 6: *De creatione*, pars tertia et quinta; resp. Lubertus Spruitius. [Also in Voetius, *Disp. selectae*.]
1656 May 24: *Disputatio theologica continens positiones aliquot miscellaneas* (resp.: Gerardus van Os). (UX)

Andreas Essenius
1654 June 24: *Disputatio theologica de infallibili fide rerum naturalium in Sacra Scriptura passim revelatarum*, author et resp.: Johannes Beusechum.
1654 Oct. 4: *Disquisitio philosophico-theologica utrum ὁ δεῖνα per propositiones et hypotheses aliquot, tam solide demonstraverit Solus quietem et terrae duplicem motum, ut S. Scripturae contrarium asserens accipienda sit, tanquam loquens secundum erroneam vulgi opinionem?*, author et resp. Henricus Troy.
(See also Niepoort, Beusechum and Troy, *Disputationes quator.*)

Jacobus Odé
1726 April 10: *Dissertatio astronomico-physica de luna habitabili*, auctor: Gulielmus Arntzenius.
1749 Sept. 24: *Dissertatio physico-astronomica de obscuritate Solis, Christo in cruce patiente facta*; auctor and resp.: Michael Kerwal.
1750 June 17: *Dissertatio philosophica de motu terrae diurno et annuo*, auctor: Petrus van Meerwyk.

Johannes Horthemels
1750 June 10: *Disputatio physico-astronomica continens examen celebratissimi hodie systematis cosmici copernicani*, author and respondens: Guilielmus Henricus s'Jacob.

Pro gradu
Andreas Demetrius, *Disputatio philosophica inauguralis, de voluntate Dei, Sole, & fortitudine*, 1640 July 2. (FS)
Gualterus de Bruyn, *Disputatio philosophica inauguralis, de malo, eo quod invitum ac spontaneum est, & motu siderum*, 1640 Dec. (H)
Bartholdus a Wesel, *Disputatio philosophica inauguralis, de materia prima, orbibus coelestibus et affectibus*, 1642 Aug. 23. (H)
Alexander de Bie, *Disputatio philosophica inauguralis, de ortu spontaneo, motu Solis, et usuris*, 1642 Aug. 25. (H)
Bernhardinus de Moor, *Disputatio philosophica inauguralis, de continuo, novis planetis, et stoicorum erroribus circa affectus*, 1643 Nov. 28. (UX)
Lambertus Velthuysen, *Disputatio philosophica inauguralis, de mundo, de vindicti, et de chalyboclisi seu deviatione versorii magnetici*, 1644 March 26.
Nicolaus Shepheard, *Disputatio philosophica inauguralis, de concursu Dei, systemate mundi, & conversionibus rerum publicarum*, 1661 April 22.
Daniel Arens, *Disputatio philosophica triplex inauguralis. Metaphysica, de natura spiritus in genere. Physica, de vero systemate mundi. Ethica, de bello*, 1671, [..] 18.
Elias à Steenbergh, *Disputatio philosophica inauguralis, de materia à facultate philosophica praefixis*, 1675 Dec. 3.
Johannes Henricus Bert, *Disputatio philosophica inauguralis de terra*, 1698.
Jacobus Leusden, *Disputatio philosophica inauguralis de mundo*, 1699.
Johannes Blomhert, *Disputatio philosophica inauguralis, de systemate mundi*, 1716 March 16.
Jacobus Odé, *Dissertatio philosophica inauguralis, de atmosphaera*, 1721 Nov. 21.
Alvaro Telles Dacosta, *Disputatio philosophica inauguralis, de maculis Solis*, 1734 June 11.
Johannes van Herwerden, *Disputatio astronomica inauguralis de motu terrae diurno atque annuo*, 1736 March 26.
Daniel van Goens, *Disputatio philosophica inauguralis comprehendens examen an Luna habitabilis sit*, 1738 June 19.
Casparus van Oort, *Dissertatio philosophica inauguralis, de meteoris quibusdam aqueis*, 1744 June 22.

Franeker

Abraham de Grau

1665: *Disputatio uranoscopia de cometis* (resp.: Hermannus Brunsfelt). (U)

1683: *Disputatio cosmologica de planetarum delatione* (resp.: Adrianus Steengracht). (L)

Tobias Andreae

1684: *Exercitatio philosophica de impossibili mundi aeternitate* (author: Hadrianus Steengracht). (A)

Pro gradu

Tjerk Nieuwenhuis, *Disputatio philosophica inauguralis de innumerabilibus mundis*, 1734 July 1. (PBL)

Groningen

Joachim Borgesius

1656 March: *Disputatio philosophica de lumine Lunae secundario* (resp.: Tob. Andreae). (FS)

1658 Sept. 25: *Disputatio philosophica de Saturno* (resp.: Johannes Petri). (FS)

Martin Schoock

1648 Jan. 26: *Disputatio historico-physica tertia expendens naturalis historiae certitudinem* (resp.: Daniel à Sanden).

1648 Feb.: *Collegii philosophici problematici disputatio septima* (resp.: Johannes Niclenius).

1648 [n.d.]: *Collegii philosophici problematici disputatio decima, continens problema physica* (resp.: Engelbertus Alberts).

1648 June 21: *Disputationum philosophicarum de convenientia orbis coelestis ac elementaris prima* (resp.: Anthonius a Dorth).

Samuel Maresius

1667 Feb. 19: *Disputatio theologica prior refutatoria libelli de Philosophia Interprete Scripturae, ac opposita Prologo anonymi* (resp.: Jacobus Berchius).

1667 July: *Disputatio theologica sexta refutatoria libelli de Philosophia Interprete Scripturae, & prior opposita ejus capita quinto* (resp.: Johannes Blencke).

Pro gradu

Johannes Ravensberg, *Disputatio philosophica inauguralis*, 1639 Feb.

Josua van Iperen, *Dissertatio philosophica inauguralis de mundi mechanismo*, 1752 Nov. 1.

Johannes Fabricius Dinckgreve, *Dissertatio philosophica inauguralis exponens nonnulla memorabilia systematis solaris*, 1777 Dec. 10.

Harderwijk

Antonius Deusing
1640 June: *Disputationum physicarum secunda. De principiis corporum naturalium internis* (resp.: Samuel Neomagus). (H)
1641 Feb. 25: *Disputationum physicarum quinta. Continens continuationem creationis. De operibus quarti diei* (resp.: Joh. Achterkerken). (H)

Amsterdam

Arnold Senguerd
1650: *Disputatio physica de mundo* (resp.: Gerardus Bicker a Swieten).

Alexander de Bie
1653 Nov. 26: *Disputationum astronomicarum secunda, de cauda cometarum* (resp.: Christianus Wittewrongel).
1653 Dec. 22: *Disputationum astronomicarum tertia, quae est ultima, de motu cometarum* (resp.: Antonius Haselbroeck).
1657 Dec. 15: *Exercitatio astronomica de terra, ac solis, et lunae maculis* (resp.: Abraham Boddens).
1659 Jan.. 21: *Disputatio astronomica de systemate mundi* (resp.: Ferdinandus Achilles a Kuffeler).
1659 April 12. *Quaestionum illustrium tertia de eclipsi Paschali prior* (resp.: Lucas Hooft).
1662 April 1: *Προγνοστικον* (resp.: Nicolaus Witsen).
1666 Sept. 11: *Disputatio de cometarum significatio* (resp.: Moesmannus Dop).

Deventer

Gisbertus ab Isendoorn
1640: *Quaestionum philosophicarum de mundo* (resp.: Andreas Petri).

Duisburg

Christophorus Wittichius
1655 June 5: *Disputatio theologica de stylo Scripturae quem adhibet cum de rebus naturalibus sermonem instituit* (resp.: Jacobus Lehnhof). (H)

Rostock

Daniel Lipstorp

1651 Nov.: *Disputatio physica de coelo & ejus partibus* (resp.: Vincentius Arnoldi). (w)

De systemate mundi copernicano, discursus physico mathematicus (w). Series of six disputations:

1652 Jan. 17: Συν θεω *De Copernicano mundi systemate, et terrae motu triplici, diurno, annuo, et illo, qui fit in seipsum, quique terrae axem semper sibi ipsi parallelum servat, exercitatio philosophica prima. Qua ostenditur, illam pythagoricorum opinionem non esse adeo improbabilem & paradoxam, sicut vulgo appaparentiis* [sic!] *coelestibus desumptis, & physicis rationibus propugnatur, suo modo Ptolemaicae praeferendam, ac probabilem esse.* (resp.: Henricus Loff).

1652 Jan: idem, *secunda* (resp.: Balthasar Jawert).

1652 Jan: idem, *tertia* (resp.: Gerardus Elers).

1652 Feb.: idem, *quarta* (resp.: Christophorus Prein).

1652 Feb.: idem, *quinta* (resp.: Balthasar Ripenow).

1652 Feb.: idem, *sexta & ultima* (resp.: Wernerus Releffs).

b. *Others*

N.N., *Epistola apologetica medicorum Mittelburgensium, qua & nothis P. Lansbergij literis respondent et a calumnia Palinodiae ipsius ab eo & filio inusta purgantur* (Middelburg, s.a.).

N.N., *Cartesius renatus, ofte naakte ontblooting der nijdige domme-kracht: onder den schijn van Gods-ijver, tegen de leere van Descartes verborgen: en eerst in 't Hollands de gemeente bekend gemaakt door J. du Bois* (s.l., 1656).

N.N., *Magni Cartesii manes ab ipsomet defensi; sive N.V. Renati Descartes Querela apologetica... Opusculum antea ineditum, nunc vero opponendum qoutidianiis Voetii & Voetianorum criminationibus, iis nominatim quas sub Theologiae Naturalis Reformatae titulo haud ita pridem emiserunt* (s.l., 1656).

N.N., *Over de woorden van vader Lodesteen van dat vervloekte, hels, duyvels, en ziel-verderflick slapen...* (s.l., 1656).

N.N., *De beweeging der aarde, in twee korte en bondige betoogen. Waar in, in het eerste beweezen, en met de Schriftuur volkomen over een gebragt word, dat de Aarde zig om zynen as wenteld. In het tweede dat de Aarde jaarlyks om de zon loopt, en niet om de aarde, zoo dat niet de aarde, maar de zon het middelpunt onzes weereldstelzel is. Waar by nog een kort vertoog. Waar uit men, door beschouwing van't geheelal, leeren kan des Scheppers grootheid en des menschen geringheid* (Rotterdam 1772). (A)

'Alaricus', *Zonnenloop en stand der aarde. Volgens de zekerheid der leere Salomons Pred. 1.4. De aarde staat in bestendigheyt. vs. 6 (De zonne) zy gaat naa het zuiden, ende zy gaat omme naa't noorden enz: Tegens over het Systema Copernicus, als dat van Pthólomie*

diende tot een confessionair berigt (mogt zyn tot overtuiginge) van den Hr: Stuid: opsteller van het vermaarde tractaatje, pseud: stuid: omtrent desselfs spot-vonnis over Mr: Alaricus, Boeren-Philosophie, of ongeleerde aanmerkingen anthi. Renatus Dubitatie of twyffelleer: zonder hem ooyt gelezen, of onderzogt te hebben enz. (Groningen 1738). (G)
'Domingo Gonzales', *Het tweede deel van de man in de maen: ofte een kort verhael, van een reyse, ghedaen in de planeet Marcurio* (The Hague 1651). (W)
'Irenaeus Philalethius' [pseud. of Abraham Heidanus], *Bedenckingen, op den staat des geschils, over de cartesiaensche philosophie, en op de nader openinghe over eenige stucken de theologie raeckende* (Rotterdam 1656) [a].
–, *De overtuigde quaetwilligheidt van Suetonius Tranquillus, blijkende uit seker boeckjen genaemt Den overtuigden cartesiaan* (Leiden 1656) [b].
'Irenaeus Philalethius, de tweede van die naem', *Den gebolden astronomus I.G.H. swymelende in syn mis-konstich bewys van wegen des Aertrijcx draey* (Leiden 1656).
'Irenaeus Philalethius, de derde van die naem' [= Johannes Hudde?], *De verstrickte astronomus Jacobus du Boys, die, terwijl hy poocht den strick van't wiskonstigh bewys van syne onnoselheyt te ontworstelen, sich hoe langer hoe vaster daer in verwert, dienende tot antwoort op den gebolden astronomus* (Leiden 1656).
'Liberius Modestinus Philosophus', *Specimen philosophiae cartesianae, thesibus aliquot expressum* (Utrecht 1656).
'Suetonius Tranquillus' [= Johannes Hoornbeek ?], *Staat des geschils, over de cartesiaansche philosophie* (Utrecht 1656) [a].
–, *Nader openinge van eenighe stucken in de Cartesiaensche philosophie raeckende de H. Theologie* (Leiden 1656) [b].
–, *Den overtyghden cartesiaen, ofte clare aenwysinghe uyt de bedenckingen van Irenaeus Philalethius, dat de stellingen en allegatien inde nader openinge tot lastte der cartesianen, in saecken de theologie raeckende, nae waerheyt en ter goeder trouwe zijn by een gebracht* (Leiden 1656) [c].
–, *Verdedichde oprechticheyt van Suetonius Tranquillus, gestelt tegen de overtuyghde quaetwilligheyt van Irenaeus Philalethius* (Leiden 1656) [d].
I.S. [Jacob Schuts], *Vermakelyke boere-praat, over het loopen of niet loopen van de son, en het draeijen en niet draeijen van de aarde. Tusschen Jaap, Kees, en Piet* (Amsterdam 1714).
J.S.K.B. [Jacob Schuts], *De vermakelyke boere-praat over het lopen van de son, en het stilstaan van de aarde, tegen C. Mennink verdedigt: nevens een ernstige waarschouwing voor desself sielverderffelykste ketterye* (Amsterdam 1714).
J.J., *Brief aan een vrint, tegens het nieuw uitgekomen boekje, getiteleert Het schriftuurbestorment gevoelen der cartesianen, omtrent het draeien of omloop des aertryks, en de stilstant der zonne, door Petrus de Laet, S.S.Th.St.* (Rotterdam 1723).
Acta der particuliere synoden van Zuid-Holland 1621-1700, W.P.C. Knuttel ed. (6 vol., The Hague 1908-1916).
Aelbert Hendricksz, see Haeyen, Aelbert.

Aitzema, Lieuwe van, *Historie of verhael van saken van staet en oorlogh, in, ende omtrent de Vereenigde Nederlanden*, IX [1657-1660] (The Hague 1664).
Amstel, Salomon van [pseud. of Daniel van Henghel], *Vredes-drangh, ofte accuraet ende leersaem discours aengaende de dispuyten ende onlusten in de kercke tot Utrecht* (Harlingen 1662).
Anderson, Alexander, *Vindiciae Archimedi. Sive elenchus cyclometriae novae, a Philippo Lansbergio nuper editae* (Paris 1616). (w)
Baillet, Adrien, *La vie de Monsieur Des-Cartes* (2 vol., Paris 1691).
Barlaeus, Caspar, *Oratio, de coeli admirandis, habita in Illustri Amstelodamensium schola, xi Feb. MDCXXXVI* (Amsterdam 1636).
Bartholin, Petrus, *Apologia pro observationibus, et hypothesibus... Tychonis Brahe... Contra... Martini Hortensii Delfensis criminationes et calumnias, quas in praefationem commentationum praeceptoris sui Philippi Lansbergii Middelburgensis, de motu terrae diurno et annuo etc. consarcinnavit* (Hafnia [Copenhagen] 1632).
[Batelier, Jacobus Johannes], *Examen accuratum disputationis primae & quasi inauguralis D. Gisberti Voetii, quam proposuit in illustri gymnasio Ultrajecti die 3. Sept. stylo vet. Anno 1634* (s.l. 1634).
Beaufort, Ludovicus, *Cosmopoea divina, seu fabrica mundi explicata* (Leiden 1656).
Beeckman, Isaac, *Journal tenu par Isaac Beeckman de 1604 à 1634*, C. de Waard ed. (4 vol., The Hague 1939-1953).
Bekker, Balthasar, *Friesche godgeleerdheid* (Amsterdam 1693).
Bibliothèque ancienne et moderne 1703-1713.
Blaeu, Willem Jansz., *Tweevoudigh onderwiis van de hemelsche en aerdsche globen; het een na de meyning van Ptolemaeus met een vasten aerdkloot; het andere na de natuerlijcke stelling van N. Copernicus met een loopenden aerdkloot* (Amsterdam 1634).
–, *Institutio astronomica de usu globorum & sphaerarum caelestium ac terrestrium*, Latin translation by Martinus Hortensius (Amsterdam 1634).
–, *Institution astronomique de l'usage des globes et sphères celestes & terrestres* (Amsterdam 1642).
Boerhaave, Herman, *Boerhaave's orations*, E. Kegel-Brinkgreve & A.M. Luyendijk-Elshout eds. (Leiden 1983).
Bois, Jacobus du, *Oude-tyds tyt-thresoor ende kerkelikke historie* (Leiden 1650).
–, *Dialogus theologico-astronomicus, in quo ventilatur quaestio astronomica, an Terra in centro universi quiescat, & Sol alaque luminaria coelestia circa eam moveantur: an vero, Sole quiescente, Terra circa eam feratur: et ex Sacris Literis Terrae quietem, Soli vero motum competere probatur: adjuncta refutatione argumentorum astronomicorum, quae in contrarium proferri solent* (Leiden 1653).
–, *Veritas et authoritas sacra, in naturalibus & astronomicis asserta & vindicata. Contra Christophori Wittichii... dissertationes duas, de S. Scripturae in philosophicis abusu. Adjuncta refutatione argumenti ab eodem authore ex Cartesii Principiis pro motu terrae desumpti* (Utrecht 1655).[a]

–, *Naecktheyt van de cartesiaensche philosophie, ontbloot in een antwoort op een cartesiaensch libel, genaemt Bewys, dat het gevoelen van die geene, die leeren der Sonne-stilstandt, en des Aerdtrijcks beweging niet strydig is met Gods Woort* (Utrecht 1655).[b]

–, *Schadelickheyt van de Cartesiaensche philosophie, ofte klaer bewijs, hoe schadelick die philosophie is, soo in het los maecken van Godes H. Woordt, als in het invoeren van nieuwe schadelicke leeringen. Tot antwoort op de tweede en vermeerderde druck van Doct. Velthuysens Bewys* (Utrecht 1656).[a]

[-], *Den ingetoomden Cartesiaen, ofte korte antwoordt op een Cartesiaensch libel, genaemt wiskonstich bewys van d'onnoselheyt, &c* (Leiden 1656).[b]

Bontekoe, Cornelis, *Alle de philosophische, medicinale en chymische werken* (Amsterdam 1689).

Brahe, Tycho, *Tychonis Brahe opera omnia*, J.L.E. Dreyer ed., 14 vol. (Copenhagen 1913-1928, repr. Amsterdam 1972).

Brakel, Wilhelmus à, *Λογικη λατρεία, dat is redelyke godts-dienst*, I (Rotterdam 1707).

Brandt, Geeraert, *Historie der reformatie, en andere kerkelyke geschiedenissen, in en ontrent de Nederlanden*, IV (Rotterdam 1704).

Braun, Johannes, *Doctrina foederum, sive systema theologicae didacticae & eclencticae* (Amsterdam 1688).

–, *Leere der verbonden ofte kort begrip der onderwysige en wederleggige godgeleerdheid* (Amsterdam 1694).

Briefe G.M. Lingelsheims, M. Berneggers und ihre Freunde, A. Reifferscheid ed. (Heilbronn 1889).

Bronnen tot de geschiedenis der Leidsche universiteit, P.C. Molhuysen ed., (7 vol., The Hague 1913-1924) (Rijks geschiedkundige publicatiën, large series, 20, 29, 38, 45, 48, 53, 56).

Burgersdijk, Franco, *Idea philosophiae naturalis sive methodus definitionum & controversiarum physicarum* (second edition, Leiden 1635).

–, *Collegium physicum, disputationibus XXXII absolutum; totam naturalem philosophiam compendiose proponens* (second edition, Leiden 1637).

Burman, Franciscus, *Synopsis theologiae, & speciatim oeconomiae foederum Dei, ab initio saeculorum usque ad consummationem eorum* (2nd edition, Utrecht 1681).

–, *Synopsis dat is kort begrip der Heilige Godgeleerdheid...* (2nd edition, Utrecht 1697).

Cardinael, Sybrand Hansz., *Mathematische ofte wisconstige bewijs-redenen, waer mede bewesen wort, dat de aerdcloot stil staet, en de sonne daghelijcx en jaerlijcx sijnen loop doet; naer de leeringhe van Ptolomaeus. Teghens het gevoelen van N. Copernicus, dat den aerdcloot sijnen daghelijckschen en jaerlijckschen loop soude doen, en de sonne stil staen in't midden vande vaste sterren-hemel* (Amsterdam 1635).

Cartesius renatus, ofte naakte ontblooting der nijdige domme-kracht: onder den schijn van Gods-ijver, tegen de leere van Descartes verborgen: en eerst in't Hollands de gemeente bekend gemaakt door J. du Bois (s.l. 1656).

Catalogus librorum qui excusi sunt typis Gisberti à Zyll. Catalogus vande boecken die Gysbert van Zyll heeft doen drucken, en die by hem te vinden zijn (Utrecht 1667). (w)
Catalogus, variorum & insignorum librorum, in quavis facultate materia & lingua praecipue vero mathematicorum & miscellaneorum: clarissimi viri Jacobi Coccaei, matheseos & linguarum oriental. ac Latinae peritissimi. Quorum auctio habebitur Haerlemi... 2 Martii, 1672 (Haarlem [1672]). (w)
Cellarius, Andreas, *Harmonica macrocosmica seu atlas universalis et novus, totius universi creati cosmographiam generalem, et novam exhibens (...)* (Amsterdam 1661).
[Cheyne, Georges], *Den schepper en Zyn bestier te kennen in Zyn schepselen; volgens het licht der reden en wiskonst. Tot opbouw van eerbiedigen godsdienst, en vernietiging van alle grondslag van atheïstery; als mede tot een regtzinnig gebruyk van de philosophie*, (Lambert ten Kate ed.) (Amsterdam 1716).
Clerc, Jean Le, *Opera philosophica*, III, *Physica*, I (Amsterdam 1700).
–, *Veteris testamenti libri historici (...); ex translatione Joannis Clerici; cum ejusdem commentario philologico, dissertationibus criticis, et tabulis chronologicis* (Amsterdam 1708)
–, *Veteris testamenti libri hagiographi...* (Amsterdam 1731).
Coccaeus, Jacobus, *Epistola de mundi, quae circumferuntur, systematis et novo alio illis certiore dialogismum paradoxum complexa* (Amsterdam 1660)a.
–, *Brief over de t'samen-stellinghen des werelts welcke in swangh gaen, ende over een nieuwe stellingh sekerder als de selve, een onghemeene bedenckinghe behelsende*, translated by I.K.V.W. [Johan Kies van Wissen] (Haarlem 1660)b.
Copernicus, Nicolaus, *Astronomia instaurata, libris sex comprehensa, qui de revolutionibus orbium coelestium inscribuntur*, Nic. Mulerius ed. (Amsterdam 1617).
– , *On the revolutions*, E. Rosen transl. (London etc. 1978).
Dee, John, 'A translation of John Dee's 'Monas hieroglyphica' (Antwerp, 1564), with an introduction and annotations', C.H. Josten ed., in: *Ambix* 12 (1964) 84-221.
Descartes, René, *Oeuvres*, C. Adam and P. Tannery eds. (2nd edition, 11 vol., Paris 1973-1978).
– *Verantwoordingh van Renatus Descartes aen d'achtbare overigheit van Uitrecht*, E.-J. Bos ed. (Amsterdam 1996).
Deurhoff, Willem, *Het voorbeeld der verdraagzaamheid onder de goddelyke bezoekingen, vertoond in de uitlegging en verklaaring van het boek Jobs* (2 vol., Amsterdam 1741).
Deusing, Anthonius, *Cosmographia catholica & astronomia secundum hypotheses Ptolemaei in concinnum, brevem, & perspicuum ordinem digesta* (Amsterdam 1642). (w)
–, *Dissertatio de vero systemate mundi, qua Copernici systema reformatur: sublatis interim infinitis pene orbibus, quibus in systemate Ptolemaico humana mens distrahitur* (Amsterdam 1643).
–, *Hexaemeron recognitum, seu de creatione meditationes, explicationibus christiano-philosophicis, et animadversionibus necessariis illustratae* (Harderwijk 1645).

Dodonaeus, Rembertus, *Cosmographica in astronomiam et geographiam isagoge* (Antwerp 1548, repr. Nieuwkoop 1963).
Doorman, G., *Octrooien voor uitvindingen in de Nederlanden uit de 16e-18e eeuw* (The Hague 1940).
Dousa, Janus (jr.), *Rerum caelestium liber primus* (Leiden 1591).
Drebbel, Cornelis, *Een kort tractaet van de natuere der elementen, ende hoe sy veroorsaecken, den wint, regen, blixem, donder, ende waeromme dienstich zijn* (Rotterdam 1621).
Effigies & vitae professorum academiae Groningae & Omlandiae; cum historiola fundationis ejusdem Acad: (Groningen 1654; repr. s.l. s.a.).
Duyn, Nicolaas, *Nutte en nodige oeffening; of kort vertoog, dat de natuur- en sterrekunde, en meer andere weetenschappen, den mensch opleiden en aanzetten tot de kenisse Gods en zyn dienst: als mede dat een meenigte schriftuurplaatsen door deeze weetenschappen worden opgeheldert* (Haarlem 1743).
Emmius, Ubbo, *Briefwechsel des Ubbo Emmius* H. Brugmans & F. Wachter eds. (2 dln., Aurich 1911 – The Hague 1923).
Engelhard, Nicolaas, *Kort doch ernstig ondersoek nopens het verschil op wat wyze de heilige Schriftuir met de onderstelling van de beweging der aarde moete overeen gebragt worden* (Groningen 1737).
Essenius, Andreas, *Synopsis controversiarum theologicarum recognita. Et index locorum totius S. Scripturae; quibus adversarii praecipue ,ad errores suos confirmandos, & Veritatem impugnandam vel declinandam abuti solent* (Amsterdam 1661).
Froidmont, Libert, *Anti-Aristarchus sive orbis terrae immobilis liber I* (Antwerp 1631).
–, *Vesta sive Ant-Aristarchi vindex. Adversus Jac. Lansbergium, Philippi f. medicum Mittelburgensem* (Antwerp 1634).
Fruytier, Jacobus, *Sions worstelingen, of historische samenspraken over de verscheide en zeer bittere wederwaardigheden van Christus kerke, met openbare en verborgene vyanden* (second edition, Rotterdam 1715).
Gassendi, Pierre, *Institutio astronomica iuxta hypotheseis tam veterum, quam Copernici, et Tychonis* (Paris 1647).
–, idem, second impression (The Hague 1656).
Gaukes, Yvo, *Dissertatio de medicina ad certitudinem mathematicam evehenda: continens certa hujus artis principia; & quomodo ex iis omnia mechanice, & methodo mathematica demonstrari possint. In ea quoque habentur diversae, cum aliorum, tum maxime Cartesii & Newtoni de rebus philosophicis sententiae sic, ut justo cuilibet veritatis arbitro apparere queat, uter ex his viris acutissimis felicius rem acu tetigerit* (Amsterdam 1712)
Geulincx, Arnold, *Opera philosophica*, J.P.N. Land ed. (3 vol., The Hague 1891-1893; repr. Stuttgart/Bad Cannstatt 1968).
Graaf, Abraham de, *De starre-kunst, leerende de hoedanigheden der beweginge van alle zichtbare hemelsche lighamen, en 't berekenen haarder zichtbare plaatzen. Mitsgaders,*

de hoedanigheden der verduistering van Zon en Maan, en de berekeninge van dien (Amsterdam 1659).
–, *Ontleding van de bril voor de Amsterdamze belachelijke geometristen* (Amsterdam 1663).
Graaf, Lieuwe Willemsz, *Korte, algemeene en zeedige wederlegging der schriften, driften, gedachten, stellingen en verstanden van veel sterre-kundigen, die al voor geruime tijd, (en nog al hedendaags) door den onvasten en menschelyke konst, het licht van Gods klare en ware woord trachten te benevelen, bysonder in der zonne stil-stand, en aartkloots beweeginge* (Amsterdam 1696).
Gravesande, Willem Jacob 's, *Physices elementa mathematica, experimenti confirmata. Sive introductio ad philosophiam Newtonianam* (2 vol., Leiden 1721).
–, *Oeuvres philosophiques et mathématiques*, J.N.S. Allamand ed. (2 vol., Amsterdam 1774).
–, *Welzijn, wijsbegeerte en wetenschap*, C. de Pater ed. (Baarn 1988).
Groenewegen, *Oefeningen over den Heidelbergschen Catechismus ofte gronden van de christelike God-geleerdheid, tot bevorderinge van de kennisse der waarheid die na de godzaligheid is* (seventh edition, Leiden 1706).
Grotius, Hugo, *Opera omnia theologica*, I, *Annotationes ad Vetus Testamentum* (Basel 1732).
Haeyen, Aelbert, *Een corte onderrichtinge belanghende die kunst vander zeevaert* (Amsterdam 1600).
Hartsoeker, Nicolaas, *Principes de physique* (Paris 1696).
–, *Beginsselen der natuur-kunde* (Amsterdam 1700).
–, *Eclaircissemens sur les conjectures physiques* (Amsterdam 1710).
Heereboord, Adrianus, *Philosophia naturalis, cum commentariis peripateticis antehac edita. Nunc vero hac posthuma editione mediam partem aucta, & novis commentariis, partim è Nob. Cartesio, Cl. Berigardo, H. Regio, aliisque praestantioribus philosophis, petitis, partim ex propria opinione dictatis, explicata* (Leiden 1663).
Heidanus, Abraham, *Corpus theologiae Christianae. In quindecim locos digestum* (Leiden 1686).
–, see (ps.) Irenaeus Philalethius.
Herbinius, Johannes, *Famosae, de Solis vel telluris motu, controversiae examen, theol.-phil., ad S. Script. normam* (Utrecht 1655).
–, *Dissertationes de admiranddis mundi cataractis supra & subterraneis* (Amsterdam 1678).
Herls, Cornelius, *Examen tractatus medici de moscho a Davide Ultralaeo in lucem editi* (Middelburg 1613).
Holwarda, Johannes Phocylides, *Πανσεληνος ἐκλειπτικη διαυγαζουσα. Id est dissertatio astronomica quae occasione ultimi Lunaris anni 1638 deliquii manuductio sit ad cognoscendum I. Statum astronomiae, praesertim Lansbergianae II. Novorum phenomenωn exortum & interitum* (Franeker 1640).

–, *Friesche sterre-konst, ofte een korte, doch volmaeckte astronomia, met de nuttigheden van dien* (Harlingen 1651-1652).
Hooft, Cornelis, *Memoriën en adviezen van Cornelis Pieterszoon Hooft*, H.A.E. van Gelder ed. (2 vol., Utrecht 1871-1925) (Werken uitgegeven door het Historisch Genootschap, gevestigd te Utrecht, second series 16, third series 48).
Hooft, Pieter, *De briefwisseling van Pieter Corneliszoon Hooft*, H.W. van Tricht ed., III, *1638-1647* (Culemborg 1979).
Hoornbeek, see (ps.) Suetonius Tranquillus.
Horrebow, Petrus, *Operum mathematico-physicorum tomus tertius* (Copenhagen 1741).
–, *De zegepralende Copernicus of eene verhandelinge over het verschilzicht des jaarlykschen loopkrings, waar in uit een menigte van sterrekundige waarnemingen de beweginge des aardkloots rondom de zon betoogt word*, translated and commented by Johannes Lulofs (Zutphen 1741).
Horrocks, Jeremias, *Opera posthuma*, John Wallis ed. (London 1673).
Hortensius, Martinus, *Responsio ad additiunculam D. Ioannis Kepleri, Caesarei mathematici, praefixam ephemeridi eius in annum 1624. In quae cum de totius astronomiae restitutione, tum imprimis de observatione diametri Solis, fide tubi dioptrici, eclipsibus utriusque luminaris, luculenter agitur* (Leiden 1631).
–, *Dissertatio de Mercurio in Sole viso et Venere invisa: Instituta cum ... D. Petro Gassendo...* (Leiden 1633).
–, 'Dissertatio de studio mathematico recte instituendo', in: H. Grotius et al., *Dissertationes de studiis instituendis* (Amsterdam 1645).
[Hudde, Johannes], *Demonstratio mathematica ineptiarum & ignorantiae Jacobi du Bois, ecclesiastae Leydensis, in oppugnanda hypothesi Copernicana, et philosophia Cartesiana* (Rotterdam 1656) [a].
[-], *Wiskonstigh-bewys der onnoselheyt van Jacobus du Bois, predicant tot Leyden: in het bestrijden van de hypothesis van Copernicus, en de philosophie van Des Cartes* (Rotterdam 1656) [b].
[-], *Den hollenden astronomus J.D.B. gecapuchont. Zijnde een antwoord op een blauw boexken genaemt Den ingetoomden Cartesiaen* (Leiden 1656) [c].
[See also Irenaeus Philalethius, *de derde van die naam*.]
Hues, Robert, *Tractatus de globis, coelesti et terrestri eorumque usu*, Johannes Isacius Pontanus ed. (Amsterdam 1617).
Huygens, Christiaan, *Wereldbeschouwer, of gissingen over de hemelsche aardkloten, en derzelver cieraad* (third impression, Amsterdam 1754; repr. Utrecht 1989).
Huygens, Constantijn, *Gedichten*, J.A. Worp ed., IV, *1644-1652* (Groningen 1894).
Jacchaeus, Gilbertus, *Institutiones physicae* (Leiden 1624).
–, idem, Amsterdam 1644.
Jacob Jansz., *Bewys redenen, verthoonende door onfeilbare gronden, dat niet den aerdtkloot*

door den hemel loopt maer de sonne, ende dat niet de sonne in 't centrum der werelt leyt, maer den aerdtkloot, krachtelijck bewesen door de heylige Schrifture, door de wercken der Nature, ende door ons eygen gesicht (Dordrecht 1688).

Jungius, Joachim, *Des Dr. Joachim Jungius aus Lübeck Briefwechsel mit seinen Schülern und Freunden. Ein Beitrag zur Kenntniss des grossen Jungius und der wissenschaftlichen wie socialen Zustände zur Zeit des dreissigjährigen Krieges*, R.C.B. Avé-Lallemant ed. (Lübeck 1863).

Keill, John, *Inleidinge tot de waare natuur- en sterrekunde, of de natuur- en sterrekundige lessen...*, translated and annotated by Johannes Lulofs (Leiden 1741).

Kepler, Johannes, *Opera omnia*, Ch. Frisch ed. (8 parts in 9 vol., Frankfurt and Erlangen 1858-1871)

–, *Gesammelte Werke*, M. Caspar etc. ed. (München, 1937-).

Koelman, Jacobus, *Het vergift van de cartesiaansche philosophie grondig ontdekt* (Amsterdam 1692).

Kyper, Albert, *Institutiones physicae* (2 vol., Leiden 1645-1646).

Laat, Petrus de, *Het schriftuur-bestormend gevoelen der cartesianen, omtrent het draaien of omloop des aardryks, en de stilstand der sonne* (Rotterdam 1723).

Langendijk, Pieter, *De wiskunstenaars of 't gevlugte juffertje*, G.W. Wolthuis ed. (Amsterdam s.a.)

Lange, Jan Jansz. de, *'t Samen-spreeckinge gehouden tusschen een schipper ende stierman, aengaende van't beweghend' firmament ende de son, of de selve stil staet, dan of dese aerdekloot haer beweecht, ende alle 24. uyren eens omgewent wert. Seer vermaeckelijck om lesen. Bewesen deur astronomische fondament-wiskonstige argumenten ende natuyrlijkcke redenen, tot lof vande navigacy ofte zeevaert* (The Hague 1636).

Lansbergen, Jacob, *Apologia, pro commentationibus Philippi Lansbergii in motum terrae diurnum & annuum: adversus Libertum Fromondum theologum Lovaniensem; & Joan. Baptistam Morinum, Doct.Med. & Parisiis mathematum professorem regium* (Middelburg 1633).

Lansbergen, Philips, *Triangulorum geometriae libri quatuor; in quibus nova & perspicua methodo, & αποδειξει, tota ipsorum triangulorum doctrina explicatur* (Leiden 1591). (w)

–, idem, second edition (Amsterdam 1631).

–, *Catechesis religionis Christianae quae in Belgii et Palatinatus ecclesijs docetur, sermonibus LII. explicatur* (Neostadii Palatinorum [= Neustadt an der Weinstrasse] 1595).

–, *Catechismus ofte onderwijsinghe inde Christelijcke religie...*, Johannes Gys transl. (third edition, Amstelredam 1645).

–, *Disputatio epistolaris... cum Carolo Franco... de moscho (...)* (Middelburg 1613)

–, *Cyclometriae novae libri duo* (Middelburg 1616)

–, idem (Middelburg 1628).

–, *Progymnasmatum astronomiae restitutae liber I. De motu solis* (Middelburg 1619).

–, idem (Middelburg 1628).
–, *Bedenckingen op den dagelyckschen, ende iaerlijckschen loop van den Aerdt-kloot. Mitsgaders op de ware af-beeldinge des sienelijcken hemels; daerin de wonderbare wercken Godts worden ontdeckt* (Middelburg 1629).
–, idem, second edition (Middelburg 1650).
–, idem, third edition, (Middelburg 1666).
–, *Commentationes in motum terrae diurnum, & annuum; et in verum adspectabilis caeli typum. In quibus επιστημονικως ostenditur, diurnum, annuumque motum, qui apparet in Sole, & Caelo, non deberi Soli, aut Caelo, sed soli Terrae: simulque adspectabilis primi Coeli typus, ad vivum exprimitur. Ex Belgico sermone in Latinum versae, a Martino Hortensio Delfensis: una cum ipsius praefatione, in qua astronomiae Brahaeanae fundamenta examinantur; & cum Lansbergiana astronomiae restitutione conferuntur* (Middelburg 1630).
–, *Uranometria libri III* (Middelburg 1631).
–, *Tabulae motuum coelestium perpetuae; ex omnium temporum observationibus constructae, temporumque omnium observationibus consentientes. Item novae et genuinae motuum coelestium theoricae et astronomicarum observationum thesaurus* (Middelburg 1632).
–, *Les tables perpétuelles de Philippe Lansbergue, des mouvemens celestes, construictes suivant les observations de tous temps. Ensemble les théories urayes & nouvelles des mouvemens celestes. Et le thresor d'observations astronomiques*, D. Goubard transl. (Middelbourg 1633).
–, *Opera omnia* (Middelburg 1663).
Leenhof, Frederik van, *De keten der bybelsche godgeleerdheit, soo als die in haar draad en samen-hang van de eerste waarheid af, door alle de wegen Gods aan een geschakeld is* (third edition, Amsterdam 1684).
Leeuwen, Simon van, *Batavia illustrata, ofte verhandelinge vanden oorspronk, voortgank, zeden, eere, staat en godtsdienst van oud Batavien, mitsgaders van den adel en regeringe van Hollandt* ('s Gravenhage 1685).
Leeuwenhoek, Antoni van, *Alle de brieven van Antoni van Leeuwenhoek / The collected letters of Antoni van Leeuwenhoek*, XI, *(1695-1696)*, L.C. Palm ed. (Lisse 1983).
Leydekker, Melchior, *Fax veritatis, seu exercitationes ad nonnullas controversias quae hodie in Belgio potissimum moventur, multa ex parte theologico-philosophicae* (Leiden 1677).
–, *Idea theologiae reformatae, ex scriptis theologorum, maxime Ultrajectensium delineata* (Utrecht 1696).
Liens, Cornelis, *Cum adversariis D.P. Lansbergij amica consertatio epistolica.* (Zierikzee 1614). [a]
–, *Mittelburgensium medicorum responsi postliminij & epistolae apologeticae ex superabundanti cautela refutatio pro D.P. Lansbergio* (Zierikzee 1614). [b]
–, *Eerste en tweede deel van de kleyne werelt: vervatende de verborgen oorsaack der minne,*

onderhandelinge van de ziele, en waare proef der selviger onsterffelijkheyt (Amstelredam 1655).
Lipsius, Justus, *Justi Lipsi Epistolae*, A. Gerlo etc. ed. (6 vol. appeared, Brussels 1978-).
Lipstorp, Daniel, *Specimina philosophiae Cartesianae. Quibus accedit ejusdem authoris Copernicus redivivus* (Leiden 1653).
Lulofs, Johan, *Onzijdige, dog vrijmoedige aanmerkingen, op het kortbondig onderzoek en berigt. Dienende tot een rechtelyk bewys dat de hedendaagsche zamenkomsten ende oeffeningen van de zogenaamde fynen ongeoorloft zyn enz.: Voor eenige dagen uitgegeven door mr. Johan Schrassert rechtsgeleerde* (Zutphen 1741).
–, *Inleiding tot eene natuur- en wiskundige beschouwinge des aardkloots* (Leiden and Zutphen 1750).
Luyts, Joannes, *Astronomica institutio. In qua doctrina sphaerica atque theorica, intermixto usu sphaere coelestis, & variis chronologicis, pertractantur* (Utrecht 1689).
Makreel, Dirck, *De lichtende leydt-starre der grote zee-vaert. Zynde een klare, en fondamentele verhandelinge, van alle de voornaemste stucken der navigatie* (Amsterdam 1671).
–, *Waernemingen en consideratien; aengaende de staert-starre, of vuurbalck, soo sich t'zedert eenige tijdt herwaert aen den Hemel heeft vertoont* [Amsterdam? 1681] (= Knuttel nr. 11795).
Maresius, Samuel, *Oartio inauguralis de usu et abusa rationis in rebus theologicis & fidei* (Groningen 1643).
–, *Hydra Socinianismi expugnata: sive Johannis Volkelii Misnici, De vera religione, (ut falso inscribuntur) libri quinque (...) cum eorundem refutatione...*, I (Groningen 1651).
–, *Sylloge disputationum aliquot selectiorum* (2 vol., Groningen 1660-1663).
–, *De abusu philosophiae Cartesianae, surrepente & vitando in rebus theologicis & fidei, dissertatio theologica* (Groningen 1670).
Mastricht, Petrus van, *Vindiciae veritatis et authoritatis Sacrae Scripturae in rebus philosophicis adversus dissertationes D. Christophori Wittichii* (Utrecht 1655).
–, *Novitatum Cartesianarum gangraena, nobiliores plerasque corporis theologici partes arrodens & exedens. Seu theologia Cartesiana, detecta* (Amsterdam 1677).
–, *Theoretico-practica theologia, qua, per singula capita theologica, pars exegetica, dogmatica, elenchtica & practica, perpetua successione conjugantur* (expanded second edition, Utrecht 1698).
Mauricius, Joan Jakob, *Onledige ouderdom. Behelzende taallievende en historische uitspanningen over 't gebruik en overneeming van vreemde woorden* (2 vol., Amsterdam 1765-1766).
Megerlin, Petrus, *Systema mundi Copernicanum, argumentis invictis demonstratum, & conciliatum theologiae* (Amsterdam 1682).

Meij, Johannes de, *Sacra physiologia, sive expositio locorum Sacrae Scripturae, in quibus agitur de rebus naturalibus* (Middelburg 1655).
–, *Halelu-jah. Of lof des heeren verbreidt op het natuir-toneel, kortelijk en grondelijk vertoonende den ganschen aerdt, oirspronck, gestalte, eigenschappen, werkingen, en uitnemenste geheimenissen des weerelds, ende deszelfs bysonderste deelen, in 't gemeen aen-gemerkt* (Middelburg 1667). [a]
–, *Het tweede halelu-jah, ofte lof des Heeren, verbreyt op het genaden-toneel* (Middelburg 1667). [b]
–, *Alle de Neder-duitsche wercken* (Middelburg 1681).
Mersenne, Marin, *Quaestiones celeberrimae in Genesim, cum accurata textus explicatione* (Paris 1623).
– *Correspondance du P. Marin Mersenne, religieux minime*, P. Tannery, C. de Waard & A. Beaulieu eds., (16 vol., Paris 1932-1986).
Merula, Paullus, *Cosmographiae generalis libri tres: item geographiae particularis libri quator: quibus Europa in genere; speciatim Hispania, Gallia, Italia, describuntur* (Leiden 1605).
Metius, Adriaan, *Institutiones astronomicae & geographicae, fondamentale ende grondelijcke onderwysinge van de sterrekonst, ende beschryvinghe der aerden, door het ghebruyck van de hemelsche ende aerdtsche globen* (Franeker 1614).
–, *Primum mobile, astronomice, sciographice, geometrice, & hydrographice, nova methodo explicatum. Editio nova* (Amsterdam 1633).
Montanus, Janus, *Den oorlog der philosophen, of de beroemde veltslag geslaagen tussen de vermaarde Aristoteles en den grooten Descartes* (s.l., 1701).
Moolen, Simon van der, *Astronomia of hemel-loop-kunde, leerende de hoedanigheden der beweginge van alle zigtbare hemel-ligten (...)* (Amsterdam 1702).
Mulerius, Nicolaus, *Tabulae Frisicae Lunae-Solares quadruplices: e fontibus Cl. Ptolemaei, Regis Alfonsi, Nic. Copernici, & Tychonis Brahe... Quibus accessere Solis tabulae totidem; hypotheses Tychonis illustratae: Kalendarium Rom. vetus cum methodo Paschali emendata* (Alkmaar 1611).
–, *Institutionum astronomicarum libri duo: quibus etiam continentur geographiae principia, nec non pleraque ad artem navigandi facientia* (Groningen 1616).
–, idem, reprint Groningen 1649.
–, *Hemelsche trompet morgenwecker ofte comeet met een langebaart* (Groningen 1618).
–, *Iudaeorum annus lunae-solaris: et Turc-Arabum annus mere lunaris. Recens uterque e suis fontibus deductus, & cum anno Romano facili methodo connexus* (Groningen 1630).
Musschenbroek, Petrus van, *Beginselen der natuurkunde, beschreven ten dienste der landgenooten* (Leiden 1736).
Muys, Weyer Willem, *Elementa physices methodo mathematica demonstrata* (Amsterdam 1711).

Newton, Isaac, *Mathematical principles of natural philosophy*, A. Motte & F. Cajori transl. and ed. (2 vol., Berkeley etc. 1962).

–, 'Further Newton correspondence', A.R. Hall ed., *Notes and records of the Royal Society of London* 37 part 1 (1982), 7-34.

Niepoort, Arnoldus, Johannes Beusechum and Henricus Troy, *Disputationes theologicae quator, de usu Sacrae Scripturae in rebus philosophicis, contra Christophori Wittichii dissertationes. Accedunt vindiciae pro iisdem disputationibus adversus illius considerationem theologicam de stylo Scripturae* (Utrecht 1656).

Nierop, Dirk Rembrandtsz. van, *Nederduytsche astronomia, dat is: onderwys van den loop des heemels...* (second edition, Amsterdam 1658).

–, *Des aertrycks beweging, en der sonne stilstant, bewysende dat dit geensins met de Christelyke religie is strydende. (...)* (Amsterdam 1661). [a]

–, *Idem*, second impression (Amsterdam 1683).

–, *Antwoort op een brief van Jacobus Coccaeus, over de 't samenstellingen des werrelds. Waer in dat wederleyt wort zijn vreemde ongefondeerde stellingh, nu onlanghs by hem voortgebracht: waer meê dat hy alle oude fondamenten gaet verwerpen, om het zijne staende te houden* (Amsterdam 1661). [b]

–, *Eenige oeffeningen, in god-lijcke, wis-konstige, en natuerlijcke dingen (...)* (Amsterdam 1674).

–, *By-voeghsel op de Nederduytse Astronomia en sonne stilstant, welcke zijn eenige aanteyckeningen dienende tot verbeteringh en vermeerdering des selfden, en dat so wel op d'eerste als tweede Druck deser Astronomia. Waer mee dat die vermeerdert zijn met een nieuwe uytreeckening op de planete Mercurius, voortkomende uyt de twee waergenomen samenstanden in de Sonne, gedaen by P. Gassendi en J. Hevelius* (Amsterdam 1677). [a]

–, *Bijvoeghsel op des aertryks beweging, of de sonne stilstant, waer in dat eenige dingen my soo nu ende dan voorgevallen sijn, dienende tot nader verklaringhe: als oock eenige tegenwerpingen beantwoordt* (Amsterdam 1677). [b]

–, *Tweede deel, op de wis-konstige rekening, welke sijn eenige aenteykeningen, dienende tot verbetering en vermeerdering des selfden (...)* (Amsterdam 1680).

–, 'Dirk Rembrantsz brieven', in: *Maandelijke mathematische liefhebberij*, 68 letters in the issues 1755-1762.

Nieuwentijt, Bernard, *Het regt gebruik der wereltbeschouwingen, ter overtuiginge van ongodisten en ongelovigen* (Amsterdam 1715).

Odé, Jacob, *Oratio de laudabili priscorum hominum philosophandi methodo* (Utrecht 1723).

–, *Principia philosophiae naturalis in usum scolarum privatarum conscripta, et captui studiosae juventutis accomodata* (2 vol., Utrecht 1727).

–, *Oratio de novis per observationes coelestes recentioris aevi inventis* (Utrecht 1743).

Outrein, Johannes d', *Proef-stukken van heilige sinne-beelden* (2 vol., Amsterdam 1700-1701).

Ouwens, Rutgerus, *Redenvoering, betogende, dat de stelling van Ptolemaeus, aangaande het lopen der sonne strydig is tegen de H. Schrift, of dat men het lopen der sonne, na de stelling van Ptolemaeus, verkeerdelyk poogd te bewyzen uit Josua x: v. 12.13.* ('s Gravenhage 1779). (L)

[Poest, Andries van der], *Aanmerkingen over het gestelde van Copernicus wegens de jaarlykse beweeging van den aardkloot om de zon* (Middelburg 1778). (KB)

Polacco, Giorgio, *Anticopernicus catholicus, seu de terrae statione, et de solis motu, contra systema copernicanum, catholicae assertiones* (Venice 1644).

Pontanus, Johannes Isacius (ed.), *Aur. Theodosii Macrobii opera* (Leiden 1597).

–, *Brieven van en aan Jo. Is. Pontanus*, P.N. van Doornick and P.C. Molhuysen ed. (Haarlem 1909).

Ptolemy, *Ptolemy's Almagest*, G.J. Toomer ed., (London 1984).

Querelle d'Utrecht, La, Th. Verbeek ed. (Paris 1988).

Raei, Johannes de, *Oratio inauguralis, de gradibus & vitiis notitiae vulgaris circa contemplationem Naturae, & officio Philosophi circa eandem* (Leiden 1651).

–, *Clavis philosophiae naturalis seu introductio ad naturae contemplationem, Aristotelico-Cartesiana* (Leiden 1654).

Reformierten klevischen Synoden im 17. Jahrhundert, Die, II, *1649-1672*, W. Petri ed. (Cologne 1979).

Ranouw, Willem van, *Kabinet der natuurlijke historiën, wetenschappen, konsten en handwerken* (Amsterdam 1719-1723).

Register van Holland en Westvriesland van de jaaren 1655 en 1656. (s.l., s.a.).

Regius, Henricus, *Fundamenta physices* (Amsterdam 1646).

–, *Fundamenta medica* (Utrecht 1647).

–, *Philosophia naturalis* (Amsterdam 1654).

–, [idem, new edition] *Philosophia naturalis, in qua tota rerum universitas, per clara & facilia principia, explanatur* (Amsterdam 1661).

–, *Conciliatio locorum S.Scripturae cum diurna et annua telluris circumrotatione* (Utrecht 1658).

Reland, Adriaan, *Oratio de incremento, quod philosophia cepit hoc saeculo* (Amsterdam 1700).

Revius, Jacob, *Methodi cartesianae consideratio theologica* (Leiden 1648).

–, *Statera philosophiae Cartesianae, qua principiorum ejus falsitas, & dogmatum impuritas expenditur ac castigatur, et ad virulentam epistolam, praefixam selectarum disputationum Adriani Heereboortii volumini primo, respondetur* (Leiden 1650).

–, *Anti-Wittichius. Sive, adversus infinitatem mundi, a Cartesio assertam, a Christophoro Wittichio theologo Duisbergensi defensam diatribe* (Leiden 1655).

Riccioli, Giovanni Battista, *Almagestum novum astronomiam veterem novamque complectens* (2 vol., Bologna 1651).

Ridderus, Franciscus, *Schriftuerlyck licht, over schijn-strijdende, duystere, en mis-duyde texten der Heylige Schrifture* (5 vol., Rotterdam and Amsterdam 1675-1680).

Roggeveen, Arent, *Het nieuwe droevige nacht-licht, ontsteken door Godts toren, ende ver-*

toont op den aerdt-kloot, in een comeet ofte staert-starre (...) waerin wort aegewesen den wonderlijcken loop der selve (Middelburg 1665).

Ryssenius, Leonardus, *Doot-stuypen der cartesianen en coccejanen* (Utrecht 1676).

[Santvoort, Dirk], *De oorsaak van de beweeging en de beginsselen der vaste lichamen* (Utrecht 1703).

[-], *Vervolg van het boek genaamt de oorsaak van de beweeging en de beginsselen der vaste lichamen* (Utrecht 1707).

Saurin, Jaques, *Discours historiques, critiques, theologiques, et moraux, sur les evenemens les plus memorables du vieux et du nouveau testament*, vol. I-II (The Hague 1728).

Scaliger, Joseph Justus, *De aequinoctiorum anticipatione diatriba* (Paris 1613)

–, *De re nummaria dissertatio, liber posthumus* (Leiden 1616).

Schagen, Gerrit Pietersz (ed.), *Wonder-vondt van de eeuwighe bewegingh die den Alckmaersche philosooph Cornelis Drebbel door een eeuwigh bewegende gheest in een cloot besloten te weghe ghebrocht heeft, welckers toeeygeningh (in 't vereeren des selvigen aen den grootmachtigen Coningh Jacob van groot Brittangen) alhier naecktelijck vertoont wordt. (...) Ooc mede by gevoeght een Boeck Pymander beschreven van Mercurius driemael de grootste die oock een philosooph en priester en Coningh soude in Egypten gheweest hebben in de tijdt van Moses* (Alkmaar 1607).

Schoock, Martin, *De scepticismo pars prior, sive libiri quator: quibus, qua antiquorum, qua recentiorum scepticorum deliria, ex suis principiis solide discutiuntur, atque certitudo non minus disciplinarum universalium, quam philosophiae theoreticae asseritur* (Groningen 1652).

–, *Physica caelestis. In qua omnia, quae in coelo siderio occurrunt, copiose discutiuntur. Non modo juxta antiquorum philosophorum placita, verum etiam recentiorum astronomorum accuratiores observationes* (Amsterdam 1663).

Schotanus à Sterringa, Joh., *Physica coelestis et terrestris. Inb quae secundum fundamenta melioris philosophiae, facili & jucunda methodo, coeli terraeque affestiones, et φαινομενα explicantur* (Franeker 1700).

Semeyns, Meindert, *Kortbondige redeneering over de gesteltheid van den aardkloot van binnen, betoogt uit de magneetische werkinge* (The Hague 1760).

Senguerd, Arnold, *Physicae exercitationes* (Amsterdam 1658).

Senguerd, Wolferd, *Philosophia naturalis, quator partibus primarias corporum species, affectiones, vicissitudines, & differentias, exhibens* (Leiden 1680).

–, *idem*, editio secunda, priore auctior (Leiden 1685).

Smetius, Henricus, *Miscellanea medica, cum praestantissimis quinque medicis (...) communicata* (Frankfurt 1611).

Smit, Pieter Maasz, *Cosmographia, of verdeelinge van de geheele wereld, als mede het maken van de hemelsche en aardsche globe: (...)* (Amsterdam, 1698) (UvA)

Smith, Writser Roelefs de, *Naturlikke ontdekking of demonstratif bewys I. Van de grootheid, naturlikke plaats, en beweeging der aard-kloot. II. Van de grootheid, staat, naturlikke plaats, snelle loop en declinatie van 't lichaam der maan. III. Van de grootheid, naturlikke plaats, snelle loop en declinatie van 't lichaam der sonne.* IV. *Van de hooge distantie des firmaments of sternheemels, van de aard-kloot, en derselver snelle beweeging. Seer naakeurig ondersogt, maat-kundig beweesen en t'saamen gestelt* (Groningen 1722). (L)

Snellius, Willebrord, *Περι λογου ἀποτομες και περι χοριου ἀποτομες resuscitata geometria* (Lugodini [Leiden] 1607).

[-], *De maculis in sole animadversis, & tanquam ab Apelle, in tabula spectandum in publica luce expositis, Batavi dissertatiuncula* ([Leiden] 1612).

–, *Eratosthenes Batavus. De terrae ambitus vera quantitate* (Leiden 1617).

–, *Descriptio cometae, qui anno 1618 mense Novembri primum effulsit* (Lugduni Batavorum 1619).

Spinoza, Baruch, *Opera*, C. Gebhardt ed. (5 vol., Heidelberg 1925–1938).

Spoors, Jacob, *Oratie van de nieuwe wonderen des wereldts, de nuttigheyd, de waerdigheyd, der wis ende meet-konsten* (Delft 1638).

Stevin, Simon, *Wisconstige gedachtenissen, inhoudende t'ghene daer hem in gheoeffent heeft den doorluchtichsten hoochgheboren vorst ende heere, Maurits Prince van Oraengen* (Leyden 1605-1608).

–, *Les oevres mathematiques de Simon Stevin de Bruges. Ou sont inserées les memoires mathematiques, esquelles s'est exercé le Très-haut & Tres-illustre Prince Maurice de Nassau, Prince d'Aurenge*, translated by Albert Girard (Leiden 1634).

–, *The principal works of Simon Stevin* (5 vol., Amsterdam 1955-1966).

Stevin, Hendric, *Wisconstich filosofisch bedryf* (Leiden 1667).

–, *Plaetboec. Vervangende de figuren of formen gehorig tottet Wisconstich filosofisch bedryf* (s.l. 1663).

[Streso, Caspar], *Korte aenmerckingen op het onbewesen bewys dat het gevoelen vander Sonne stillestant ende des aertrijckx beweginghe niet strijdigh is met Godts-Woort, Aengeteykent tot onderdrukkinge van een opspringende wortel der bitterheyt tegens de autoriteyt vande H. Schrifture* (The Hague, 1656). [a]

–, *Dissertatio de usu philosophiae inter Christianos* (The Hague 1656). [b]

Ultralaeus, David, *Tractatus medicus per-brevis. In quo succincte ac disposite disputatur an puerperae liceat exhibere moschum* (Middelburg 1613).

'Témoignages sur l'ancienne université de Nimègue (1655-1671)', H. Bots ed., in: *Lias* 19 (1992) 215-253.

Ursinus, Zacharias, *Het schat-boeck der christelycke leere, ofte uytlegginge over de catechismus der gereformeerde kercken in Nederlandt*, David Pareus ed., Festus Hommius transl., 2nd edition, *'inde welcke bij gedaen is uyt Balt. Copius, Hier. Bastingius, Phil. Lansbergius, Georg Spindlerus, ende andere die over de Catechismum geschreven hebben, wat dient tot breeder verclaringhe'*, Festus Hommius ed. (2 vol., Leiden 1606).

Varenius, Bernhard, *Geographia generalis, in qua affectiones generales telluris explicantur* (Amsterdam 1650).

[Velthuysen, Lambert van], *Bewys, dat het gevoelen van die genen, die leeren der sonne stilstandt, en des aertrycks beweging niet strydich is met Godts-woort* (s.l. 1655).
–, *Bewys dat noch de leere van der sonne stilstant, en des aertryx bewegingh, noch de gronden vande philsophie van Renatus des Cartes strijdig sijn met Godts Woort. Gestelt tegen een tractaet van J. du Bois, predikant tot Leyden; genaemt Naectheyt vande Cartesiaensche Philosophie ontbloot, (Utrecht 1656).*
– *Nader bewys dat noch de leere van der sonne stilstant, en des aertryx beweging, noch de gronden van de philosophie van Renatus des Cartes strijdig sijn met Godts woort. Gestelt tegen een tractaet van J. du Bois, predikant tot Leyden; genaemt Schadelickheyt van de Cartesiaensche philsophie (Utrecht 1657).*
Verqualje, Jan, *Uytvindingh der langhte van oost en west (...)* (Amsterdam 1661).
–, *Klaer bewys, by 't welcke seckerlijck blijckt: ...* (Amsterdam 1663).
Voet, Daniel, *Physiologia*, second edition, Gerard de Vries ed. (Utrecht 1678).
Voet, Paulus, *Theologia naturalis reformata* (Utrecht 1656).
Voetius, Gisbertus, *Thersites heautontimorumenos. Hoc est, remonstrantium hyperaspistes, catechesi, et liturgiae Germanicae, Gallicae, & Belgicae denuo insultans, retusus* (Utrecht 1635).
–, *Sermoen van de nutticheydt der academien ende scholen/ mitsgaders der wetenschappen ende consten die in de selve gheleert werden* (Utrecht 1636).
–, *Selectarum disputationum theologicarum pars prima-quinta* (5 vol., Utrecht 1648-1696).
Vossius, Gerardus Joannes, *Universalis philosophiae κρωτηριασμος. Dispute soutenue à l'université de Leyde le 23 février 1598. Thèses et défense*, M. van Straaten ed. (Leiden 1955)
–, *De theologia gentili, et physiologia christiana, sive de origine ac progressu idololatriae* (2 vol., Amsterdam 1641).
Witt, Johan de, *Brieven van Johan de Witt*, 1, *1650-1657 (1658)*, R. Fruin and G.W. Kernkamp ed. (Werken van het Historisch genootschap (gevestigd te Utrecht), third series vol. 18) (Amsterdam 1906).
Wittichius, Christophorus, *Dissertationes duae quarum prior de S.Scripturae in rebus philosophicis abusu, examinat, 1, An physicae genuinam principium sit Scriptura? 2. An haec de rebus naturalibus loquens accuratam semper veritatem, an potius sensum & opinionem vulgi saepius sequatur? Altera dispositionem & ordinem totius universi & principalium ejus corporum tradit, sententiamque nobilissimi Cartesii, de vera quiete & vero motu terrae defendit* (Amsterdam 1653).
–, *Consideratio theologica de stylo Scripturae quem adhibet cum de rebus naturalibus sermonem instituit* (Leiden 1656).
– *Consensus veritatis in Scriptura Divina et infallibili revelatae cum veritate philosophica a Renato des Cartes detecta, cujus occasione liber II. & III. Principiorum Philosophiae dicti des-Cartes maximam partem illustrantur* (Nijmegen 1659).
–, idem, second edition (Amsterdam 1688).

–, *Theologia pacifica, in que varia problemata theologica inter reformatos theologos agatari solita ventilantur, simul usus philosophiae cartesianae in diversis theologiae partibus demonstratur, & ad dissertationem celeberrimi viri Samuelis Maresii, de abusu philsophiae cartesianae in rebus theologicis & fidei, modeste respondetur* (third edition, Leiden 1683).

Wyermars, Hendrik, *Den ingebeelde chaos, en gewaande werels-wording der oude, en hedendaagze wysgeeren, veridelt en weerlegt, byzonder de gevoelens hier omtrent, van T. Lucretius Carus en Dirk Santvoort* (Amsterdam 1710).

Secondary literature

Aarsen, P.J., *Tot Middelburgh by Richard Schilders*, in: *Zeeuws tijdschrift* 19 (1969) 104-108.

Aiton, E.J., *The vortex theory of planetary motions* (London and New York 1972).

Andersen, K., Stevin's theory of perspective: the origin of a Dutch academic approach to perspective, in: *Tractrix* 2 (1990) 25-62.

Andriesse, C.D., *Titan kan niet slapen. Een biografie van Christiaan Huygens* (Amsterdam and Antwerp 1993).

Applebaum and Hatch, Boulliau, Mercator and Horrocks's Venus in Sole visa, in: JHA 14 (1983)

Armogathe, J.-R., *La vérité des Ecritures et la nouvelle physique*, in: J.-R. Armogathe ed., *Le grand siècle et la bible* (Paris 1989) 49-60.

Asselt, W.J. van, *Amicitia Dei. Een onderzoek naar de structuur van de theologie van Johannes Coccejus (1603-1669)* (dissertation Utrecht 1988).

–, *Ultimum tempus nobis imminet. Eschatologische structuren van de theologie van Johannes Coccejus*, in: *Nederlands archief voor kerkgeschiedenis* 76 (1996) 189-226.

Averdunk, H, and J. Müller-Reinhard, *Gerhard Mercator und die Geographen unter seinen Nachkommen* (Gotha 1914; repr. Amsterdam 1969).

Baan, J. van der, Philippus en Petrus Lansbergen. *Eene bijdrage, als toevoegsel aan die van Dr. A.A. Fokker (...)* in: *Archief. Vroegere en latere mededeelingen voornamelijk in betrekking tot Zeeland*, VII (1869) 205-227.

–, *Kerkelijke procedure over de zaak van Philippus en Petrus Lansbergen*, in: *Kerkhistorisch archief* 4 (1866) 339-384.

Baldwin, M., 'Magnetism and the anti-Copernican polemic', in: JHA 16 (1985) 155-174.

Bailly, J.-S., *Histoire de l'astronomie moderne depuis la fondation de l'école d'Alexandrie, jusqu'à l'époque de 1730*, II (Paris 1779).

Bauer, B., *Die Rolle des Hofastrologen und Hofmathematicus als fürstlicher Berater*, in: A. Buck ed., *Höfischer Humanismus* (Weinheim 1989) 93-117.

Baumgartner, F.J., 'Scepticism and French interest in Copernicanism to 1630', in: JHA 17 (1986) 77-88.

Bennett, J.A., 'Cosmology and the magnetical philosophy, 1640-1680', in: *JHA* 12 (1981) 165-177.
Berg, J. van den, *Het stroomlandschap van de Gereformeerde Kerk in Nederland tussen 1650 en 1750*, in: Broeyer and Van der Wall (1994) 9-27.
Berkel, K. van, *Spiegheling en daet bij Beeckman en Stevin*, in: *TGGNWT* 2 (1979) 89-100.
–, *Universiteit en natuurwetenschap in de 17e eeuw, in het bijzonder in de republiek*, in: H.A.M. Snelders and K. van Berkel ed., *Natuurwetenschappen van Renaissance tot Darwin. Thema's uit de wetenschapsgeschiedenis* (The Hague 1981).
–, *Isaac Beeckman (1588-1637) en de mechanisering van het wereldbeeld* (Amsterdam 1983). [a]
–, *De geschriften van Rudolf Snellius. Een bijdrage tot de geschiedenis van het ramisme in Nederland*, in: *TGGNWT* 6 (1983) 185-194. [b]
–, *Descartes in debat met Voetius. De mislukte introductie van het cartesianisme aan de Utrechtse universiteit (1639-1645)*, *TGGNWT* 7 (1984) 4-18.
–, *In het voetspoor van Stevin. Geschiedenis van de natuurwetenschap in Nederland 1580-1940* (Meppel 1985).
–, 'Galileo in Holland before the *Discorsi*: Isaac Beeckman's reaction to Galileo's work',in: C.S. Maffioli and L.C. Palm ed., *Italian scientists in the Low Countries in the XVIIth & XVIIIth centuries* (Amsterdam/Atlanta 1989) 101-110.
–, *Alexandrië aan de Amstel? De illusies van Martinus Hortensius (1605-1639), eerste hoogleraar in de wiskunde in Amsterdam*, in: E.O.G. Haitsma Mulier et al. ed., *Athenaeum illustre. Elf studies over de Amsterdamse doorluchtige school 1632-1877* (Amsterdam 1997) 200-225.
– Berkel, K. van, L.C. Palm and A. Van Helden ed., *A history of science in the Netherlands. Survey, themes and reference* (Leiden 1999).
Berkvens-Stevelinck, C., *Nicolaas Hartsoeker contre Isaac Newton ou pourquoi les planètes se meuvent-elles?*, in *Lias* 2 (1975) 313-328.
Biagioli, M., 'The social status of Italian mathematicians, 1450-1600', in: *History of Science* 27 (1989) 41-95.
–, *Galileo courtier. The practice of science in the culture of absolutism* (Chicago and London 1993).
Bierens de Haan, D., *Bouwstoffen voor de geschiedenis der wis- en natuurkundige wetenschappen in de Nederlanden* (2 vol., Amsterdam 1878-1887).
–, *Bibliographie néerlandaise historique-scientifique des ouvrages importants dont les auteurs sont nés aux 16e, 17e et 18e siècles, sur les sciences mathématiques et physiques, avec leurs applications* (Rome 1883; reprint Nieuwkoop 1960).
Bisschop, R., *Sions vorst en volk. Het tweede-Israëlidee als theocratisch concept in de gereformeerde kerk van de Republiek tussen ca. 1650 en ca. 1750* (Veenendaal 1993).
Bijl, M. van der, *De tweedracht van voetianen en cocejanen in politiek perspectief*, in: Broeyer and Van der Wall (1994) 74-94.

Bizer, E., *Die reformierte Orthodoxie und der Cartesianismus*, in: *Zeitschrift für Theologie und Kirche* 55 (1958) 306-372.

Blackwell, R.J., *Galileo, Bellarmine and the Bible* (Notre Dame and London 1991).

Blok, F.F., 'Isaac Vossius and the Blaeus', in: *Quaerendo* 26 (1996) 77-84, 87-93.

Blotevogel, H.H., and R.H. Vermij, 'Einleitung: Wer war Gerhard Mercator?', in: idem, *Gerhard Mercator und die geistigen Strömingen des 16 und 17. Jahrhunderts* (Bochum 1995), DMS 3, 1-15.

Bos, E.P., and H.A. Krop (ed.), *Franco Burgersdijk (1590-1635). Neo-Aristotelianism in Leiden* (Amsterdam and Atlanta 1992).

Bodel Nyenhuis, J.T., 'Levensbijzonderheden van den Nederlandschen geschiedschrijver, Johannes Isacius Pontanus, meest geput uit deszelfs album', in: *Bijdragen voor vaderlandsche geschiedenis en oudheidkunde* 2 (1840) 81-109.

Boeles, J.J.G., *Vondel en de sterrenkunde* (Nijmegen 1906).

Boogaard, M. Th. uit den, *De gereformeerden en Oranje tijdens het eerste stadhouderloze tijdperk* (Groningen and Djakarta 1955).

Bots, H., 'Ismaël Boulliau, ses voyages en Hollande', in: *Lias* 1 (1974) 29-62.

Bots, J., *Tussen Descartes en Darwin. Geloof en natuurwetenschap in de achttiende eeuw in Nederland* (Assen 1972).

Bouman, H., *Geschiedenis van de voormalige Geldersche hoogeschool en hare hoogleeraren* (2 vol., Utrecht 1844-1847).

Braekman, W.L., 'Joos Balbiaen, dokter, alchemist en legerkapitein tijdens de Gentse calvinistische republiek (1576-1584)', in: *Handelingen der Maatschappij voor geschiedenis en oudheidkunde te Gent*, nieuwe reeks 40 (1986) 85-95.

Breugelmans, R., and E. Dekker, 'Adriaan Anthonisz and the Gregorian calendar', in: T. Croiset van Uchelen, K. van der Horst and G. Schilder, ed., *Theatrum orbis librorum. Liber amicorum presented to Nico Israel on the occasion of his seventieth bitrhday* (Utrecht 1989) 137-157.

Brockliss, L.W.B., *French higher education in the seventeenth and eighteenth centuries. A cultural history* (Oxford 1987).

– , 'Copernicus in the university: the French experience', in: J. Henry and S. Hutton ed., *New perspectives on Renaissance thought: Essays... in memory of Charles B. Smith* (London 1990) 190-213.

Broeyer, F.G.M., 'William III and the Church in Utrecht after the French occupation (1672-1673)', in: J. van de Berg and P.G. Hoftijzer ed., *Church, change and revolution* (Publications of the Sir Thomas Browne institute, Leiden, n.s., 12) (Leiden 1991) 179-191.

–, 'Franciscus Burman, een coccejaan in voetiaans vaarwater', in: Broeyer and Van der Wall ed. (1994) 104-130.

Broeyer, F.G.M., and E.G.E. van der Wall ed., *Een richtingenstrijd in de gereformeerde kerk. Voetianen en coccejanen 1650-1750* (Zoetermeer 1994).

Brotóns, V.N., 'The reception of Copernicus in sixteenth-century Spain. The case of Diego de Zúñiga', in: *Isis* 86 (1995) 52-78.
Brugman, J., 'Arabic scolarship', in: Th.H. Lunsingh Scheurleer and G.H.M. Posthumus Meyjes ed., *Leiden university in the seventeenth century: an exchange of learning* (Leiden 1975) 202-215.
Buck, A., *Die humanistische Tradition in der Romania* (Bad Homburg etc. 1968).
Buisman, J.W., *Tussen vroomheid en Verlichting. Een cultuurhistorisch en -sociologisch onderzoek naar enkele aspecten van de Verlichting in Nederland (1755-1810)* (Zwolle 1992).
Bunge, W. van, 'Van Velthuysen, Batelier and Bredenburg on Spinoza's interpretation of the Scriptures', in: P. Cristofolini ed., *The Spinozistic heresy. The debate on the Tractatus theologico-politicus, 1670-1677 of Benedictus de Spinoza and the immediate reception of Spinozism* (Amsterdam and Maarssen 1995) 49-65.
–, 'Filosofie', in: W. Frijhoff and M. Spiess, *1650. Bevochten eendracht* (The Hague 1999) 280-349.
–, *From Stevin to Spinoza. An essay on philosophy in the seventeenth-century Dutch Republic* (Leiden etc. 2001).
Burns, R.M., *The great debate on miracles. From Joseph Glanvill to David Hume* (Lewisburgh 1981).
Burstyn, H.L., 'Early explanations of the role of the earth's rotation in the circulation of the atmosphere and the ocean', in: *Isis* 57 (1966) 167-187.
Byvanck, 'De platen in de Aratea van Hugo de Groot', in: *Mededelingen der Koninklijke Nederlandsche Akademie van Wetenschappen, afd. letterkunde*, new series, 12 (1949) 169-235.
Cameron, J.K., 'Humanism in the Low Countries', in: A. Goodman and A. MacKay, *The impact of humanism on western Europe* (London and New York 1990) 137-163.
Clulee, N.H., *John Dee's natural philosophy. Between science and religion* (London and New York 1988).
Cohen, I.B., 'Perfect numbers in the Copernican system: Rheticus and Huygens', in: *Science and history* (Wroclaw etc. 1978) (Studia Copernicana 16) 419-425.
Cohen, G., *Écrivains francais en Hollande dans la première moitié du XVIIe siècle* (Paris 1920).
Cohen, H.F., *Quantifying music. The science of music at the first stage of the scientific revolution, 1580-1650* (Dordrecht etc. 1984).
Conring, E., *Kirche und Staat nach der Lehre der niederländischen Calvinisten in der ersten Hälfte des 17. Jahrhunderts* (Neukirchen-Vluyn 1965).
Cragg, G.R., *From puritanism to the age of reason. A study of changes in religious thought within the church of England 1660 to 1700* (Cambridge 1950).
Cramer, J.A., *Abraham Heidanus en zijn cartesianisme* (Utrecht 1889).

–, *De theologische faculteit te Utrecht ten tijde van Voetius* (Utrecht 1932).
Crone, E., 'Afgunst en ruzie tussen leermeesters in de stuurmanskunst', in: *Vereeniging Nederlandsch Historisch Scheepvaartmuseum Amsterdam, jaarverslag 1959, 1960 en 1961* (Amsterdam 1962) 37-49.
Crosland, M., *Historical studies in the language of chemistry* (London 1962).
Dahm, J.J., 'Science and apologetics in the early Boyle lectures', in: *Church history* 39 (1970) 172-186.
Dam, H.J. van, 'Filoloog en dichter in Leiden', in: H.J.M. Nellen and J. Trapman ed., *De Hollandse jaren van Hugo de Groot (1583-1621)* (Hilversum 1996) 67-86.
Dankbaar, W.F., 'De stichting van de Leidse universiteit en de eerste decennia van haar bestaan', in: idem, *Hoogtepunten uit het Nederlandsche calvinisme in de zestiende eeuw* (Haarlem 1946) 126-161.
Dauphiné, *Guillaume de Saluste du Bartas, poète scientifique* (Paris 1983).
Davids, C.A., 'Van Anthonisz tot Lastman. Navigatieboeken in Nederland in de zestiende en het begin van de zeventiende eeuw', in: *Lucas Jansz. Waghenaer van Enckhuysen. De maritieme cartografie in de Nederlanden in de zestiende en het begin van de zeventiende eeuw* (Enkhuizen 1984) 73-88.
– *Zeewezen en wetenschap. De wetenschap en de ontwikkeling van de navigatietechniek in Nederland tussen 1585 en 1815* (Amsterdam and Dieren 1985).
–, 'Ondernemers in kennis. Het zeevaartkundig onderwijs in de Republiek gedurende de zeventiende eeuw', in: *De zeventiende eeuw* 7 (1991) 37-48.
Dear, P. 'Mersenne's suggestion: Cartesian meditation and the mathematical model of knowledge in the seventeenth century', in: R. Ariew and M. Grene, *Descartes and his contemporaries. Meditations, objections and replies* (Chicago and London 1995) 44-62.
Deason, G.B., 'John Wilkins and Galileo Galilei: Copernicanism and biblical interpretation in the protestant and catholic traditions', in: E.A. McKee and B.G. Armstrong ed., *Probing the reformed tradition* (Louisville 1989) 313-338.
Decavele, J. (ed.), *Het eind van een rebelse droom. Opstellen over het calvinistisch bewind te Gent (1577-1584) en de terugkeer van de stad onder de gehoorzaamheid van de koning van Spanje (17 september 1584)* (Gent 1984).
Dekker, E., 'Sterrenkunde in de zeventiende eeuw', in: *De zeventiende eeuw* 2 (1986) no.2, 84-100.
Delambre, J.B.J., *Histoire de l'astronomie moderne* (2 vol., Paris 1821).
D'Elia, A., *Christiaan Huygens. Una biografia intelletuale* (Milan 1985).
Depau, R., *Simon Stevin* (Bruxelles 1942).
Dibon, P.A.G., *L'enseignement philosophique dans les universités néerlandaises à l'époque pré-cartésienne (1575-1650)* (diss. Leiden, 1954).
–, *Regards sur la Hollande du siècle d'or* (Napoli 1990).

Dijck, G.C.M. van, *De Bossche optimaten. Geschiedenis van de Illustere Lieve Vrouwebroederschap te 's Hertogenbosch, 1318-1973* (Tilburg 1973).
Dijksterhuis, E.J., *Simon Stevin* ('s-Gravenhage 1943).
–, *Simon Stevin. Science in the Netherlands around 1600* (The Hague 1970).
Dobbs, B.J., and M.C. Jacob, *Newton and the culture of Newtonianism* (New Jersey 1995).
Dobrzycki, J., (ed.), *The reception of Copernicus' heliocentric theory. Proceedings from a symposium organised by the Nicolas Copernicus committee of the International Union of the History and Philosophy of Science, Torún, Poland, 1973* (Dordrecht and Boston 1972) (also published as *Colloquia Copernicana*, 1).
Donahue, W.H., *The dissolution of the celestial spheres 1595-1650* (New York 1981).
Dorsten, J.A. van, *Poets, patrons and professors. Sir Philip Sydney, Daniel Roger and the Leiden humanists* (Leiden 1962).
Drake-Brockman, J.A., 'The perpetuum mobile of Cornelis Drebbel', in: W.D. Hackmann and J.A. Turner, *Learning, language and invention: essays presented to Francis Maddison* (Aldershot and Paris 1994) 124-147.
Dreyer, J.L.E., *A history of astronomy from Thales to Kepler* (second edition, New York 1953).
Duhem, P., *Le système du monde. Histoire des doctrines cosmologiques de Platon à Copernic* (10 vol., Paris 1913-1959).
Duker, A.C., *Gisbertus Voetius* (4 vol., Leiden 1897-1915).
Dunin Borkowski, S. von, *Spinoza* (4 vol., Münster 1933-1936).
Eastwood, B.S., 'Kepler as historian of science: precursors of Copernicus heliocentrism according to *De revolutionibus*, 1, 10', in: *Proceedings of the American Philosophical Society held at Philadelphia for Promoting Useful Knowledge* 126 (1982) 367-394.
–, 'Origins and contents of the Leiden planetary configuration (Ms. Voss. q. 79, fol 93v), an artistic astronomical schema of the early Middle Ages', in: *Viator* 14 (1983) 1-40.
Ebels-Hoving, B., 'Theorie en therapie. Antonius Deusing (1612-1666) en de koorts', in: F. Huisman and C. Santing ed., *Medische geschiedenis in regionaal perspectief: Groningen 1500-1900* (Rotterdam 1997) 49-68.
Eeghen, I.H. van., *De Amsterdamse boekhandel 1680-1725*, (6 vol., Amsterdam 1960-1978).
– 'Frederik de Wit, Amsterdams uitgever', in: *Jacob Aertz. Colom's kaart van Holland in 1681* (Alphen aan den Rijn 1990).
Ellinger, G., *Geschichte der neulateinischen Literatur Deutschlands im sechzehnten Jahrhundert*, III-1, *Geschichte der neulateinischen Lyrik in den Niederlanden vom Ausgang des fünfzehnten bis zum Beginn des siebzehnten Jahrhunderts* (Berlin 1933).
Eschweiler, K., 'Die Philosophie der spanischen Spätscholastik auf der

deutschen Universitäten des 17. Jahrhunderts', in: *Spanischen Forschungen der Görres-Gesellschaft* 1 (1928) 251-325.

Evans, R.J.W., *Rudolf II and his world. A study in intellectual history 1576-1612* (Oxford 1973).

Evenhuis, *Ook dat was Amsterdam*, III, *De kerk der hervorming in de tweede helft van de zeventiende eeuw: nabloei en inzinking* (Baarn 1971); IV, *De kerk der hervorming in de achttiende eeuw: de grote crisis* (Baarn 1974).

Evers, M., 'Pro Newtone et religione: de receptie van Newton en de Engelse fysicotheologen in de 'Bibliothèque ancienne et moderne' (1714-1727), *Documentatieblad werkgroep achttiende eeuw* 20 (1988) pp. 247-67.

Eyffinger, A., 'The Dutch period in the life of Hugo Grotius', in: *Hugo Grotius: a great European 1583-1645. Contributions concerning his activities as a humanist scholar* (Delft 1983) 5-23.

Feingold, M., *The mathematicians' apprenticeship. Science, universities and society in England, 1560-1640* (Cambridge etc. 1984).

Ferguson, J., *Bibliotheca chemica. A bibliography of books on alchemy chemistry and pharmaceutics* (1906; repr. 2 vol. London 1954).

Fix, A.C., 'Willem Deurhoff (1650-1717): merchant and philosopher', in: *Geschiedenis van de wijsbegeerte in Nederland* 1 (1990) 153-164.

Fokker, A.A., 'Philippus Lansbergen en zijne zonen Pieter en Jacob. Bijdrage tot hun leven', in: *Archief. Vroegere en latere mededeelingen voornamelijk in betrekking tot Zeeland*, 1, part 5 (1862) 52-100.

Fontaine Verwey, H. de la, *Uit de wereld van het boek*, III, *In en om de 'Vergulde Sonnewyser'* (Amsterdam 1979).

Frijhoff, W. Th.M., *La société néerlandaise et ses gradués, 1575-1814. Une recherche sérielle sur le statut des intellectuels à partir des registres universitaires* (Amsterdam 1981).

Fruin, R., 'Een brief van Philippus van Lansbergen', in: *Archief. Vroegere en latere mededeelingen voornamelijk in betrekking tot Zeeland* (1903) 143-148.

Fuchs, Th., *Die Mechanisierung des Herzens. Harvey und Descartes – der vitale und der mechanische Aspekt des Kreislaufs* (Frankfurt am Main 1992).

Galama, S., *Het wijsgerig onderwijs aan de hogeschool te Franeker 1585-1811* (Franeker 1954).

Galuzzi, P. (ed.), *Novità celesti e crisi di sapere. Atti del convegno internazionale di studi Galileiani* (Firenze 1984).

Gascoigne, J., 'From Bentley to the Victorians: the rise and fall of British Newtonian natural theology', in: *Science in context* 2 (1988) 219-256.

Gaukroger, S., *Descartes. An intellectual biography* (Oxford 1995).

Gent, R.H. van, 'De hemelatlas van Andreas Cellarius: het meesterwerk van een vergeten Hollandse kosmograaf', in: *Caert-thresoor* 19 (2000) 9-25.

Gerlo, A., 'Copernic et Simon Stevin', in: *Ciel et terre* 69 (1953) 277-288.

Geyer-Kordesch, J., 'Deutschsprachiche Bücher aus Holland: medizinische Kontroversen und ihr Publikum in Deutschland 1680-1730', in: M.J. van Lieburg and R. Toellner ed., *Deutsch-niederländische Beziehungen in der Medizin des 17. Jahrhunderts* (Amsterdam 1982).

Gibbs, C.G., 'The role of the Dutch Republic as the intellectual entrepot of Europe in the seventeenth and eighteenth centuries, c. 1680-1730', in: *BMGN* 86 (1971) 255-287.

Gillespie, N.C., 'Natural history, natural theology and social order: John Ray and the 'Newtonian ideology'', in: *Journal for the history of biology* 20 (1987) 1-49.

Gingerich, O, *The eye of heaven. Ptolemy, Copernicus, Kepler* (New York 1993).

Glazemaker 1682-1982. Catalogus bij een tentoonstelling over de vertaler Jan Hendriksz Glazemaker in de universiteitsbibliotheek van Amsterdam (Amsterdam 1982).

Graafland, C., 'Structuurverschillen tussen Voetiaanse en coccejaanse geloofsleer', in: Broeyer and Van der Wall ed. (1994) 28-53.

Grafton, A., 'Civic humanism and scientific scholarship at Leiden', in: Th. Bender (ed.), *The university and the city. From medieval origins to the present* (New York 1988) 59-78.

–, 'Humanism, magic and science', in: A. Goodman and A. MacKay, *The impact of humanism on western Europe* (London and New York 1990) 99-117.

–, 'Humanism and science in Rudolphine Prague: Kepler in context', in: idem, *Defenders of the text. The traditions of scholarship in an age of science, 1450-1800* (Cambridge (Mass.) and London 1991) 178-203.

–, *Joseph Scaliger. A study in the history of classical scholarschip*, II, *Historical chronology* (Oxford 1993).

Grant, E., *Planets, stars and orbs. The medieval cosmos 1200-1687* (Cambridge 1994).

Griffin, M.I.J., *Latitudinarianism in the seventeenth-century Church of England* (Leiden 1992).

Grössing, H., *Humanistische naturwissenschaft. Zur Geschichte der Wiener mathematischen Schulen des 15. und 16. Jahrhunderts* (Baden-Baden 1983).

Guhrauer, G.E., *Joachim Jungius und sein Zeitalter* (Stuttgart and Tübingen 1850).

Günther, S., 'Die Kompromiss-Weltsysteme des XVI, XVII und XVIII Jahrhunderts', in: *Annales internationales d'histoire. Congrès de Paris 1900. 5e section. Histoire des sciences* (Paris 1901) 121-145.

–, *Varenius* (Leipzig 1905) (Klassiker der Naturwissenschaften, IV).

Guerrini, A., 'The Tory Newtonians: Gregory, Pitcairne, and their circle', in: *Journal of British studies* 25 (1986) 288-311.

Haak, S.P., *Paullus Merula 1558-1607* (Zutphen 1901).

Haan, A.A.M. de, 'Geschiedenis van het wijsgerig onderwijs te Deventer', in: H.W. Blom, H.A. Krop and M.R. Wielema ed., *Deventer denkers. De geschiedenis van het wijsgerig onderwijs te Deventer* (Hilversum 1993) 29-122.

Haasbroek, N.D., *Gemma Frisius, Tycho Brahe and Snellius and their triangulations* (Delft 1968).
Hallyn, F., 'Un poème inédit de Philippe Lansbergen sur l'étoile nouvelle de 1604', *Humanistica Lovaniensia* 46 (1997) 258-265.
–, 'La préface de Gemma Frisius aux Ephemerides de Stadius (1556)', in: *Scientiarum historia* 24 (1998) 3-15.
Hannaway, O., *The chemists and the word. The didactic origins of chemistry* (Baltimore and London, 1975).
Hartfelder, K., *Philipp Melanchthon als praeceptor Germaniae* (Berlin 1889; repr. Nieuwkoop 1964).
Harris, L.E., *The two Netherlanders. Humphrey Bradley and Cornelis Drebbel* (Leiden 1961).
Heckel, J., *Cura religionis, jus in sacra, jus circa sacra* (Darmstad 1962).
Heesakkers, Ch., and W. Reinders, *Genoeglijk bovenal zijn mij de Muzen. De Leidse neolatijnse dichter Janus Dousa (1545-1604)* (Leiden 1993).
Helden, A. van, 'Huygens and the astronomers', in: H.J.M. Bos a.o. ed., *Studies on Christiaan Huygens* (Lisse 1980) 147-165.
–, *Measuring the universe. Cosmic dimensions from Aristarchus to Halley* (Chicago and London 1985).
Heuvel, C.M.J.M. van den, *'Papiere bolwercken'. De introductie van de Italiaanse stede- en vestingbouw in de Nederlanden (1540-1609) en het gebruik van tekeningen* (Alphen aan den Rijn 1991).
–, 'Architectuurmodellen en voorstellingen van een logische orde: Stevin en Ramus', in: *Feit & fictie. Tijdschrift voor de geschiedenis van de representatie* 2 (1994) 95-111.
–, 'De huysbou, de crychconst en de wysentijt. Stevins teksten over architectuur, stede- en vestingbouw in het licht van zijn wetenschappelijk oeuvre', in: V. Logghe ed., *Spiegeling en daet. Simon Stevin van Brugghe (1548-1620)* (Catalogue exposition Brugge) (Brugge 1995) 44-69.
Hooykaas, R., *Humanisme, science et réforme. Pierre de la Ramée (1515-1572)* (Leiden 1958).
–, 'The reception of Copernicanism in England and the Netherlands', in: C. Wilson e.a., *The Anglo-Dutch contribution to early modern society. An Anglo-Netherlands symposium* (Oxford 1976) 33-44.
–, *G.J. Rheticus' treatise on Holy Scripture and the motion of the earth* (Amsterdam etc. 1984).
Hoppe, B., 'Die Vernetzung der mathematisch ausgerichteten Anwendungsgebiete mit den Fächern des Quadriviums in der Frühen Neuzeit', in: I. Hantsche ed., *Der 'mathematicus'. Zur Entwicklung und Bedeutung einer neuen Berufsgruppe in der Zeit Gerhard Mercators* (Bochum 1996) (DSM 4) 1-40.

Huisman, C., *Neerlands Israël. Het natiebesef der traditioneel-gereformeerden in de achttiende eeuw* (Dordrecht 1983).
Huisman, F., *Stadsbelang en standsbesef. Gezondheidszorg en medisch beroep in Groningen 1500-1730* (Rotterdam 1992).
Huisman, G.C., 'Boeken en brieven van Nicolaus Mulerius', in: *Biblotheek, wetenschap en cultuur. Opstellen aangeboden aan mr. W.R.H. Koops bij zijn afscheid als bibliothecaris der Rijksuniversiteit te Groningen* (Groningen 1990).
Ihsanoglu, E., 'Introduction of western science to the Ottoman world: a case study of modern astronomy (1660-1860)', in: E. Ihsanoglu (ed.) *Transfer of modern science and technology to the Muslim world* (Istanbul 1992) 67-120.
Israel, J.I., *The Dutch Republic. Its rise, greatness and fall 1477-1806* (Oxford 1995).
Jaeger, F.M., *Cornelis Drebbel en zijne tijdgenoten* (Groningen 1922).
Janssen, A., 'Wirkung eines Romans als Inspirationsquelle: Francis Godwins The Man in the Moon', in: *Arcadia. Zeitschrift für vergleichende Literaturwissenschaft* 20 (1985) 20-46.
Janssen, F.A., 'Dutch translations of the Corpus Hermeticum', in: T. Croiset van Uchelen, K. van der Horst and G. Schilder, ed., *Theatrum orbis librorum. Liber amicorum presented to Nico Israel on the occasion of his seventieth bitrhday* (Utrecht 1989) 137-157.
Johnson, F.R., 'Astronomical text-books in the sixteenth century', in: E.A. Underwood ed., *Science medicine and history. Essays on the evolution of scientific thought and medical practice writtenin honour of Charles Singer*, 1 (London etc. 1953) 285-302.
Johnston, S., 'The identity of the mathematical practitioner in 16th-century England', in: I. Hantsche ed., *Der 'mathematicus'. Zur Entwicklung und Bedeutung einer neuen Berufsgruppe in der Zeit Gerhard Mercators* (Bochum 1996) (DSM 4) 93-120.
Jorink, H.G.M., 'Tussen Aristoteles en Copernicus. De natuurfilosofische opvattingen van Nicolaus Mulerius (1564-1630), in: H.A. Krop, J.A. van Ruler, A.J. Vanderjagt, *Zeer kundige professoren. Beoefening van de filosofie in Groningen van 1614 tot 1996* (Hilversum 1997) 69-83.
Juynboll, W.M.C., *Zeventiende-eeuwsche beoefenaars van het Arabisch in Nederland* (Utrecht 1931).
Kästner, A.G., *Geschichte der Mathematik seit der Wiederherstellung der Wissenschaften bis an das Ende des achtzehnten Jahrhunderts* 4 vol. (Göttingen 1796-1800, repr. Hildesheim/New York 1970).
Kallinen, M., *Change and stability. Natural philosophy at the academy of Turku (1640-1713)* (Helsinki 1995).
Karpp, H., 'Der Beitrag Keplers und Galileis zum neuzeitlichen Schriftverständnis', in: *Zeitschrift für Theologie und Kirche* 67 (1970) 40-55.

Keuning, J., and M. Donkersloot-de Vrij, *Willem Jansz. Blaeu. A biography and history of his works as a cartographer and publisher* (Amsterdam 1973).
Kleerkooper, M.M., and W.P. van Stockum, *De boekhandel te Amsterdam voornamelijk in de 17e eeuw* (2 vol., The Hague 1914-1916).
Korninger, S., *Die Naturauffassung in der englischen Dichtung des 17. Jahrhunderts* (Wien-Stuttgart 1956) (Wiener Beiträge zur englischen Philologie, 64).
Kossmann, E.F., *De boekhandel te 's Gravenhage tot het eind van de 18de eeuw* (The Hague 1937).
Krogt, P. van der, *Globi Neerlandici. The production of globes in the Low Countries* (Utrecht 1993).
–, 'Commercial cartography in the Netherlands with particular reference to atlas production (16th-18th centuries)', in: *La cartografia dels Països Baixos* (Barcelona 1995) 71-140.
Kuhn, Th., *The Copernican revolution. Planetary astronomy in the development of western thought* (Cambridge, Mass. 1957).
Kusukawa, S., *The transformation of natural philosophy. The case of Philip Melanchthon* (Cambridge 1995).
Laplanche, F., *L'écriture, le sacré et l'histoire. Érudits et politiques protestants devant la Bible en France au XVIIe siècle* (Amsterdam and Maarssen 1986).
–, 'Débats et combats autour de la Bible dans l'orthodoxie réformée', in: J.-R. Armogathe ed., *Le grand siècle et la Bible* (Paris 1989) 117-140.
Laursen, J.C., and J.P. McDonald, 'The beneficial lies controversy in the Huguenot Netherlands, 1705-1731: an unpublished manuscript at the root of the *cas Saurin*', in: *Studies on Voltaire and the eighteenth century*, CCCXIX (Oxford 1994) 67-103.
Lerner, M.-P., 'L' 'hérésie' héliocentrique: du soupcon à la condamnation', in: *Sciences et religion de Copernic à Galilée (1540-1610)* (Rome 1999).
Lieburg, F.A., *De nadere reformatie in Utrecht ten tijde van Voetius. Sporen in de gereformeerde kerkeraadsacta* (Rotterdam 1989).
Lindeboom, J., *Frederik Adolf van der Marck, Een achttiende-eeuwsch leeraar van het natuurrecht* (The Hague 1947).
Maanen, J.A. van, 'Korrespondenten von G.W. Leibniz: Joachim Nieuwstadt, geb. um 1624 – gest. nach 1684', in: *Studia Leibnitiana* 15 (1983) 115-119.
–, *Facets of seventeenth century mathematics in the Netherlands* (dissertation Utrecht 1987).
McAdoo, H.R., *The spirit of Anglicanism. A survey of Anglican theological method in the seventeenth century* (London 1965).
McColley, G., 'An early friend of the Copernican theory: Gemma Frisius', in: *Isis* 26 (1937) 322-325.
McGahagan, T.A., *Cartesianism in the Netherlands, 1639-1676; the new science and the*

Calvinist counter-reformation (unpublished dissertation, Universiteit of Pennsylvania, 1976).
Marnef, G., *Het calvinistisch bewind te Mechelen 1580-1585* (Kortrijk 1987).
Matthey, I., 'De betekenis van de natuur en de natuurwetenschappen voor Constantijn Huygens', in: H. Bots ed., *Constantijn Huygens. Zijn plaats in geleerd Europa* (Amsterdam 1973) 334-459.
Meertens, P.J., *Letterkundig leven in Zeeland in de zestiende en de eerste helft der zeventiende eeuw* (Amsterdam 1943).
Meinsma, K.O., *Spinoza et son cercle. Etude critique historique sur les hétérodoxes hollandais* (Paris 1983).
Menk, G., *Die Hohe Schule Herborn in ihrer Frühzeit (1584-1660). Ein Beitrag zum Hochschulwesen des deutschen Kalvinismus im Zeitalter der Gegenreformation* (Wiesbaden 1981; Veröffentlichungen der Historischen Kommission für Nassau, 30).
Meskens, A., *Wiskunde tussen Renaissance en barok. Aspecten van wiskunde-beoefening te Antwerpen 1550-1620* (Antwerp 1994) (Publikaties van de stadsbibliotheek en het archief en museum voor het Vlaamse cultuurleven, 41-43).
Methuen, C., 'The role of the heavens in the thought of Philip Melanchthon', in: *Journal of the history of ideas* 57 (1996) 385-403.
Metzger, H., *Attraction universelle et religion naturelle chez quelques commentateurs anglais de Newton* (3 vol., Paris 1938).
Michel, H., 'Le mouvement perpetuel de Drebbel', in: *Physis* 13 (1971) 289-294.
Moesgaard, K.P., 'How Copernicanism took root in Denmark and Norway', in: *Reception* (1972), 117-151.
–, 'Cosmology in the wake of Tycho Brahe's astronomy', in: W. Yourgrau and A.D. Beck ed., *Cosmology, history and theology* (New York and London 1977) 293-305.
Monchamp, G., *Galilée et la Belgique. Essai historique sur les vicissitudes du système de Copernic en Belgique* (Saint Trond and Brussels, 1892).
Mormino, G., *Penetralia motus. La fondazione relativistica della meccanica in Christiaan Huygens, con l'edizione del Codex Hugeniorum 7A* (Florence 1993).
Mostert, R., and M. Mostert, 'Using astronomy as an aid to dating manuscripts. The example of the Leiden Aratea planetarium', in: *Quaerendo* 20 (1990) 248-261.
Muller, E. and K. Zandvliet (ed.), *Admissies als landmeter in Nederland voor 1811* (Alphen aan den Rijn 1987).
Muller, R.A., *Post-reformation reformed dogmatics*, 1, *Prolegomena to theology* (Grand Rapids 1987); 11, *Holy Scripture: the cognitive foundation of theology* (Grand Rapids 1993).
Nakayama, S., 'Diffusion of Copernicanism in Japan', in: *Reception* (1972) 153-188.

Nauta, D., *Samuel Maresius* (Amsterdam 1935).
Nellen, H.J.M., *Ismaël Boulliau (1605-1694. Astronome, épistolier, nouvelliste et intermédiaire scientifique. Ses rapports avec les milieux du 'libertinage érudit'* (Amsterdam and Maarssen 1994).
Nelson, A., 'Hugo Grotius. Quelques observations sur ses débuts comme philologue, sur ses études de droit romain et sur ses relations avec J.-A. de Thou, historien et président au Parlement de Paris', in: *Kungl. humanistiska vetenskaps-samfundet i Uppsala. Årsbok, 1952*, 33-63.
Nobbs, D., *Theocracy and toleration. A study of the disputes in Dutch Calvinism from 1600 to 1650* (Cambridge 1938).
Nouhuys, T. van, *The age of two-faced Janus. The comets of 1577 and 1618 and the decline of the Aristotelian world view in the Netherlands* (Leiden 1998).
North, J.D., 'The reluctant revolutionaries: astronomy after Copernicus', in: *Colloquia copernicana*, III, *Astronomy of Copernicus and its background* (Wroclaw 1975) 169-184.
Obbema, P.F.J., 'Die Handschrift', in: B. Bischoff e.a., *Aratea. Kommentar zum Aratus des Germanicus ms. Voss. lat. Q. 79. Bibliotheek der Rijksuniversiteit Leiden* (Luzern 1989; companion volume to a facsimile edition of the manuscript).
Oeser, E., 'Copernicus und die ägyptische Hypothese. Ein Beitrag zur Frage nach der Struktur wissenschaftlicher Revolutionen', in: *Philosophia naturalis* 14 (1973) 276-308.
Ortroy, F. van, *Bio-bibliographie de Gemma Frisius fondateur de l'école Belge de géographie, de son fils Corneille et de ses neveux les Arsenius* (Bruxelles 1920, repr. Amsterdam 1966).
Pagnoni-Sturlese, M.R., 'Lazar Zetzner, bibliopola Argentinensis. Alchimia e lullismo a Strasburgo agli inizi dell'età moderna', in: M.J.F.M. Hoenen and A. de Libera ed., *Albertus Magnus und der Albertismus. Deutsche philosophische Kultur des Mittelalters* (Leiden 1995) 353-379.
Pater, C. de, 'Experimental physics', in: Th.H. Lunsingh Scheurleer and G.H.M. Posthumus Meyjes ed., *Leiden university in the seventeenth century: an exchange of learning* (Leiden 1975) 309-327.
–, *Petrus van Musschenbroek (1692-1761), een newtoniaans natuuronderzoeker* (diss. Utrecht 1979).
–, Ds. Hermanus Johannes Krom: worstelen met 'geloof en wetenschap', in: *Worstelende wetenschap. Aspecten van wetenschapsbeoefening in Zeeland van de zestiende tot in de negentiende eeuw* (Separate issue from: *Archief. Mededelingen van het Koninklijk Zeeuwsch Genootschap der Wetenschappen*, 1987) 182-205.
–, 'The textbooks of 's Gravesande and Van Musschenbroek in Italy', in: C.S. Maffioli and L.C. Palm ed., *Italian scientists in the Low Countries in the XVIIth and XVIIIth centuries* (Amsterdam 1989) 231-241.
–, 'Nicolaus Engelhard (1696-1765) en zijn kritiek op de Beginselen der na-

tuurkunde van Petrus van Musschenbroek (1692-1761): wolffianisme versus newtonianisme' in: TGGNWT 13 (1990) 141-162.
Pedersen, O., 'In quest of Sacrobosco', in: JHA 16 (1985) 175-221.
Petersen, P., *Geschichte der aristotelischen Philosophie im protestantischen Deutschland* (Leipzig 1921, repr. Stuttgart-Bad Cannstatt 1964).
Philipp, W., *Das Werden der Aufklärung in theologiegeschichtlicher Sicht* (Göttingen 1957).
Pingré, A.-G., *Annales célestes du dix-septième siècle*, M.G. Bigourdan ed. (Paris 1901).
Poortman, W.C. and J. Augusteijn, *Kaarten in bijbels (16e-18e eeuw)* (Zoetermeer 1995).
Popkin, R.H., 'Cartesianism and biblical criticism', in: Th. M. Lennon and J.M. Nicholas (ed.), *Problems of Cartesianism* (Kingston and Montreal 1982) 61-81.
Posthumus Meyjes, E.J.W., *Jacobus Revius, zijn leven en werken* (Amsterdam 1895).
Price, D.J. de S., 'Contra-Copernicus: a critical re-estimation of the mathematical planetary theory of Ptolemy, Copernicus and Kepler', in: M. Clagett ed., *Critical problems in the history of science* (Madison 1959) 197-218.
Pumphrey, S., 'Neo-Aristotelianism and the magnetic philosophy', in: J. Henry and S. Hutton ed., *New perspectives on Renaissance thought* (London 1990) 177-189.
Radelet-De Grave, P., 'La reception du Copernicanisme dans les milieux jesuites belges', in: Opsomer 89-100.
Rademaker, C.S.M., *Life and work of Gerardus Joannes Vossius (1577-1649)* (Assen 1981).
Rappard, F.A. van, *Ernst Brinck, eerst secretaris van het Nederlandsche gezantschap te Konstantinopel, later burgemeester van Harderwijk* (Utrecht 1868).
Redondi, P., 'Libert Froidmont 'opposant' et ami de Galilée', in: A.-C. Bernès ed., *Libert Froidmont et les resistances aux revolutions scientifiques* (Haccourt 1988) 83-103.
H. Richter, Willem Janszoon Blaeu with Tycho Brahe on Hven, and his map of the island: some new facts', in: *Imago mundi* 3 (1939) 53-60.
Ridder-Symoens, H. de, 'Enkele gegevens over Joos Balbiaens universitaire studies', in: *Handelingen der Maatschappij voor geschiedenis en oudheidkunde te Gent*, nieuwe reeks 40 (1986) 97-100.
Ridderbos, S., *De philologie aan de Leidsche universiteit gedurende de eerste vijfentwintig jaren van haar bestaan* (Leiden 1906).
Roorda, D.J., 'Prins Willem III en het Utrechtse regeringsreglement. Een schets van gebeurtenissen, achtergronden en problemen', in: *Van standen tot Staten. 600 jaar Staten van Utrecht 1375-1975* (Utrecht 1975) 91-133.
Rothschuh, K.E., 'Henricus Regius und Descartes. Neue Einblicke in die

frühe Physiologie (1640-1641) des Regius', in: *Archives internationales d'histoire des sciences* 21 (1968) 39-66.
Rudolph, H., 'Kosmosspekulation und Trinitätslehre. Ein Beitrag zur Beziehung zwischen Weltbild und Theologie bei Paracelsus', in: *Paracelsus in der Tradition* (Salzburger Beiträge zur Paracelsusforschung 21) (Wien 1980) 32-47.
Ruestow, E.G., *Physics at 17th and 18th century Leiden* (The Hague 1973).
Ruler, J.A. van, *The crisis of causality. Voetius and Descartes on God, nature and change* (Leiden etc. 1995).
–, 'Waren er muilezels op de zesde dag? Descartes, Voetius en de zeventiende-eeuwse methodenstrijd', in: F. Egmond, E. Jorink and R. Vermij ed., *Kometen, monsters en muilezels. Het veranderende natuurbeeld en de natuurwetenschap in de zeventiende eeuw* (Haarlem 1999) 120-132.
Ruska, J., *Tabula smaragdina. Ein Beitrag zur Geschichte der hermetischen Literatur* (Heidelberg 1926).
Russel, J.L., 'Kepler's laws of planetary motions, 1609-1666', in: BJHS 2 (1964) 1-24.
–, 'The Copernican system in Great Britain', in: *Reception* (1972), 189-239.
–, 'Catholic astronomers and the Copernican system after the condemnation of Galileo', in: *Annals of science* 46 (1989) 365-386.
Salman, J., *Populair drukwerk in de Gouden Eeuw. De almanak als lectuur en handelswaar* (Zutphen 1999).
–, 'Van sodanige almanacken, die gevult zijn met ergerlijcke bijvoegselen en oncuyse en onstigtelijcke grillen', in: *Literatuur* 10 (1993) 74-80.
Sandblad, H., 'The reception of the Copernican system in Sweden', in: *Reception* (1972), 241-270.
Sandler, Chr., *Johann Baptista Homann, Matthäus Seutter und ihre Landkarten. Ein Beitrag zur Geschichte der Kartographie* (Amsterdam s.a. [1962]).
Sassen, F., *Het oudste wijsgerig onderwijs te Leiden (1575-1619)* (Amsterdam 1941) (Mededeelingen der Nederlandsche Akademie van Wetenschappen, afdeeling letterkunde, new series vol. IV no. 1). [a]
–, *Henricus Renerius, de eerste 'cartesiaansche' hoogleeraar te Utrecht* (Amsterdam 1941) (Mededeelingen der Nederlandsche Akademie van Wetenschappen, afdeeling letterkunde, new series vol. IV no. 20). [b]
–, 'Adriaan Heereboord (1614-1661). De opkomst van het cartesianisme te Leiden', in: *Algemeen Nederlandsch tijdschrift voor wijsbegeerte en psychologie* 36 (1942-1943) 12-22.
–, *Het wijsgerig onderwijs aan de illustere school te Breda (1646-1669)* (Amsterdam 1962) (Mededelingen der Koninklijke Nederlandse Akademie van Wetenschappen, afd. letterkunde, nieuwe reeks XXV no. 7).
–, *Johan Lulofs (1711-1768) en de reformatorische Verlichting in de Nederlanden* (Amster-

dam 1965) (Mededelingen der Koninklijke Nederlandse Akademie van Wetenschappen, afd. letterkunde, nieuwe reeks XXVIII nr. 7).
Saunders, J.L. *Justus Lipsius. The philosophy of Renaissance Stoicism* (New York 1955).
Savage-Smith, E., and C. Wakefield, 'Jacob Golius and celestial cartography', in: W.D. Hackmann and J.A. Turner, *Learning, language and invention: essays presented to Francis Maddison* (Aldershot and Paris 1994) 238-260.
Schaffer, S., 'Newton's comets and the transformation of astrology', in: P.Curry ed., *Astrology, science and society. Historical essays* (Woodbridge 1987) 219-243.
Schama, S., *The embarrassment of riches. An interpretation of Dutch culture in the Golden Age* (London 1987).
Schatzberg, W., 'Gottsched as a populariser of science', in: *Modern language notes* 83 (1968) 753-755.
Schenkeveld-Van der Dussen, M.H., 'Een dominee als kosmoloog', in: H. van Dijk a.o. ed., *In de zevende hemel. Opstellen voor P.E.L. Verkuyl over literatuur en kosmos* (Groningen 1993) 177-181.
Schilder, G., *The world map of 1669 by Jodocus Hondius the elder and Nicolaas Visscher* (Amsterdam 1978).
–, *Monumenta cartographica Neerlandica*, III, *Twee wandkaarten van de wereld van Blaeu: 1619 en 1645/46. Two wall-maps of the world by Blaeu: 1619 and 1645-1646* (Alphen aan den Rijn 1990).
–, *Monumenta cartographica Neerlandica*, IV, *Losse kaarten en topografische prenten uitgegeven door Willem Jansz. Blaeu. Single-sheet maps and topographical prints published by Willem Jansz. Blaeu* (Alphen aan den Rijn 1993).
Schilder, G., and J. Weelu, *The world map of 1611 by Pieter van der Keere* (Amsterdam 1980).
Schilling, H., 'Afkeer van domineesheerschappij: ein neuzeitlicher Typus des Antiklerikalismus', in: P.A. Dykema and H.A. Oberman ed., *Anticlericalism in late medieval and early modern Europe* (Leiden 1993) 655-668.
Schmitt, C.B., 'Philosophy and science in sixteenth-century universities: some preliminary comments', in: J.E. Murdoch and E.D. Schneider, I., *Johannes Faulhaber 1580-1635. Rechenmeister in einer Welt des Umbruchs* (Basel etc. 1993).
Schneppen, H., *Niederländische Universitäten und deutsches Geistesleben. Von der Gründung der Universität Leiden bis ins späte 18. Jahrhundert* (Münster 1960).
Schofield, C.J. *Tychonic and semi-Tuchonic world systems* (New York 1981).
Scholder, K., *Ursprünge und Probleme der Bibelkritik im 17. Jahrhundert. Ein Beitrag zur Entstehung der historisch-kritischen Theologie* (München 1966). [English translation: *The birth of modern critical theology. Origins and problems of biblical criticism in the seventeenth century* (London and Philadelphia 1990)].
Schotel, G.D.J., *Bijdrage tot de geschiedenis der kerkelijke en wereldlijke kleeding* (2 vol., The Hague 1854-1856).

–, *De openbare eeredienst der Nederl. Hervormde Kerk in de zestiende, zeventiende en achttiende eeuw*, second edition, H.C. Rogge ed., (Leiden s.a.).
Schuster, J.A., *Descartes and the scientific revolution 1618-1634. An interpretation* (unpublished PhD dissertation, Princeton university, 1977).
Secret, F., 'Réforme et alchemie', in: *Bulletin de la Société de l'histoire du protestantisme français* 124 (1978) 173-186.
Seidengart, J., 'Les théories cosmologiques de Christiaan Huygens', in: *Huygens et la France* (Paris 1982) 209-222.
Shapiro, B., 'Latitudinarianism and science in seventeenth-century England', in: *Past and present* 40 (1968) 16-41.
–, *John Wilkins 1614-1672. An intellectual biography* (Berkeley and Los Angeles 1969).
Shirley, R.W., *The mapping of the world. Early printed world maps 1472-1700* (London 1983).
Smet, A. de, 'Gemma Frisius et Nicolas Copernic', in: *Der Globusfreund* 21-23 (1972-1974) 72-80.
–, 'Copernic et les Pays-Bas', in: *Janus* 60 (1973) 13-23.
–, 'Rembertus Dodoens (Dodonaeus) en Nicolaus Copernicus', in: *Studia Mechliniensia. Bijdragen aangeboden aan dr. Henry Joosen t.g.v. zijn 65e verjaardag* (Malines/Mechelen 1976) 185-189.
Smit, J., *Dirck Rembrantsz van Nierop 1610-1682* (2 vol., private edition [Winkel 1992]).
Snelders, H.A.M., 'Alkmaarse natuurwetenschappers uit de 16de en 17de eeuw', in: *Van Spaans beleg tot Bataafse tijd. Alkmaars stedelijk leven in de 17de en 18de eeuw* (Alkmaarse historische reeks, 4) (Zutphen 1980) 101-122.
–, 'Antoni van Leeuwenhoek's mechanistic view of the world', in: L.C. Palm and H.A.M. Snelders ed., *Antoni van Leeuwenhoek 1632-1723. Studies on the life and work of the Delft scientist commemoratin the 350th anniversary of his birthday* (Amsterdam 1982) 57-78.
–, 'Science and religion in the seventeenth century: the case of the northern Netherlands', in: C.S. Maffioli and L.C. Palm ed., *Italian scientists in the Low Countries in the XVIIth and XVIIIth centuries* (Amsterdam 1989) 65-77.
Stephenson, B., *Kepler's physical astronomy* (Princeton 1994).
Stewart, M.A., 'Libertas philosophandi: from natural to speculative philosophy', in: *Australian journal of politics and history* 40 (994) 29-46.
Stimson, D., *The gradual acceptance of the Copernican theory of the universe* (New York 1917).
Strien, T. van, 'De zeven planeten volgens Jeremias de Decker', in: H. van Dijk, M.H. Schenkeveld-van der Dussen and J.M.J. Sicking ed., *In de zevende hemel. Opstellen voor P.E.L. Verkuyl over literatuur en kosmos* (Groningen 1993) 92-96.

Struik, D.J., *Het land van Stevin en Huygens* (Nijmegen 1979).
Stumpel, Jeroen, 'On painting and planets. A note on art theory and the Copernican revolution', in: id., *The province of painting. Theories of Italian Renaissance art* (Utrecht 1990) 231-258.
Taub, L.C., *Ptolemy's universe. The natural philosophical and ethical foundations of Ptolemy's astronomy* (Chicago and LaSalle, 1993).
Telle, J., 'Astrologie und Alchemie im 16. Jahrhundert. Zu den astroalchemischen Lehrdichtungen von Christoph von Hirschenberg und Basilius Valentinus', in: A. Buck (ed.), *Die okkulten Wissenschaften in der Renaissance* (Wiesbaden 1992).
Terpstra, H., *Friesche sterrekonst. Geschiedenis van de Friesche sterrenkunde en aanverwante wetenschappen door de eeuwen heen* (Franeker 1981).
Thiele, R., 'Breves in sphaeram meditatiunculae. Die Vorlesungsausarbeitung des Bartholomäus Mercator im Spiegel der zeitgenössischen kosmographischen Literatur', in: H.H. Blotevogel and R.H. Vermij ed., *Gerhard Mercator und die geistigen Strömungen des 16. und 17. Jahrhunderts* (Bochum 1995) (DMS 3) 147-174.
Thijssen-Schoute, C.L., *Nederlands cartesianisme* (Amsterdam 1954; repr. Utrecht 1989).
–, *Uit de Republiek der Letteren. Elf studiën op het gebied der ideeëngeschiedenis van de Gouden Eeuw* (The Hague 1967).
Thomassen, K. (ed.), *Alba amicorum. Vijf eeuwen vriendschap op papier gezet; het album amicorum en het poëziealbum in de Nederlanden* (Maarssen and 's Gravenhage 1990).
Thoren, V.E., *The Lord of Uraniborg. A biography of Tycho Brahe* (Cambridge etc. 1990).
Thorndike, L., *The sphere of Sacrobosco and its commentators* (Chicago 1949).
–, *A history of magic and experimental science* (8 vol., New York 1923-1958) (several editions).
Tierie, G., *Cornelis Drebbel (1572-1633)* (diss. Leiden 1932) (Amsterdam 1932) [in English].
Topper, D., 'Galileo, sunspots and the motions of the earth', in: *Isis* 90 (1999) 757-767.
Trevisani, F., *Descartes in Germania. La ricenzione del cartesianismo nella Facoltà filosofica e medica di Duisburg (1652-1703)* (Milan 1992)
Trevor-Roper, H., 'The Paracelsian movement', in: idem, *Renaissance essays* (London 1985) 149-199.
Trimp, J.C., *Jodocus van Lodensteyn, predikant en dichter* (Kampen 1987).
Tuzet, H., *Le cosmos et l'imagination* (Paris 1965).
Valk, J. van der, 'Jan van Hout', in: *Tijdschrift voor Nederlandsche taal- en letterkunde* 25 (1906) 75-77.

Vandenbossche, H., 'Hendrik Wyermars' ingebeelde chaos (1710): een 'tractatus de emendatione Spinozae'', in: *Tijdschrift voor de studie van de Verlichting* 2 (1974) 321-369.

Vanpaemel, G.H.W., *Echo's van een wetenschappelijke revolutie. De mechanistische natuurwetenschap aan de Leuvense artesfaculteit (1650-1797)* (Brussel 1986).

–, 'Science disdained: Galileo and the problem of longitude', in: C.S. Maffioli and L.C. Palm ed., *Italian scientists in the Low Countries in the XVIIth and XVIIIth centuries* (Amsterdam/Atlanta 1989) 111-129.

–, 'Gemengde wiskunde en experimentele fysica', in: *De zeventiende eeuw* 7 (1991) 49-57.

–, 'Mercator and the scientific Renaissance at the university of Leuven', in: H.H. Blotevogel and R.H. Vermij ed., *Gerhard Mercator und die geistigen Strömungen des 16. und 17. Jahrhunderts* (Bochum 1995) (DMS 3) 33-48. [a]

–, 'Het copernicanisme aan de oude Leuvense universiteit', in: Opsomer, 101-122. [1995b]

Veen, S.D. van, *Zondagsrust en zondagsheiliging in de zeventiende eeuw* (Nijkerk 1889).

Verbeek, Th., 'Descartes and atheism: the Utrecht crisis', in: *Nederlands archief voor kerkgeschiedenis* 71 (1991) 211-233.

–, *Descartes and the Dutch. Early reactions to Cartesian philosophy 1637-1650* (Carbondale and Edwardsville, 1992).

–, 'From 'learned ignorance' to scepticism. Descartes and Calvinist orthodoxy', in: R.H. Popkin and A. Vanderjagt (ed.), *Scepticism and irreligion in the seventeenth and eighteenth centuries* (Leiden 1993) 31-45.

–, 'Henricus Reneri (1593-1639)', in: H.W. Blom, H.A. Krop and M.R. Wielema ed., *Deventer denkers. De geschiedenis van het wijsgerig onderwijs te Deventer* (Hilversum 1993) 123-134.

–, 'Regius' Fundamenta physices', in: *Journal of the history of ideas* 55 (1994) 533-551.

Verdonk, J.J., *Petrus Ramus en de wiskunde* (Assen 1966).

– , 'Vom Einfluss des P. de la Ramée auf Simon Stevin', in: *Centaurus* 13 (1969) 251-262.

Verkerk, C.L., '*Aratea*: a review of the literature concerning MS. Vossianus lat. q. 79 in Leiden university library', in: *Journal of medieval history* 6 (1980) 245-287.

Verkuyl, P.E.L., 'Sterren-konst-licht bij Huygens-lectuur', in: *Nieuwe taalgids* 58 (1965) 234-243.

–, 'Huygens ludens. Een en ander over 'Meteorologia' (1636)', in: *Tijdschrift voor Nederlandse taal- en letterkunde* 100 (1984) 81-128.

– 'Kosmos-beelden in Hoofts lyriek',in: *Tijdschrift voor Nederlandse taal-en letterkunde* 101 (1985) 1-33.

–, 'Sterren-konst en dichtkonst bij Huygens', in: N.F. Streekstra and P.E.L.

Verkuyl ed., *Huygens in noorder licht. Lezingen van het Groningse Huygens-symposium* (Groningen 1987) 69-98.
Vermij, R.H., *Secularisering en natuurwetenschap in de zeventiende en achttiende eeuw: Bernard Nieuwentijt* (Amsterdam 1991).
–, 'Het copernicanisme in de Republiek: een verkenning', in: *Tijdschrift voor geschiedenis* 106 (1993) 349-367.
– 'The beginnings of physico-theology: England, Holland, Germany', in: H. Kattenstedt (ed.), *'Grenz-Überschreitung'. Wandlungen der Geisteshaltung, dargestellt an Beispielen aus Geographie und Wissenschaftshistorie, Theologie, Religions- und Erziehungswissenschaft, Philosophie, Musikwissenschaft und Liturgie. Festschrift zum 70. Geburtstag von Manfred Büttner* (Bochum 1993) 173-184.
–, 'De typus universitatis', in: M. Watelet ed., *Gerardus Mercator Rupelmundanus* (Antwerpen 1994) 235-239.
– 'Bijdrage tot de bio-bibliografie van Johannes Hudde', in: *Gewina* 18 (1995) 25-35.
–, 'Matter and motion: Toland and Spinoza', in: W. van Bunge and W. Klever, *Disguised and overt Spinozism around 1700* (Leiden etc. 1996) 275-288.
–, 'Mathematics at Leiden: Stevin, Snellius, Scaliger', in: I. Hantsche ed., *Der 'mathematicus'. Zur Entwicklung und Bedeutung einer neuen Berufsgruppe in der Zeit Gerhard Mercators* (Bochum 1996) (DSM 4) 75-92. [a]
–, 'Correspondence of Dirk Rembrandtszoon van Nierop (1610-1682)', in: *Lias* 23 (1996) 49-68. [b]
–, 'Nederlandse wereldkaarten en het stelsel van Copernicus in de 17de eeuw', in: *Caert-thresoor* 16 (1997) 33-40.
–, 'Waarom werd Philips Lansbergen copernicaan?', in: *Scientiarum historia* 24 (1998) 39-64.
–, 'Johannes Lulofs als vertegenwoordiger van het newtonianisme in de Republiek', in: *Gewina* 22 (1999) 136-150.
–, 'Dirk Santvoort, een achttiende-eeuws materialist', in: *Geschiedenis van de wijsbegeerte in Nederland* 11 (2000) 61-80.
Vermij, R.H., and J.A. van Maanen, 'An unpublished autograph by Christiaan Huygens: His letter to David Gregory of 18 January 1694', in: *Annals of science* 49 (1992) 507-523.
Vernet, J., 'Copernicus in Spain', in: *Reception* (1972), 271-291.
Visser, H.B., *De geschiedenis van den sabbatsstrijd onder de Gereformeerden in de zeventiende eeuw* (Utrecht 1939).
Visser, J.Th. de, *Kerk en staat*, II, *Nederland (vóór en tijdens de Republiek* (Leiden 1926).
Vos, P.D. de, *De vroedschap van Zierikzee van de tweede helft der 16de eeuw tot 1795* (Middelburg 1931, repr. Alphen aan den Rijn 1982).

Vrijer, M.J.A. de, *Henricus Regius. Een 'cartesiaansch' hoogleeraar aan de Utrechtse hoogeschool* (The Hague 1917).

Waard, C. de, 'Nog twee brieven van Philips Lansbergen', in: *Archief. Vroegere en latere mededeelingen voornamelijk in betrekking tot Zeeland* (1915) 93-99.

Walker, D.P., *The ancient theology. Studies in Christian Platonism from the fifteenth to the eighteenth century* (London 1972).

Wall, B.E., 'Anatomy of a precursor. The historiography of Aristarchos of Samos', in: *Studies in history and philosophy of science* 6 (1975) 201-228.

Wall, E. van der, 'Profetie en providentie: de coccejanen en de vroege Verlichting', in: P. Bange a.o. (ed.), *Kerk en Verlichting* (Zwolle 1990).

–, 'Orthodoxy and scepticism in the early Dutch Enlightenment', in: R.H. Popkin and A. Vanderjagt (ed.), *Scepticism and irreligion in the seventeenth and eighteenth centuries* (Leiden 1993) 121-141.

–, 'Cartesianism and Cocceianism: a natural alliance?', in: M. Magdelaine a.o. (ed.), *De l'humanisme aux Lumières, Bayle et le protestantisme* (Paris and Oxford 1996) 445-455.

Walther, Ch. F., 'Catalogue méthodique des dissertations ou thèses académiques imprimées par les Elzeviers de 1616 à 1712, recueillies pour la première fois dans la Bibliothèque Impériale publique à Saint-Petersbourg', in: *Bulletin du bibliophile Belge* 17 (1861) 347-360; 18 (1862) 1-25, 89-105, 217-228, 313-324, 409-415; 19 (1863) 1-11, 149-154.

Warner, D.J., *The sky explored. Celestial cartography 1500-1800* (New York and Amsterdam 1979).

Waterbolk, E.H., 'The 'reception' of Copernicus' teachings by Gemma Frisius (1508-1555)', in: *Lias* 1 (1974) 225-242.

Westman, R.S., 'Kepler's theory of hypothesis and the 'realist dilemma'', in: *Studies in history and philosophy of science* 3 (1972) 233-264.

– 'The Melanchthon circle, Rheticus and the Wittenberg interpretation of the Copernican theory', in: *Isis* 66 (1975) 165-193. [a]

–, 'Three responses to the Copernican theory: Johannes Praetorius, Tycho Brahe, and Michael Maestlin', in: R.S. Westman (ed.), *The Copernican achievement* (Berkeley etc. 1975) 285-345. [b]

–, 'Magical reform and astronomical reform: the Yates thesis reconsidered', in: R.S. Westman and J.E. McGuire, *Hermeticism and the scientific revolution. Papers read at a Clark Library seminar, March 9, 1974* (Los Angeles 1977) 1-91.

–, 'The astronomer's role in the 16th century: a preliminary study', in: *History of science* 18 (1980) 105-147. [a]

–, 'Humanism and scientific roles in the 16th century', in: R. Schmitz and F. Krafft ed., *Humanismus und Naturwissenschaften* (Boppard am Rhein 1980) 83-99. [b]

–, 'The Copernicans and the churches', in: D.C. Lindberg and R.L. Numbers ed., *God and nature. Historical essays on the encounter between Christianity and science* (Berkeley etc. 1986) 76-113.

Westra, F., *Nederlandse ingenieurs en de fortificatiewerken in het eerste tijdperk van de Tachtigjarige Oorlog, 1573-1604* (Alphen aan den Rijn, 1992).

Wielema, M.R., 'Nicolaas Hartsoeker (1656-1725): van mechanisme naar vitalisme', in: *Gewina* 15 (1992) 243-261.

–, 'Nicolaus Engelhard (1696-1765). De Leibniz-Wolffiaanse metafysica te Groningen', in: H.A. Krop, J.A. van Ruler, A.J. Vanderjagt, *Zeer kundige professoren. Beoefening van de filosofie in Groningen van 1614 tot 1996* (Hilversum 1997) 149-161.

Wiersum, E., 'Alchemie te Middelburg in 1604', in: *Archief. Vroegere en latere mededeelingen voornamelijk in betrekking tot Zeeland* (1918) 119-120.

Wiesenfeldt, G.B., *Leerer Raum in Minervas Haus. Experimentelle Naturlehre an der Universität Leiden, 1675-1715* (diss. Hamburg 1999).

Wightman, W.P.D., *Science and the Renaissance. An introduction to the study of the emergence of the sciences in the sixteenth century* (2 vol., Edinburgh 1962).

Wijk, W.E., 'De quadratuur van den cirkel', in: *Natuur en mens* (1938) (separate).

Wijnman, H.F., 'De Amsterdamsche rekenmeester Sybrandt Hansz. Cardinael', in: *Het boek* 22 (1933/1934) 73-94.

Wilson, C., 'Predictive astronomy in the century after Kepler', in: R. Taton and C. Wilson (ed.), *Planetary astronomy from the Renaissance to the rise of astrophysics: Tycho Brahe to Newton* (The general history of asronomy, II-A) (Cambridge etc. 1989) 161-206.

Winter, P.J. van, *Hoger beroepsonderwijs avant-la-lettre. Bemoeiingen met de vorming van landmeters en ingenieurs bij de Nederlandse universiteiten van de 17e en 18e eeuw* (Amsterdam enz. 1988) (Verhandelingen der Koninklijke Nederlandse Akademie van Wetenschappen, afd. letterkunde, nieuwe reeks, 137).

Wolthuis, G.W., 'Pieter Langendijk en de wiskunstenaars. Een bijdrage tot de kennis der zeden van onze wiskundige voorvaderen', in: *Tijdschrift voor taal en letteren* 24 (1936) 163-183, 238-261.

Wortel, Th.P.H., 'Adriaen Anthonisz., de vestingbouwer', in: idem, *Uit de geschiedenis van Alkmaar en omgeving* (Bergen (N.H.) 1990) 175-218.

Yates, F.A., *Giordano Bruno and the Hermetic tradition* (Chicago etc. 1964).

Zemplén, J., 'The reception of Copernicanism in Hungary. A contribution to the history of natural philosophy and physics in the XVIIth and XVIIIth centuries', in: *Reception*, 311-356.

Zinner, E., *Entstehung und Ausbreitung der Coppernicanischen Lehre* (Erlangen 1943).

Zuidervaart, H.J., 'Astronomische waarnemingen en wetenschappelijke contacten van Jan de Munck (1687-1768), stadsarchitect van Middelburg', in: *Worstelende wetenschap. Aspecten van wetenschapsbeoefening in Zeeland van de zestiende*

tot in de negentiende eeuw (Separate issue from: *Archief. Mededelingen van het Koninklijk Zeeuwsch Genootschap der Wetenschappen*, 1987) 103-170.
–, *Speculatie, wetenschap en vernuft. Fysica en astronomie volgens Wytze Foppes Dongjuma (1707-1778), instrumentmaker te Leeuwarden* (Leeuwarden 1995).
–, (with assistance of H. Hoitsma), 'Een Zeeuws planetarium uit de tweede helft van de 18e eeuw', *Archief van het Koninklijk Zeeuwsch Genootschap der Wetenschappen* (1982) 69-148.
–, *Van 'konstgenoten' en hemelse fenomenen. Nederlandse sterrenkunde in de achttiende eeuw* (Rotterdam 1999).

Index